Advances in Image Enhancement

Advances in Image Enhancement

Editors

Chunwei Tian
Wenqi Ren
Yudong Liang

MDPI • Basel • Beijing • Wuhan • Barcelona • Belgrade • Manchester • Tokyo • Cluj • Tianjin

Editors

Chunwei Tian
School of Software
Northwestern Polytechnical University
Xi'an
China

Wenqi Ren
School of Cyber Science and Technology
Sun YAT-SEN University
Shenzhen
China

Yudong Liang
School of Computer and Information Technology
Shanxi University
Taiyuan
China

Editorial Office
MDPI
St. Alban-Anlage 66
4052 Basel, Switzerland

This is a reprint of articles from the Special Issue published online in the open access journal *Electronics* (ISSN 2079-9292) (available at: www.mdpi.com/journal/electronics/special_issues/Image_Enhancement_1).

For citation purposes, cite each article independently as indicated on the article page online and as indicated below:

LastName, A.A.; LastName, B.B.; LastName, C.C. Article Title. *Journal Name* **Year**, *Volume Number*, Page Range.

ISBN 978-3-0365-7941-2 (Hbk)
ISBN 978-3-0365-7940-5 (PDF)

Contents

About the Editors

Chunwei Tian

Chunwei Tian is currently an Associate Professor with the School of Software, Northwestern Polytechnical University, China. Moreover, he is a member of the National Engineering Laboratory for Integrated Aerospace Ground-Ocean Big Data Application Technology. He became one of the World's Top 2% of Scientists in 2022. He has obtained a 2021 Shenzhen CCF Excellent Doctoral Dissertation and 2022 Harbin Institute of Technology. His research interests include image restoration and deep learning. He has published over 50 papers in academic journals and conferences, including *IEEE TNNLS, IEEE TMM, IEEE TSMC, Pattern Recognition, Neural Networks, Information Sciences,* and *ICASSP*. He has four ESI highly cited papers, three homepage papers of *Neural Networks*, one homepage paper of *TMM*, and published one excellent paper in 2020 for *CAAI Transactions on Intelligence Technology*. Moreover, he has three codes which are rated as the contribution codes of GitHub 2020. He has two papers which are integrated on iHub and Profillic. He has obtained the awards of open science excellent author program from Wiley and an excellent paper from Taicang. He has one paper code which has been collected by OSCS. His one paper technique has been used by MetronMind. His three paper techniques have resulted in him being invited to act as a benchmark for image super-resolution. Additionally, he is an Associate Editor/Young Editor of *CAAI Transactions on Intelligence Technology, Defence Technology, Intelligent Data Analysis, Mathematics, International Journal of Image and Graphics, Data Science and Management, Ordnance Industry Automation, Frontiers in Robotics and AI*, etc. Moreover, he is also a reviewer of some journals and conferences, such as *IEEE TIP, IEEE TII, IEEE TNNLS, IEEE TCYB, IEEE TSMC, NN, Information Sciences, CVIU, Information Fusion, AAAI,* and *ICASSP*.

Wenqi Ren

Wenqi Ren is an Associate Professor at Sun Yat-sen University. His primary research interests include computer vision and artificial intelligence. He has published over 60 top-tier papers in international journals and conferences in the field, with more than 8500 citations on Google Scholar. Six of his papers have been selected as highly cited papers in ESI. He has served as Aera Chair and Senior Program Committee member for several international conferences in the field of artificial intelligence and computer vision. He has led multiple national-level and industry projects. He has received the Outstanding Ph.D. Thesis Award from the China Computer Federation and the Wu Wenjun Artificial Intelligence Excellent Youth Award. He was also recognized as a Highly Cited Researcher in China by Elsevier in 2022.

Yudong Liang

Yudong Liang received his Ph.D. from Xi'an Jiaotong University, Xi'an, China, in 2017, and he is currently an Associate Professor with the School of Computer and Information Technology, Shanxi University. His primary research interests include computer vision, image processing, and artificial intelligence, especially focusing on image enhancement tasks, image quality assessment, and high-level vision tasks on low-quality images. In recent years, he has published papers in top conferences and SCI journals such as *IJCAI, ACM MM, ECCV, ICME, IEEE TIP*, and *Pattern Recognition*. He serves as a reviewer and PC member for top conferences and journals in the relevant communities. He has led or participated in multiple national-level and industry projects. He is currently a member of the Hybrid Artificial Intelligence Expert Committee of the Chinese Association of Automation.

Preface to "Advances in Image Enhancement"

In the era of the Internet of Things, images have played important roles in human–computer interactions, and with the arrival of big data technology, people have higher requirements of image qualities, especially ones collected in dark light. This can be addressed through the development of camera hardware quality, i.e., the resolution and exposure time of cameras, which may require high computational costs. As an alternative, image enhancement techniques can exact salient features to improve the quality of captured images according to the differences in diverse features, although they suffer from some challenges, i.e., a low contrast, artifacts, and overexposure, thus making it decidedly necessary to determine how to use advanced image enhancement techniques.

To address these issues, we investigated advances in image enhancement on electronics. This investigation includes low- and high-level vision. That is, a convolutional capsule network, retinex theory, and histogram equalization can be used for image enhancement. Improving network architectures can enhance the effects of image super-resolution. Traditional machine learning, i.e., wavelet, non-local moment mean filtering, CNN, and GAN, can be used to remove noise to obtain clear images. Moreover, using multi-scale technique and cascading architecture can be utilized for image watermark removal and dehazing. Additionally, few samples, wavelet, and deep networks can mine more semantic segmentation for image segmentation. Fusing different features from CNNs and traditional machine learning can improve the accuracy of image recognition.

The topic of advances in image enhancement on electronics is here presented as a reprint, which brings together the research accomplishments of researchers from academia and industry. The secondary goal of this reprint is to show the latest research results of advances in image enhancement.

Chunwei Tian, Wenqi Ren, and Yudong Liang
Editors

electronics

Article

Power Line Scene Recognition Based on Convolutional Capsule Network with Image Enhancement

Kuansheng Zou *, Shuaiqiang Zhao and Zhenbang Jiang

School of Electrical Engineering and Automation, Jiangsu Normal University, Xuzhou 221116, China
* Correspondence: zoukuansheng@jsnu.edu.cn

Abstract: With the popularization of unmanned aerial vehicle (UAV) applications and the continuous development of the power grid network, identifying power line scenarios in advance is very important for the safety of low-altitude flight. The power line scene recognition (PLSR) under complex background environments is particularly important. The complex background environment of power lines is usually mixed by forests, rivers, mountains, buildings, and so on. In these environments, the detection of slender power lines is particularly difficult. In this paper, a PLSR method of complex backgrounds based on the convolutional capsule network with image enhancement is proposed. The enhancement edge features of power line scenes based on the guided filter are fused with the convolutional capsule network framework. First, the guided filter is used to enhance the power line features in order to improve the recognition of the power line in the complex background. Second, the convolutional capsule network is used to extract the depth hierarchical features of the scene image of power lines. Finally, the output layer of the convolutional capsule network identifies the power line and non-power line scenes, and through the decoding layer, the power lines are reconstructed in the power line scene. Experimental results show that the accuracy of the proposed method obtains 97.43% on the public dataset. Robustness and generalization test results show that it has a good application prospect. Furthermore, the power lines can be accurately extracted from the complex backgrounds based on the reconstructed module.

Keywords: capsule network; image enhancement; power line scene recognition; complex background

Citation: Zou, K.; Zhao, S.; Jiang, Z. Power Line Scene Recognition Based on Convolutional Capsule Network with Image Enhancement. *Electronics* **2022**, *11*, 2834. https://doi.org/10.3390/electronics11182834

Academic Editor: Byung-Gyu Kim

Received: 31 July 2022
Accepted: 6 September 2022
Published: 8 September 2022

1. Introduction

With the continuous development of the modern power grid system, the demand for electricity is also greatly increased, and transmission lines spread to all parts of the world in a complex network. It is also of great significance for low-altitude flight to detect the power lines and implement obstacle avoidance. The Australian transport safety report shows that between 1994 and 2004, there were 119 helicopter crashes into power lines, of which 45 caused fatal injuries and 22 caused serious injury [1]. Hitting power lines will cause serious damage to the helicopter. The U.S. military data report shows that 54 power line collisions occurred between 1997 and 2006, resulting in 13 deaths and economic losses of up to USD 224 million [2]. Flight safety accidents threaten people's lives and cause huge economic losses.

Flight obstacle avoidance mainly depends on the pilot's reaction and experience. They can avoid large obstacles, but small obstacles, especially power lines, they often fail to dodge, which in turn leads to disasters. The power line scene recognition (PLSR) is mainly used for the flight obstacle avoidance of power lines, which can identify the presence or absence of power lines in advance, and use this as a judgment basis for reminding the driver. Thus, it is a meaningful research work and has a huge market prospect.

Although there were many publications in scene recognition of remote sensing images [3–8], little research focused on the PLSR. The leading cause for this is that the public dataset of the power lines is very scarce; only three types of power line data sets could be easily downloaded on the internet [9–12]. Among them, only one type can be used for the classification and recognition of power line scenarios [9,10]. Due to the inherent characteristics of power lines and the low resolution of datasets, it is difficult to obtain good recognition results. There were still state-of-the-art PLSR methods presented in recent years. Yetgin et al. [13] presented a PLSR framework based on the discrete cosine transform (DCT) of scenes obtained from aircraft-based cameras. This work attacked the problem of extracting signatures from the DCT coefficients by systematically changing the DCT matrix sizes and applying known classifiers to the DCT sub-matrices. The details were given in [14]. First, the image filtering was used to reduce the interference of noise and normalize the amplitude. Second, different types of image features of power lines were extracted through the DCT, local binary pattern (LBP), and histogram of oriented gradient (HOG), respectively. The absolute value of the logarithm of the discrete cosine coefficient in the DCT domain was taken to emphasize the dynamic range. Finally, the naive bayes (NB), random forest (RF), and support vector machine (SVM) classifier were used for the PLSR task. Although these kinds of methods were simple, it needed to manually set the feature extraction and feature matching methods. The PLSR method, based on deep learning, does not require manual feature extraction of power lines, and the established convolutional neural network (CNN) model can automatically extract effective features. Thus, some researchers tried to apply the CNNs to PLSR [15]. The VGG19 model and the ResNet50 model were fine-tuned to adapt to the power line dataset in literature [15], and an end-to-end PLSR method is proposed. The VGG19 model and the ResNet50 model were divided into five stages, and then the feature maps of these five stages were outputted. The feature maps were inputted to the NB, RF, and SVM classifiers, respectively, for the PLSR task. A fast PLSR network for the pixel-wise straight and curved power line detection method is proposed in [16]. The edge attention fusion module was combined together with a filter block, which extracts semantic and spatial information to improve the PLSR result along the boundary.

The power line extraction (PLE) is the pixel-wise PLSR method, which was paid more attention than the PLSR task. A PLE method based on the weakly supervised learning, which solved the problem of labeling large-scale datasets, was proposed in [17]. A PLE method based on pyramid patch classification, which used a CNN-based classifier to help eliminate power line pseudo-targets, was proposed in [18]. The generative adversarial network was combined with the conic and hue perturbation to enhance the datasets to reduce the model parameters and computational complexity through model pruning in [19]. Artificially synthesized power line images were used as the training data, and a fast single-shot line segment detector (LSD) was proposed in [20]. A real-time segmentation model for power lines was proposed in [21]. They used a spatial branch to capture rich spatial information and utilized classification with subnet-level skip connections. It recovered long-distance features and improved the performance of the power line extraction. Liu et al. improved the Unet model and its variants to the power line scene recognition and extraction task [22].

Since the capsule network (CapsNet) [23] is widely used in various classification tasks with its rich feature expression ability and effectiveness on small data sets and achieved good classification results [24–29]. The CapsNet is also tentatively studied in the scene recognition of remote sensing images [3–8]. Thus, in this paper, the CapsNet is selected as the backbone network, and the edge of line features of the power lines are enhanced. Finally, a novel PLSR method is proposed. The main innovation can be summarized as follows:

(1) A PLSR method based on the convolutional CapsNet fused with image enhancement is proposed. The edge structures of the power lines are enhanced by using the guided filter. The lone points and lines that are reinforced at the same time are weakened by the convolutional CapsNet. Various experiments show that it is suitable for the PLSR task with complex backgrounds.

(2) The power line scene recognition and feature extraction tasks can be performed simultaneously based on the convolutional CapsNet structure. The PLSR task is performed based on the output of the digital capsule layer, and the PLE task is performed based on

the output of a reconstructed module. The sections of this paper are arranged as follows: The CapsNet is introduced in Section 2. The proposed convolutional CapsNet with image enhancement is explained in Section 3. The scene recognition results and analysis of the proposed method are given in Section 4. The reconstruction results and analysis of the proposed method are shown in Section 5. The conclusion is obtained in Section 6.

2. Capsule Network

The CapsNet is used to maintain the location information and the inherent attributes of objects in the image, which can model the spatial relationship of the image [23]. In the CapsNet structure, the scalar output of the feature detector in the CNN is replaced with a vector output, and the maximum pooling is replaced with a protocol routing simultaneously. Meanwhile, all the capsules, except the last capsule layer, maintain the convolutional structure. By doing this, the advantages of the CNN in copying the learned knowledge across space is retained. The higher-level capsules can cover a larger image area the same as the CNN. Unlike with the maximum pooling, the CapsNet can partially retain the precise location information of entities in the region through the protocol routing [30]. The CapsNet is composed of the input layer, output layer, convolutional layer, primary caps layer, and digit caps layer. The convolutional layer is used to extract the low-level features of the detect target. The primary caps layer is used to express the spatial relationship between the features, and transfers the extracted features to the digit caps layer. The dynamic routing algorithm is used to predict the classification results in the digit caps layer [31]. The coupling coefficient c, according to the similarity between the low-level capsule layer and the high-level capsule layer, is adjusted. The weight W between networks is updated. If the similarity between the i-th capsule in the lower layer and the j-th capsule in the upper layer is greater, the coupling coefficient c_{ij} is greater, and the formula is shown in Equation (1). Where the initial value of a priori coupled probability b_{ij} of the capsule i and the capsule j is set to 0, and updated as Equation (2).

$$c_{ij} = \frac{\exp(b_{ij})}{\sum_k \exp(b_{ij})} \tag{1}$$

$$b_{ij} \leftarrow b_{ij} + \hat{u}_{j|i} v_j \tag{2}$$

where the calculation method is shown as Equations (3) and (4), respectively.

$$\hat{u}_{j|i} = W_{ij} u_i \tag{3}$$

$$v_j = \frac{\|s_j\|^2}{1 + \|s_j\|^2} \frac{s_j}{\|s_j\|} \tag{4}$$

where in Equation (4), S_j is the input vector of the j-th capsule in the upper layer, and the formula is given as follows:

$$s_j = \sum_i c_{ij} \hat{u}_{j|i} \tag{5}$$

The object function of the CapsNet is defined as follows:

$$L_k = T_k \cdot max\left(0, \left(m^+ - \|v_k\|\right)\right)^2 + \lambda(1 - T_k) \cdot max\left(0, \left(\|v_k\| - m^-\right)\right)^2 \tag{6}$$

where v_k is the output of a capsule in the softmax layer. T_k represents the tag of the k-th target. If a training sample belongs to class k, T_k = 1. Otherwise, T_k = 0. m^-, and m^+ are, respectively, the upper bound for the probability of a training sample not belonging to class k and the lower bound for the probability of a training patch being an instance of class k. They are set as m^+ = 0.9 and m^- = 0.1. λ is a weight regularization factor, which is usually set as 0.5 [32].

The CapsNet was used to classify the MNIST images of 28 × 28 at first. The original network has a convolution layer, including 256 convolution cores with a scale of 9 × 9, and outputs a local feature map with a scale of 20 × 20 as the input of primary caps. The primary caps contain 32 different capsules, each with eight 9 × 9 × 256 convolution kernels. Both layers use the ReLU activation function. Moreover, the digital capsule layer outputs 16-D vector reconstruction objects contain all the required instantiation parameters [26,33].

3. Convolutional Capsule Network with Image Enhancement

3.1. Motivation

The aerial image of power lines is mainly taken by the inspected unmanned aerial vehicles (UAVs), which has its inherent characteristics. In terms of color and lustre, the brightness of power lines is uniform and is higher than the backgrounds. In terms of shape, the power line usually exists in the form of a straight line, with a pixel width of about 1~5 [23], but some power lines, in the shape of a solitary vertical curve, still exist. In terms of spatial relationship, power lines usually run parallel to each other throughout the image, except for single ones. The background of power lines is complex and changeable. It is found that the background images of power lines are mostly forest, lake, river, field, mountain, sky, white cloud, pole, tower scene, and so on. It makes the power line scene recognition and extraction task challenging.

In general, power lines account for less than 15% of pixels in power line scenes. The complex backgrounds also have good edge features. Thus, pooling operation in the CNN may lose the spatial information of power lines, or misdetect part of the edge background as power lines. Due to the excellent performance of the CapsNet in the image classification mentioned above, the CapsNet is our first choice for the PLSR task. The CapsNet also has drawbacks: (1) It is unable to handle large size input well (2) It is unable to fully extract the input features. (3) The classification accuracy decreases with the complexity of the dataset. Two additional convolutional layers are used to better extract features and reduce input size simultaneously. The guided filter can enhance the edge lines well, meanwhile, the CapsNet can preserve the spatial relationship of power lines. Thus, the convolutional CapsNet with image hencement by guided filter is proposed.

3.2. Image Enhancement with Guided Filter

Experiments show that the guided filter [34] proposed by He et al. can better enhance the edge features of power lines and increase the recognition accuracy of power lines in complex backgrounds. The guided filter [34] is an edge-preserving algorithm based on the local linear model. It uses a guided image to guide the filtering process, defines any pixel in the image as a linear relationship with some of its adjacent pixels, and performs filtering processing, respectively. Finally, all local filtering results are accumulated to derive the global filtering results, and an output image with a structure similar to the input image is obtained.

The output image f^o of the guided filter can be linearly represented by the guided image I_i in a square window i, as shown below [35].

$$f^o = a_k I_i + b_k, \ \forall i \in w_k \tag{7}$$

where w_k is a square window with a radius of r centered on the pixel k, a_k and b_k are constants in f^o, and their coefficients are solved by minimizing the following energy function:

$$E(a_k, b_k) = \sum_{i \in w_k} \left(\left(f_i^o - f_i^{in} \right)^2 + \eta a_k^2 \right) \tag{8}$$

where η is the regularization parameter to prevent it with too large a value, f_i^{in} is the input image of the filter.

Because the guided filter uses a guided image for reference, choosing a different guided image will obtain different learning tasks. It is suitable for the deep learning process. A power line scene image as input is shown in Figure 1a, and its enhanced image by the guided filter is shown in Figure 1b. Where the input image itself is selected as the guided image. It is obvious that the power lines are enhanced. Simultaneously, grass and the outline of a wheat field are also enhanced. If the CNN is used for the deep learning network, these enhanced backgrounds will represent the surrounding spaces because of several pooling operations. If the surrounding spaces are considered by the CapsNet, power lines can be easily distinguished with the enhanced backgrounds.

Figure 1. Power line scene image enhancement by using a guided filter. The input image itself is selected as the guided image. (**a**) Power line image. (**b**) Enhanced image.

If the ground truth images are selected as the guided image, the training process of the network will be sped up. A power line image as input is shown in Figure 2a, the ground truth label is shown in Figure 2b, and the output image of the guided filter by using the ground truth as the guided image is shown in Figure 2c. Obviously, the output image, by using the guided filter, is greatly enhanced. If this type of guided filter is combined with deep learning, not only will the training time be greatly reduced, the network performance will be also improved.

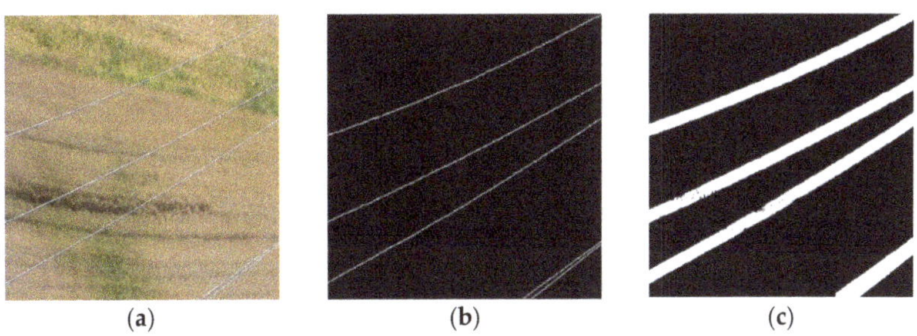

Figure 2. Power line scene image enhancement by using a guided filter. The ground truth is selected as the guided image. (**a**) Power line image. (**b**) Ground Truth. (**c**) Enhanced image.

In practice, there are no responded ground truth labels with the input images. Except for choosing the input image itself, more clarity for an image with special features can be selected as the guided image. For example, the line segment detection (LSD) [36] can better outline the power lines, it can be considered as the guided image. A power line image as input is shown in Figure 3a, the responded LSD map is shown in Figure 3b, and the output image of the guided filter by using the LSD as the guided image is shown in Figure 3c. Obviously, the output image, by using the guided filter, is greatly enhanced.

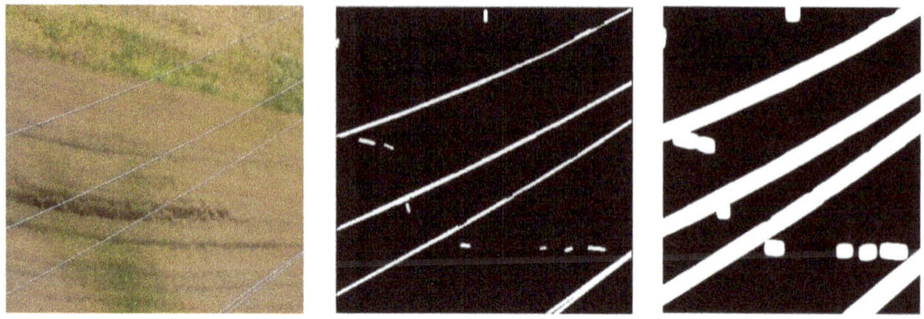

(**a**) (**b**) (**c**)

Figure 3. Power line scene image enhancement by using the guided filter. The LSD image is selected as the guided image. (**a**) Power line image. (**b**) LSD image. (**c**) Enhanced image.

In order to further improve scene recognition performance, the reconstructed image can be used as the guided image again for feedback. For feature extraction and classification, the ground truth image, feature enhanced image, or reconstructed image can be selected as the guided image to improve the accuracy. The input image itself is used as the guided image, and has a wider application. In order to make a more general comparison, the input image itself is selected as the guided image in the experiment. In brief, the guided filter with its variations [37–40], has a broad research prospect in the field of deep learning.

3.3. Convolutional CapsNet Framework

The proposed PLSR framework is shown in Figure 4. After image enhancement, the original image of 128 × 128 × 3 is grayed to an image of 128 × 128 × 1 and enters the first convolutional layer. The first convolutional layer contains 32 kernel functions with a scale of 5 × 5, and stride = 2. The output 64 × 64 × 32 feature image enters the second convolutional layer. The second convolutional layer contains 64 kernel functions with a scale of 5 × 5, stripe = 2. The output 32 × 32 × 64 feature map enters the third convolutional layer. The third convolutional layer contains 128 kernel functions with a scale of 9 × 9, and stripe = 2. The output 16 × 16 × 128 feature map enters the primary capsule layer. The primary capsule layer contains 32 different capsules, each capsule performs eight times of 9 × 9 kernel convolution, and stripe = 1. The last digital capsule layer outputs 16-D vector, which is used for binary classification tasks (power line scene or non-power line scene), and provides necessary information for image reconstruction. The ReLU activation function is applied to all layers. After the subsequent reconstruction module, the digital capsule can reconstruct the extracted power line binary image. The dimensions are 128 × 128 × 1.

The specific parameters of the convolutional CapsNet structure are shown in Table 1. In this paper, before the primary capsule layer, three convolutional layers with a stripe of 2 are selected in order to reduce the image dimension and extract more image information. The convolutional layer with a stripe of 2 can prevent the loss of spatial information caused by the pooling layer. The power line itself is very slender, and the spatial information is particularly important for the identification and extraction of power lines.

Table 1. The convolutional CapsNet structure.

	Filter	Kernel Size	Stride	Output
input				128 × 128 × 1
Conv1	32	5 × 5	2	64 × 64 × 32
Conv2	64	5 × 5	2	32 × 32 × 64
Conv3	128	9 × 9	2	16 × 16 × 128
primary capsule	32 × 8	9 × 9	1	16 × 16 × 256
digital capsule	-	-	-	16 × 2
output	-	-	-	2

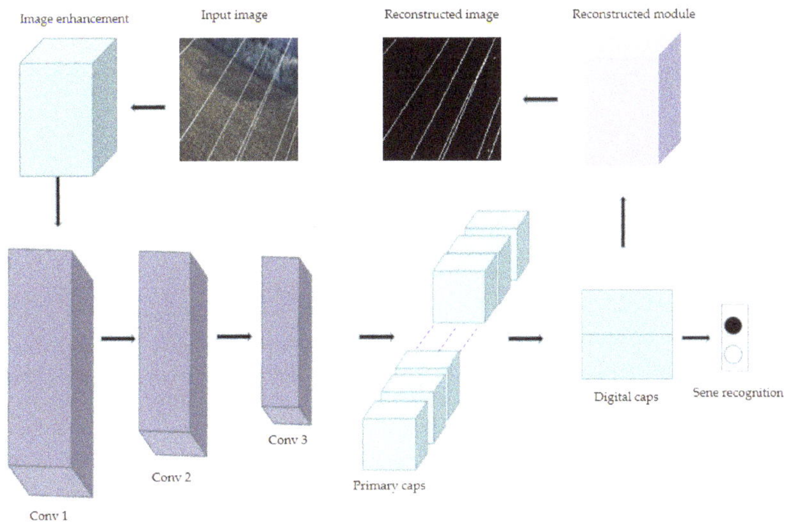

Figure 4. The proposed PLSR framework.

4. Scene Recognition Results and Analysis

4.1. Dataset and Experimental Configuration

The public data set of power line scenarios is adopted for experiment in this paper [9]. The dataset contains two subsets, infrared (IR) and visible light (VL). Each subset contains two parts, including and excluding. Each part has 2000 images of power line scenarios with 128 × 128 pixels. The subset with visible light [9] is used to carry out the experiment in this paper. The dataset is divided into training set, cross-validation set, and test set according to 3:1:1.

The configuration used in this paper, in terms of the hardware and the software platform, is shown in Table 2.

Table 2. Configuration of the experimental environment.

Platform	Configuration
Operating system	64 bit version of Windows 10
Central processing unit (CPU) Graphic	Intel(R) Core(TM) i9-10900k CPU @ 3.70 GHz
processing unit (GPU)	NVIDIA GeForce RTX 2070 8 G
Deep learning framework	PyTorch1.7
Compilers	PyCharm
Scripting language	Python 3.7
Solid state disk (SSD)	500 GB

The experimental parameters used to train the proposed network are shown in Table 3.

Table 3. Experimental parameters of the convolutional CapsNet.

Parameters	Configuration
Input Size	128 × 128 × 1
Batch size	64
Optimizer	Adam
Learning rate	0.001
Training epochs	200

4.2. Evaluation Metric

In this experiment, the accuracy rate is selected as the evaluation criteria, and the formula is given as Equation (9).

$$Accuracy = \frac{Number\ of\ correct\ predictions}{Total\ number\ of\ predictions} \tag{9}$$

The PLSR task is a binary classification problem, and the above-mentioned formula can be written as Equation (10).

$$Accuracy = \frac{(TP + TN)}{(TP + FP + TN + FN)} \tag{10}$$

where *TP* indicates that the actual case is positive, and the prediction is positive; *TN* indicates that the actual case is negative, and the prediction is negative; *FP* indicates that the actual case is negative, and the prediction is positive; *FN* indicates that the actual case is positive, and the prediction is negative.

4.3. Experimental Results and Analysis

4.3.1. Scene Recognition Results and Analysis

The visualization results of the proposed convolutional capsule network, with image enhancement on the visible light data set, are shown in Figure 5, where all the 32 images are visually displayed. The lower left part with the red font represents the real label, and the lower right part with the yellow font represents the model prediction results. Where 0 represents the scene without power lines, and 1 represents the scene containing power lines. All the 32 images are visually displayed, the presence or absence of power lines are correctly judged by using the proposed method.

Figure 5. Visualization results of power line scene recognition.

In order to verify the superiority of the method on the visible light data set, the comparative experiments with the traditional image processing based methods [13,14] are given in Figure 6. The parameters of these compared methods are given based on literature [13,14]. The detailed methods are listed as follows: SVM is used to classify local binary pattern (LBP) features; naïve bayes (NB) is used to classify LBP features; random forest is used to classify LBP features; SVM is used to classify histogram of oriented gradient (HOG) features; naïve bayes is used to classify HOG features; random forest (RF) is used to classify HOG features; SVM is used to classify classical selection DCT (CS_DCT) features; naïve bayes is used to classify CS_DCT features; random forest is used to classify CS_DCT features; SVM is used to classify reversed selection DCT (RS_DCT) features; naïve bayes is used to classify RS_DCT features; and random forest is used to classify RS_DCT features. Although a good detection result can be obtained by the DCT+RF, the feature extractor and matching method should be manually set. If the DCT+RF is tested on a larger dataset with a more complex background, the calculation will become more complicated, and the detection accuracy will not be guaranteed. The proposed model achieved the highest accuracy of 97.43%, which was 7.93% higher than the second place. It can be seen that on the visible light dataset, the proposed model has significant advantages over traditional image processing methods.

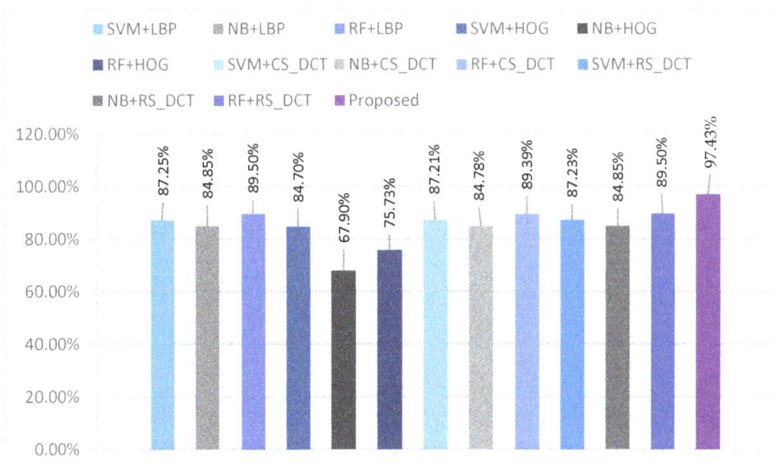

Figure 6. Comparison results with the traditional image processing based methods.

The proposed model is also compared with the deep learning methods implemented by us, and the experimental results are shown in Figure 7. The CapsNet is implemented as follows: After graying the 128 × 128 × 3 power line scene image, it is resized to the size of 28 × 28 × 1 and input into the original CapsNet network, without changing the network architecture. The accuracy is 77% by using the original CapsNet. The attention mechanism based CapsNet achieved accuracy of 78.8% [41]. Resizing the size from 128 × 128 × 1 to 28 × 28 × 1 simply results in the loss of the spatial information of power lines. Even with the attention mechanism-based CapsNet, it is hard to improve the accuracy of classification. Comparing the experimental results of the convolutional CapsNet, it can be seen that the two additional convolutional layers, without pooling operation, are effective, as the accuracy gets to 92.38% from 77%. The accuracy of the convolutional attention-based CapsNet (CA-CapsNet) reaches 93.5%. When image enhancement is added, the proposed model achieves the highest accuracy of 97.43%, and the guided convolutional attention-based CapsNet (GCA-CapsNet) obtains 97.15%. Since the power lines are very thin and run throughout the image, it is hard to design which part should be paid more attention, especially when both the edge lines of power lines and surrounding backgrounds are enhanced together.

Furthermore, U-net gets a very good classification performance, the accuracy of which is calculated by us from the result in [22]. It is verified that image enhancement with the guided filter is effective in improving the accuracy of the convolutional CapsNet and its variations. It also can be combined with other methods. It also has a further research value to improve the performance of itself by exploring more information.

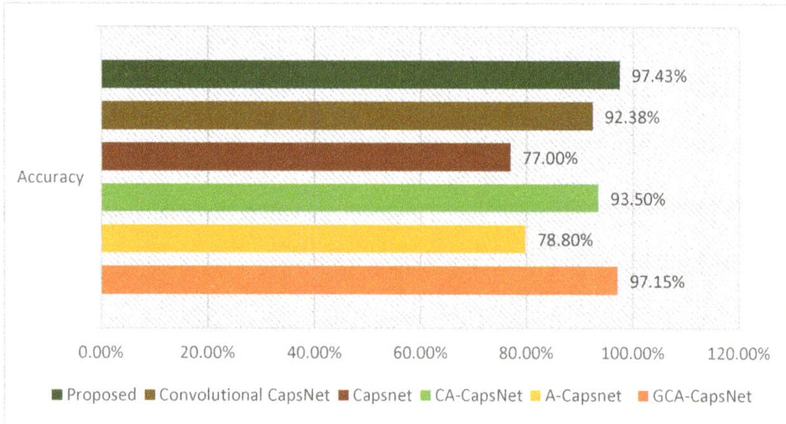

Figure 7. Comparison results with the deep learning based methods.

4.3.2. Performance Robustness Analysis

The robustness of the proposed PLSR method is tested in this section. The test dataset, containing 800 images, is selected for the experiment. The quantitative results are shown in Table 3. The accuracy of power line scene recognition in fog, snowfall, strong light, and motion-blurred scenes are 95.8%, 92.1%, 96.6%, and 88.3%, respectively. Compared with the normal scenes, the deviation of power line scene recognition accuracy in the above four scenes is −1.67%, −5.47%, −0.85%, and −9.37%, respectively. The deviation of motion-blurred scenes is slightly higher, but it is also less than 10%, and its performance is better than that of many normal scenes in Table 4. Other scenarios have a good performance robustness. Because the power line has the characteristics of small targets and weak features in aerial images, motion blur will affect the boundary response of the foreground and background. Through image feature enhancement and two additional convolution layers, the proposed method improves the robustness of power line scene recognition in the complex environments.

Table 4. Performance comparison of PLSR methods.

Scenes	Accuracy
Foggy	95.8
Strong light	92.1
Snow fall	96.6
motion blur	88.3

4.3.3. Generalization Test and Analysis

In order to evaluate the generalization performance of the proposed model more clearly, the test dataset in [12], containing 120 power line scene images with complex backgrounds, are selected, and another similar 80 images without power lines are also selected for testing. The total accuracy is 94.8%. Part of the test results of the proposed PLSR is shown in Figure 8. The lower left part with the red font represents the real label, and the lower right part with the yellow font represents the model prediction results. Where 0 represents the scene without power lines, and 1 represents the scene containing power lines. The experiment shows the recognition cases of eighteen images, of which the 13th image, the 14th image, and the 18th image are the display of false recognition cases.

Figure 8. The visual generalization test results of the proposed method.

5. Reconstruction Results and Analysis

The CapsNet uses an automatic encoder structure to reconstruct data; the automatic encoder is composed of an encoder and decoder [16]. This section discusses and analyzes the effect of power line reconstruction based on capsule network. In the proposed CapsNet, the encoder is composed of a convolution layer, primary capsule layer, and digital capsule layer. The decoder includes three full connection layers. The decoder uses the image features of the power line scene generated in the encoder to reconstruct an image with the same size as the input image. During reconstruction, the encoder uses the difference of the mean square error between the reconstructed image and the label image. Low error indicates that the reconstructed image is similar to the label image. The decoder structure of the proposed model is shown in Figure 9.

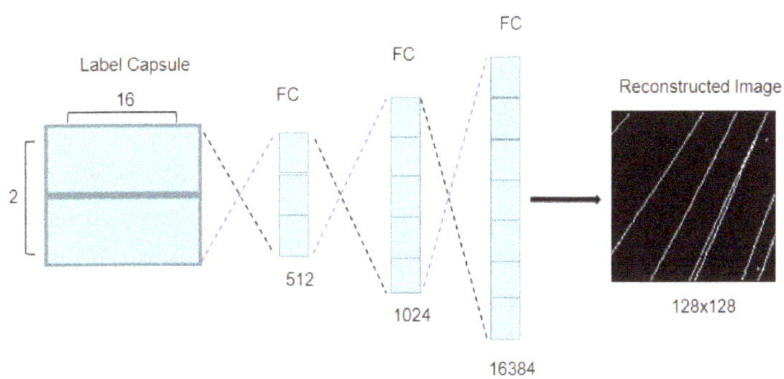

Figure 9. The decoder structure of the proposed model.

Based on the proposed method, the PLE results, with typical background, are shown in Figure 10. The six power line scene images with typical backgrounds are given in Figure 10a. The first and third pictures show the background of the tower. The second and fifth images show the field backgrounds. The fourth shows the grassland background. The sixth picture shows the road background. Figure 10b shows the real power line label corresponding to the original image, and Figure 10c shows the PLE results based on the proposed method. It can be seen that the power line can be completely extracted from the background by using the proposed model.

Figure 10. The PLE results with typical background. (**a**) Power line images. (**b**) Ground truth labels. (**c**) PLE results based on the proposed model.

In order to continue to evaluate the effect of the reconstruction model in the pixel level-recognition of power lines, Figure 11a shows six power line scene images with complex backgrounds. Due to the influence of complex backgrounds, the power lines are difficult to be found with naked eyes. The first and second pictures show the forest backgrounds. The third and fourth pictures show the mountain backgrounds. The fifth and sixth images show the backgrounds of the field. Figure 11b shows the real power line label corresponding to the original image, and Figure 11c shows the PLE results based on the proposed method. In these six images, although the power line is difficult to distinguish with naked eyes, the first image is perfectly extracted. The second and fourth images are partially bent and broken. The third and fifth pictures are partially missed, and the sixth picture has a small section of trees with multiple inspections. Overall, a good pixel level-recognition effect is achieved.

Figure 11. The power line extraction results under a complex background. (**a**) Power line images. (**b**) Ground truth labels. (**c**) PLE results based on the proposed method.

6. Conclusions

In this paper, the background of power line scene recognition is carefully analyzed at first, and the guided filter is found that can enhance the power line features effectively. Thus, the feature enhancement module with the guided filter is introduced to weaken the influence of complex background images on power line detection and extraction. A convolutional capsule network is used to design the power line scene recognition and extraction method. Experiments show that the proposed method has a high recognition accuracy and good robustness in the PLSR task. The image output from the convolutional capsule network decoder can also obtain a better power line pixel-level recognition effect. Based on the proposed method, we can not only judge whether there is a power line scene, but also extract the power line completely from the scene image of power lines. It lays a foundation for the future research of UAV tracking along the line and fault diagnosis attached to components of power lines.

For the issue of not-so-perfect performance robustness in a strong light environment, the fusion of infrared images and visible light images can be introduced in the future, since in the strong light environment, although the power lines are indistinguishable from the background, the high-temperature power lines can be distinguished from the low-temperature background environment. For the issue of not-so-good performance robustness in a motion blur environment, in the future, more stable and active disturbance rejection UAV trajectory-tracking methods can be studied to obtain a better image capture effect and reduce motion blur in aerial images.

This article makes sense despite its simplicity. The selection of guided images in the guided filter is variable, which makes the combination with deep learning have unlimited potential. New features, such as edge detection, texture preservation, and image enhancement could be used as guiding images, which will enhance the performance of the network. In addition, the design is flexible and simple, and the computational complexity is lower than that of the attention mechanism, which can be widely combined without various deep learning tasks. In supervised learning, selecting the ground truth label as the guide image can greatly improve the training performance of the network. In unsupervised learning and predictive analysis tasks, first selecting the original image or enhanced image as the guided image, and then selecting the reconstructed image as the guided image as the relevant feedback will improve the performance of the image classification task.

Author Contributions: Conceptualization, methodology, writing—review and editing, supervision, project administration, K.Z.; validation, software, writing—original draft, visualization, S.Z.; formal analysis, investigation, resources, data curation, Z.J. All authors have read and agreed to the published version of the manuscript.

Funding: This research received no external funding.

Data Availability Statement: The data that support the findings of this study are openly available in [Mendeley Data] at [https://data.mendeley.com/datasets/n6wrv4ry6v/8 (accessed on 30 July 2022)] and [Mendeley Data] at [https://data.mendeley.com/datasets/twxp8xccsw/9 (accessed on 30 July 2022)]. The data partly support for generalization test are openly available in [Github] at [https://github.com/SnorkerHeng /PLD-UAV (accessed on 30 July 2022)].

Conflicts of Interest: The authors declare no conflict of interest.

References

1. ASTB. Wire-Strike Accidents in General Aviation: Data Analysis 1994 to 2004. *ATSB Transp. Saf. Investig. Rep. Aust. Gov.* **2006**. Available online: https://www.atsb.gov.au/media/32640/wirestrikes_20050055.pdf (accessed on 23 January 2020).
2. Song, B.; Li, X. Power line detection from optical images. *Neurocomputing* **2014**, *129*, 350–361. [CrossRef]
3. Guo, Y.; Liao, J.; Shen, G. A Deep Learning Model with Capsules Embedded for High-Resolution Image Classification. *IEEE J. Sel. Top. Appl. Earth Obs. Remote Sens.* **2021**, *14*, 214–223. [CrossRef]
4. Khodadadzadeh, M.; Ding, X.; Chaurasia, P.; Coyle, D. A Hybrid Capsule Network for Hyperspectral Image Classification. *IEEE J. Sel. Top. Appl. Earth Obs. Remote Sens.* **2021**, *14*, 11824–11839. [CrossRef]
5. Mei, Z.; Yin, Z.; Kong, X.; Wang, L.; Ren, H. Cascade Residual Capsule Network for Hyperspectral Image Classification. *IEEE J. Sel. Top. Appl. Earth Obs. Remote Sens.* **2022**, *15*, 3089–3106. [CrossRef]
6. Wang, J.; Guo, S.; Huang, R.; Li, L.; Zhang, X.; Jiao, L. Dual-Channel Capsule Generation Adversarial Network for Hyperspectral Image Classification. *IEEE Trans. Geosci. Remote Sens.* **2022**, *60*, 5501016. [CrossRef]
7. Yu, Y.; Liu, C.; Guan, H.; Wang, L.; Gao, S.; Zhang, H.; Zhang, Y.; Li, J. Land Cover Classification of Multispectral LiDAR Data with an Efficient Self-Attention Capsule Network. *IEEE Trans. Geosci. Remote Sens.* **2022**, *19*, 6501505. [CrossRef]

8. Paoletti, M.; Moreno-Álvarez, S.; Haut, J. Multiple Attention-Guided Capsule Networks for Hyperspectral Image Classification. *IEEE Trans. Geosci. Remote Sens.* **2022**, *60*, 5520420. [CrossRef]
9. Yetgin, Ö.; Gerek, Ö. Powerline Image Dataset (Infrared-IR and Visible Light-VL), Mendeley Data, V8. Available online: https://data.mendeley.com/datasets/twxp8xccsw/1 (accessed on 14 June 2020).
10. Yetgin, Ö.; Gerek, Ö. Ground Truth of Powerline Dataset (Infrared-IR and Visible Light-VL), Mendeley Data, V1. Available online: https://data.mendeley.com/datasets/twxp8xccsw/9 (accessed on 14 June 2020).
11. Abdelfattah, R.; Wang, X.; Wang, S. TTPLA: An Aerial-Image Dataset for Detection and Segmentation of Transmission Towers and Power Lines. In Proceedings of the 15th Asian Conference on Computer Vision (ACCV), Kyoto, Japan, 30 November–4 December 2020; pp. 1–17.
12. Heng, S. Two Datasets for Power Line Detection in UAV Images. Available online: https://github.com/SnorkerHeng/PLD-UAV (accessed on 14 June 2020).
13. Yetgin, Ö.; Gerek, Ö. Automatic recognition of scenes with power line wires in real life aerial images using DCT-based features. *Digit. Signal Process.* **2018**, *77*, 102–119. [CrossRef]
14. Yetgin, Ö.; Gerek, Ö. Feature extraction, selection and classification code for power line scene recognition. *Softwarex* **2017**, *8*, 43–47. [CrossRef]
15. Yetgin, Ö.; Benligiray, B.; Gerek, Ö. Power Line Recognition from Aerial Images with Deep Learning. *IEEE Trans. Aerosp. Electron. Syst.* **2019**, *55*, 2241–2252.
16. Zhu, K.; Xu, C.; Cai, G.; Wei, Y. Fast-PLDN: Fast power line detection network. *J. Real-Time Image Process.* **2022**, *19*, 3–13.
17. Choi, H.; Koo, G.; Kim, B.; Kim, S. Weakly supervised power line detection algorithm using a recursive noisy label update with refined broken line segments. *Expert Syst. Appl.* **2021**, *165*, 113895.1–113895.9.
18. Li, Y.; Pan, C.; Cao, X.; Wu, D. Power Line Detection by Pyramidal Patch Classification. *IEEE Trans. Emerg. Top. Comput. Intell.* **2018**, *3*, 416–426. [CrossRef]
19. Xu, G.; Li, G. Research on lightweight neural network of aerial power line image segmentation. *J. Image Graph.* **2021**, *26*, 2605–2618.
20. Nguyen, V.; Jenssen, R.; Roverso, D. LS-Net: Fast single-shot line-segment detector. *Mach. Vis. Appl.* **2021**, *32*, 1–16.
21. Gao, Z.; Yang, G.; Li, E.; Liang, Z.; Guo, R. Efficient parallel branch network with multi-scale feature fusion for real-time overhead power line segmentation. *IEEE Sens. J.* **2021**, *21*, 12220–12227.
22. Liu, J.; Li, Y.; Gong, Z.; Liu, X.; Zhou, Y. Power line recognition method via fully convolutional network. *J. Image Graph.* **2020**, *25*, 956–966.
23. Sabour, S.; Frosst, N.; Hinton, G. Dynamic routing between capsules. In Proceedings of the 31st International Conference on Neural Information Processing Systems, Long Beach, CA, USA, 4–9 December 2017; pp. 3859–3869.
24. Zhao, Z.; Cheng, S. Capsule networks with non-iterative cluster routing. *Neural Netw.* **2021**, *143*, 690–697.
25. Kim, J.; Jang, S.; Park, E.; Choi, S. Text classification using capsules. *Neurocomputing* **2020**, *376*, 214–221.
26. Toraman, S.; Alakus, T.; Turkoglu, I. Convolutional capsnet: A novel artificial neural network approach to detect COVID-19 disease from X-ray images using capsule networks. *Chaos Solitons Fractals* **2020**, *140*, 110–122.
27. Kakillioglu, B.; Ren, A.; Wang, Y.; Velipasalar, S. 3D capsule networks for object classification with weight pruning. *IEEE Access* **2020**, *8*, 27393–27405.
28. Fahim, S.R.; Sarker, S.K.; Muyeen, S.M.; Das, S.K.; Kamwa, I. A deep learning based intelligent approach in detection and classification of transmission line faults. *Electr. Power Energy Syst.* **2021**, *133*, 102–107.
29. Moghaddam, A.; Etemad, A.; Pereira, F.; Correia, P. CapsField: Light field-based face and expression recognition in the wild using capsule routing. *IEEE Trans. Image Process.* **2021**, *30*, 2627–2642.
30. Yu, Y.; Ren, Y.; Guan, H.; Li, D.; Yu, C.; Jin, S.; Wang, L. Capsule feature pyramid network for building footprint extraction from high-resolution aerial imagery. *IEEE Geosci. Remote Sens. Lett.* **2021**, *18*, 895–899.
31. Zhang, T.; Zou, C. Study on Image Classification of Capsule Network Using Fuzzy Clustering. *Comput. Sci.* **2019**, *46*, 279–285.
32. Zhang, G.; Ding, X.; Yang, J.; Wang, H. Hyperspectral remote sensing classification based on multi-scale adaptive capsule network. *Laser Optoelectron. Prog.* **2021**, *13*, 2445.
33. Kai, Q.; Chi, Z.; Wang, L.; Jian, C.; Yan, B. Accurate reconstruction of image stimuli from human functional magnetic resonance imaging based on the decoding model with capsule network architecture. *Front. Neuroinform.* **2018**, *12*, 62.
34. He, K.; Sun, J.; Tang, X. Guided image filtering. *IEEE Trans. Pattern Anal. Mach. Intell.* **2013**, *35*, 1397–1409.
35. Badrinarayanan, V.; Kendall, A.; Cipolla, R. SegNet: A Deep convolutional encoder-decoder architecture for image segmentation. *IEEE Trans. Pattern Anal. Mach. Intell.* **2017**, *39*, 2481–2495.
36. Gu, G.; Ko, B.; Go, S.; Lee, S.; Lee, J.; Shin, M. Towards Light-weight and Real-time Line Segment Detection. In Proceedings of the 36th AAAI Conference on Artificial Intelligence, Virtual, Palo Alto, CA, USA, 22 February–1 March 2022; pp. 1–17.
37. Zeng, R.; Song, Y. A Fast Routing Capsule Network with Improved Dense Blocks. *IEEE Trans. Ind. Inform.* **2022**, *18*, 4383–4392. [CrossRef]
38. Chen, J.; Liu, Z. Mask Dynamic Routing to Combined Model of Deep Capsule Network and U-Net. *IEEE Trans. Neural Netw. Learn. Syst.* **2020**, *31*, 2653–2664. [CrossRef]
39. Pinckaers, H.; Ginneken, B.; Litjens, G. Streaming Convolutional Neural Networks for End-to-End Learning with Multi-Megapixel Images. *IEEE Trans. Pattern Anal. Mach. Intell.* **2022**, *44*, 1581–1590. [CrossRef] [PubMed]

40. Wu, H.; Zheng, S.; Zhang, J.; Huang, K. Fast End-to-End Trainable Guided Filter. In Proceedings of the 2018 IEEE/CVF Conference on Computer Vision and Pattern Recognition (CVPR), Salt Lake City, UT, USA, 18–23 June 2018; pp. 1–10.
41. Li, C.; Wang, B.; Zhang, S.; Liu, Y.; Song, R.; Cheng, J.; Chen, X. Emotion recognition from EEG based on multi-task learning with capsule network and attention mechanism. *Comput. Biol. Med.* **2022**, *143*, 105303. [CrossRef] [PubMed]

electronics

MDPI

Article

Low-Illumination Road Image Enhancement by Fusing Retinex Theory and Histogram Equalization

Yi Han [1], Xiangyong Chen [1], Yi Zhong [1,*], Yanqing Huang [2], Zhuo Li [2], Ping Han [1], Qing Li [3] and Zhenhui Yuan [4]

1 School of Information Engineering, Wuhan University of Technology, Wuhan 430070, China
2 SAIC GM Wuling Automobile Co., Ltd., Liuzhou 545007, China
3 Peng Cheng Laboratory, Shenzhen 518066, China
4 Department of Computer and Information Science, Northumbria University, Newcastle Upon Tyne NE1 8ST, UK
* Correspondence: zhongyi@whut.edu.cn; Tel.: +86-27-8785-8005

Abstract: Low-illumination image enhancement can provide more information than the original image in low-light scenarios, e.g., nighttime driving. Traditional deep-learning-based image enhancement algorithms struggle to balance the performance between the overall illumination enhancement and local edge details, due to limitations of time and computational cost. This paper proposes a histogram equalization–multiscale Retinex combination approach (HE-MSR-COM) that aims at solving the blur edge problem of HE and the uncertainty in selecting parameters for image illumination enhancement in MSR. The enhanced illumination information is extracted from the low-frequency component in the HE-enhanced image, and the enhanced edge information is obtained from the high-frequency component in the MSR-enhanced image. By designing adaptive fusion weights of HE and MSR, the proposed method effectively combines enhanced illumination and edge information. The experimental results show that HE-MSR-COM improves the image quality by 23.95% and 10.6% in two datasets, respectively, compared with HE, contrast-limited adaptive histogram equalization (CLAHE), MSR, and gamma correction (GC).

Keywords: low illumination; image enhancement; Retinex theory; histogram equalization; image fusion

Citation: Han, Y.; Chen, X.; Zhong, Y.; Huang, Y.; Li, Z.; Han, P.; Li, Q.; Yuan, Z. Low-Illumination Road Image Enhancement by Fusing Retinex Theory and Histogram Equalization. *Electronics* **2023**, *12*, 990. https://doi.org/10.3390/electronics12040990

Academic Editor: Chiman Kwan

Received: 28 January 2023
Revised: 12 February 2023
Accepted: 15 February 2023
Published: 16 February 2023

1. Introduction

With the development of automatic driving technology, computer vision methods are based on simulated human vision, and they are also used to carry out important sensing tasks in multiple automatic driving scenarios, such as object detection, semantic road segmentation, etc. Due to changes in ambient light, such as day and night, the visibility of the images varies significantly. If the computer vision algorithm needs to ensure stable performance under different lighting conditions, it should cover all lighting scenes as much as possible during training. This will undoubtedly require more time and human labor resources in collecting the dataset, as well as training based on this dataset. Image enhancement is an effective solution to solve the above problems. The night image is enhanced by the day image, which can greatly enhance the information perception of computer vision and human vision. Through image enhancement, image characteristics such as brightness, contrast, signal-to-noise ratio, edge sharpness, and color accuracy are improved [1,2], and the feature differences between night and day images are further reduced. This increases the degree of image aggregation in the feature space, which is beneficial to the training and reasoning processes of visual deep learning networks. Traditional image enhancement methods are based on mathematical computations that do not need training in advance. This can save computing power for computationally constrained automated driving applications. Such methods can be used as the data preprocessing module of the deep-learning-based automatic driving computer vision tasks for images with low illumination at night.

Land and McCann proposed and developed the Retinex theory [3,4]. Retinex theory regards the image as the superposition of two components: illumination and reflectance. Illumination is the influence of ambient light in the imaging process. The reflectance component represents the natural properties of the objects in the image and is not affected by other factors. The purpose of the Retinex algorithm is to separate the reflectance component of the object from the image, removing the effects of ambient lighting. In night image enhancement, the Retinex algorithm can obtain the reflectance component of night image objects and remove unfavorable illumination conditions.

Many image enhancement algorithms are derived from Retinex theory. These algorithms separate only the reflectance components of the object and ignore the illumination. This normally leads to poor imaging results. Reflectance components pay more attention to the high-frequency information, such as the edge texture of the image, but lack information on color and brightness. This is not conducive to enhancing contrast and producing proper brightness. In addition, better image enhancement requires more careful manual parameter adjustment to guarantee high performance [5]. This limitation makes it difficult to generalize the algorithm based on the Retinex theory in practice.

HE (histogram equalization) has been widely used for image brightness enhancement [6]. It expands the existing gray levels of the original image to the whole gray level (0–255). For example, for night images, the overall image style is dark, and the gray level is concentrated in a small gray level range. HE can significantly improve the image brightness by expanding the gray level distribution to the entire gray level. The classic HE increases brightness by evenly distributing the entire gray level. The average brightness of the enhanced image changes dramatically. However, if there are both over-light and over-dark areas in one image, HE will map the pixel brightness in the two areas to medium-level brightness. A bright pixel may be mapped to the same medium brightness as a dark pixel, resulting in the loss of image edge details [7]. Additionally, HE expands the gray level of the image from 0–50 to 0–255, meaning that the brightness of pixels with the same gray level will be different after expansion. This brings high-frequency noise to the enhanced image.

DCT (discrete cosine transform) is similar to DFT (discrete Fourier transform) but only operates with real numbers. Compared with DFT, DCT has better aggregation for certain information. In the image field, images are often processed by DCT and IDCT (inverse discrete cosine transform) in the frequency domain. The low-frequency signal of the image mainly corresponds to the slowly changing information, such as color and brightness. High-frequency signals correspond to rapidly changing information in images, such as the edges. Ordinary high-pass filters or low-pass filters can only achieve image smoothing or sharpening [8]. The image enhancement algorithm based on Retinex theory retains more edge information, but the visual effect of the image depends on fine parameter adjustment. HE enhances the lighting information better, but the edge information is lost. The image frequency domain transformation is performed via DCT, combing with the advantages of the Retinex and HE. HE enhances the image brightness, and more edge information is retained by the Retinex algorithm.

This paper proposes HE-MSR-COM, which combines the low-frequency information of HE-enhanced images with the high-frequency information of MSR-enhanced images. The low-frequency information of HE-enhanced images can provide enhanced illumination, to ensure a better visual experience. While the high-frequency information of MSR can retain more edge details, it will improve the quality of the image, e.g., contrast, mean gradient, etc. This method can filter the high-frequency noise brought by HE and achieve performance balance and optimization with overall illumination and edge details. This paper mainly focuses on enhancing low-illumination images. Images under rainy and foggy weather can be categorized as low-illumination images and can use the same processing method proposed in this paper. The salt and pepper noise introduced by these typical weather conditions needs further image processing steps, such as image noise reduction, which is not within the scope of this paper.

The structure of this paper is as follows: Section 2 introduces the development of different research directions and related work on night image enhancement. In Section 3, the relevant theoretical basis is introduced, and the research method of this paper is proposed. Section 4 describes the selection of the dataset and experimental evaluations, as well as the analysis of the experimental results. Section 5 summarizes the performance of the proposed algorithm and indicates future research directions.

2. Related Works

2.1. Retinex Theory

Many low-light image enhancement algorithms have been developed based on Retinex theory. Jobson et al. improved the Retinex theory and proposed SSR (single-scale Retinex) [9] and MSR (multiscale Retinex) [10]. These methods simply assume that the illumination is smooth and the reflectance components are unsmooth. The Gaussian low-pass filter (LPF) and logarithm operation are used to estimate the illumination of the image. The gradient and region size in different images are different. SSR needs to strike a balance between overall illumination estimation and local image detail performance. MSR uses different weights for several linear LPFs to estimate illuminance. This can ensure the balance of performance in the overall illumination and local image details. Wang et al. [11] proposed a low-illumination color image enhancement algorithm based on the Gabor filter and Retinex theory. The algorithm extracts the illumination component from the HSI (hue, saturation, intensity) color space of the original image. The authors enhanced the illumination component using MSRCR (multiscale Retinex with color restore) to obtain the enhanced illumination component and illuminated images. Additionally, the original image of the RGB space is enhanced using the SSR algorithm. Then, the illuminated image and the enhanced image are weighted and fused for better performance. Traditional Retinex-based algorithms use Gaussian filters (GSFs) to estimate illumination. However, GSFs cannot adapt themselves to different backgrounds in images, which is the main reason why they cannot accurately estimate illumination [12]. Tao et al. [13] replaced the GSF with a region covariance filter (RCF), which depends on the covariance matrix of local image features for each pixel. As a result, the RCF is adaptive to different pixels in an image and can estimate illumination more accurately than the GSF. The RCF Retinex algorithm increases contrast, cancels noise, and enhances detail compared with GSF Retinex algorithms. However, the calculation of RCF Retinex is time-consuming and impractical.

The performance of these methods often depends on the careful selection of the parameters of the filters and their corresponding weights, and most of these parameters require human-involved decisions, which are time- and human-labor-intensive and are impractical for real-time applications such as night image enhancement in autonomous driving.

2.2. Histogram Equalization

Histogram equalization (HE) is used to enhance contrast and improve image quality. Yeong Kim [14] considered that the original HE algorithm would cause the loss of edge information and, therefore, reduce the image contrast. They proposed bi-histogram equalization (BBHE) to enhance the image contrast. The average value of illumination is used as a threshold to distinguish dark and bright areas. The HE algorithm is used in both a bright area and a dark area to reduce the loss of edge information. However, this results in unbalanced overall distribution of illumination in the enhanced image. Chen et al. [15] believe that the median illumination is more appropriate as the threshold instead of the average illumination. Therefore, they proposed dualistic sub-image histogram equalization (DSIHE) to prevent over-light or over-dark areas from affecting the threshold, and their experiments proved that the median is more statistically significant. Ooi et al. [16] proposed bi-histogram equalization with a plateau level (BHEPL), which reduces the processing time compared to BBHE. Ooi et al. [17] proposed quadrant dynamic histogram equalization (QDHE), which divides the histogram into four (quadrant) sub-histograms based on the input image's median value. It reduces noise amplification and over-enhancement. Salah et al. [18] proposed a combination of gamma correction and the retinal filter (gamma-HM-COMP), which preserves the contrast between the gray levels of the original pixels, thereby preserving more edge information. Tan et al. [19] proposed a background-brightness-preserving HE (BBPHE) based on nonlinear histogram equalization. This method divides the image into background regions and non-background regions. It can enhance the brightness of the whole image and preserve the edge information of the object as much as possible. Adaptive histogram equalization (AHE) is a commonly used method that calculates the local gray histogram of images to obtain more local details and improve contrast. Shome et al. [20] proposed a contrast-limited AHE (CLAHE) to overcome the problem that AHE will overamplify the noise in the same area of the image. On the other hand, Lin et al. [21] proposed averaging histogram equalization (AVHEQ) for color images. This algorithm divides the original image into sub-images and equalizes them independently. It proposes a new mathematical algorithm to determine the optimal threshold and achieves better performance compared with conventional methods such as BBHE, DSIHE, and BHEPL. Chen et al. [22] used a fast guide filter to decompose the image into a base layer and a detail layer. The plateau equalization (PE) enhances the detail and the background separately, increasing the contrast of the detail. Kwan et al. [23] used a second-order histogram matching algorithm that enhances 16-bit infrared video contrast. This optimizes the possible information loss caused by using processed 8-bit infrared video. The performance of this method has been improved in the target detection using You Only Look Once (YOLO) and classification using a residual network (ResNet). Liao et al. [24] proposed an innovative box filtering method by combining the mean and median filtering techniques to achieve the balance between noise removal and edge preservation.

HE-based algorithms are popular because they are easy to implement and fast to process. However, these algorithms also have various limitations, such as adding noise to the output image and increasing the contrast of the background rather than the object in the image. The direct stretching on the gray level also causes the loss of edge information, resulting in a fuzzy edge. Much research has been carried out to prevent the loss of edge information. However, this issue is more complicated to solve in complex illumination scenes.

2.3. Data-Driven Methods

Recently, many image enhancement methods have been combined with deep learning. These methods use a data-driven approach to enhance the night image adaptively based on a priori trained model. CNN (convolutional neural network) is a typical approach that employs supervision training of a large number of labeled datasets and has shown good adaptability to different scenes. It is a resource-intensive task to collect the required datasets that contain a large number of paired low-light and normal-light images as sample data and label data, respectively. LLNet [25] is trained by pseudo-labels generated by random gamma correction. These unreal labels are given by the traditional image enhancement algorithm, which limits its enhancement effect. Due to the cost of the dataset and the poor generalization ability of CNNs, this method often results in artifacts and unnatural images.

Methods based on unsupervised GANs (generative adversarial networks) do not require a large number of paired images as a training set. These methods can mitigate the cost of collecting labeled datasets. EnlightenGAN [26], a low-light image enhancement algorithm based on an unsupervised GAN, uses unpaired low-light and normal-light data as the dataset. However, the performance of GAN methods is highly affected by the selection of the dataset. GAN methods can produce unpredictable outputs. Some produces features that fool the discriminator and are regarded as the correct result, which is actually an unsatisfactory result.

Qu et al. [27] adopted deep learning to compensate for the defects of traditional image enhancement methods. However, these methods rely heavily on datasets with perfect scenes for training. It is challenging to allocate adequate computing resources to image enhancement in real-time automatic driving applications.

3. Method

The structure of the proposed HE-MSR-COM Algorithm 1 contains three main parts, including the MSR enhancement module, HE enhancement module, and frequency-domain fusion module. The MSR and HE enhancement modules are responsible for obtaining the edge and illumination enhancement information of the image, respectively, which can be seen in Figure 1. The frequency-domain fusion module is used to adaptively unify the edge and illumination information by deriving weights for different scenarios.

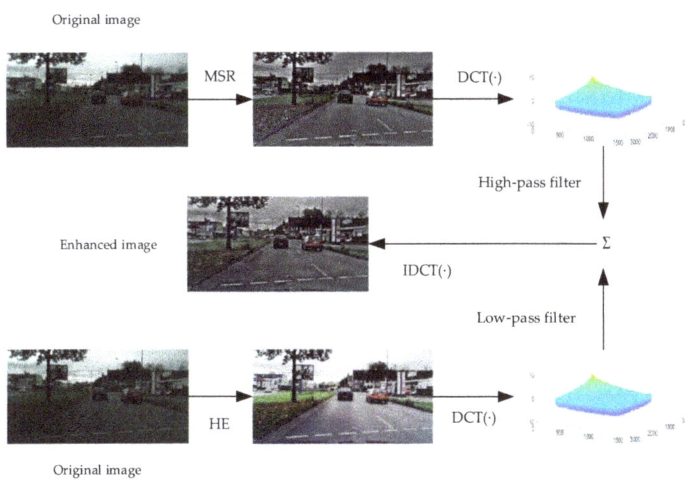

Figure 1. Overview of frequency−domain fusion based on MSR and HE.

3.1. MSR Image Enhancement

Retinex theory is based on the idea that images are a combination of illumination and reflectance. The theory of Retinex is shown in Figure 2.

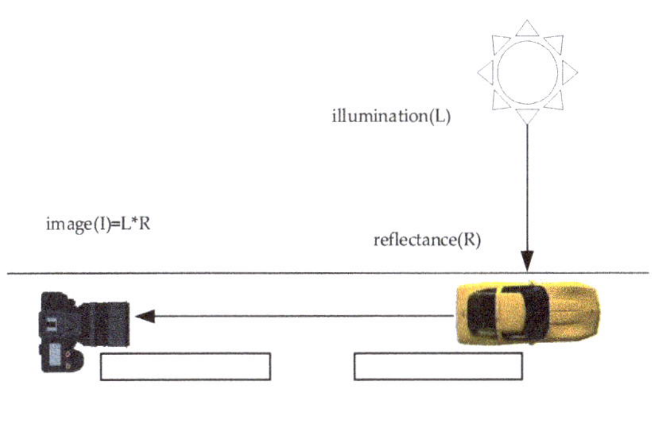

Figure 2. Sketch of Retinex theory.

Retinex theory can be defined as follows:

$$I = L * R \tag{1}$$

where I is the original image, L is a matrix of illumination, and the matrix R represents the reflectance components of the object in I. The operation $*$ is the matrix multiplication of the corresponding elements. Illumination L is a dynamic result of a series of different light sources, such as clear daytime lighting, nighttime street lighting, and other common lighting environments. The reflectance component R represents the key information for humans or computers to understand the semantics of the images. MSR separates the reflectance components to reduce the interference of the dynamic lighting environment with the image semantics. This allows the observers to better understand the image. It is difficult to calculate the reflectance component R directly. By first estimating the illumination L, R can be computed indirectly by $R = I/L$. The MSR can be defined as follows:

$$I(x, y, c) = L(x, y, c) \times R(x, y, c) \quad (2)$$

$$R(x, y, c) = I(x, y, c) \ / \ L(x, y, c) \quad (3)$$

$$\log(R(x, y, c)) = \log(I(x, y, c)) - \log(L(x, y, c)) \quad (4)$$

The image is composed of multiple pixels; x and y are the two-dimensional coordinates of the image pixels, and c is the channel of the image. If the image is gray, then c is 1, representing the gray channel. If it is a color image, c is 1, 2, or 3, representing the R, G, and B color channels, respectively. Equation (4) is the logarithmic form of Equation (3).

It is assumed that the illumination component L changes slowly on different objects, while the object reflectance R changes significantly at the edges of the objects. Therefore, the common method is to estimate the slowly changing illumination L by the Gaussian filtering method in the spatial domain. The Gaussian filter is used to estimate the illumination by calculating a weighted average of a pixel and its surrounding pixels. L can be estimated as follows:

$$L(x, y, c) = I(x, y, c) \times G(x, y, c) \quad (5)$$

$$G(x, y, c) = \frac{1}{2\pi\sigma^2} \exp\left(-\frac{x^2 + y^2}{2\sigma^2}\right) \quad (6)$$

The parameter σ is a key parameter of the Gaussian filter, which determines the filtering scale when estimating the illumination. Selecting a large value is not conducive to local illumination estimation. A small value of σ would defeat the original purpose of the hypothesis and would not be conducive to estimating the overall illumination. Therefore, MSR estimates the illumination by using three different scales: large, medium, and small. The accurate illumination is determined by the weighted average value.

3.2. HE Image Enhancement

The grayscale distribution histograms of over-light or over-dark images are concentrated in the area of high or low brightness, respectively. The grayscale distribution histograms of images with normal lighting are evenly distributed within the overall gray value range. HE mainly uses the CDF (cumulative distribution function) to shift the gray/brightness of the image to ensure that it is distributed uniformly within the overall gray value range, which is similar to that of a normal lighting image.

For the original gray image $I(x, y)$, there are N pixels whose value range is $[P_{min}, P_{max}]$. The brightness is divided into L discrete levels with a range of $[0, L-1]$. The original histogram of the image is obtained by (7). CDF is defined by (8).

$$H(k) = \frac{n_k}{N}, \quad for\ 0 \le k \le L-1 \quad (7)$$

$$CDF(k) = \sum_{i=0}^{k} H(k) \quad (8)$$

where H (k) is the PDF (probability density function) of the pixel with a brightness of k, and also the histogram height of the pixel with a brightness of k, while n_k is the number of pixels with a brightness of k.

$$P_{out} = \mathrm{CDF}(P_{in}) \times (L - 1) \tag{9}$$

HE pixel brightness mapping is defined in (9). P_{in} is the input pixel brightness, while P_{out} is the output brightness of the corresponding pixel. For the color images, the three channels (RGB) can be enhanced by the above HE. The grayscale distribution before and after HE enhancement is shown in Figure 3.

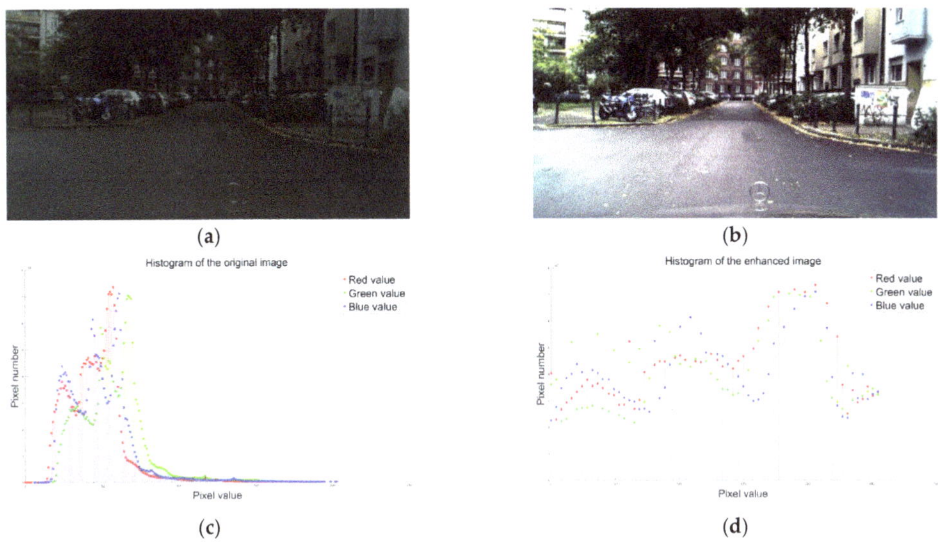

Figure 3. HE enhancement demo: (**a**) Original image. (**b**) HE-enhanced image. (**c**) RGB histogram of the original image. (**d**) RGB histogram of the HE-enhanced image.

The original night image has low brightness, and its pixel brightness is concentrated in a small range, resulting in poor visibility. After the enhancement, the image brightness is evenly distributed in the value area, and the overall image brightness increases noticeably.

3.3. Image Fusion

MSR can separate the object reflectance components of the image and, thus, retain edge information. The visibility of the MSR-enhanced image is limited, as MSR eliminates the illumination components and only keeps the reflectance components. HE directly modifies the gray value of pixels to achieve better enhancement in illumination, but it also introduces high-frequency noise to the enhanced image. Direct conversion on the gray level will also cause the loss of edge information, which is the key information for semantic segmentation in autonomous driving. The image illumination and color information are mainly in the low-frequency range, while the edge information is mainly in the high-frequency range. The enhancement effect of MSR is more remarkable in the high-frequency range, but it is not stable in the low-frequency range. Conversely, HE can effectively enhance the low-frequency information, but it also causes high-frequency noise and loss of edge information—mainly located in the high-frequency range of the image. The proposed HE-MSR-COM combines the above two methods by using DCT to generate high-quality images that include the high-frequency information from MSR and the low-frequency information from HE.

The proposed HE-MSR-COM uses the high-frequency information of the MSR-enhanced image to obtain the clear edge information and uses the low-frequency information of the HE-enhanced image to obtain the enhanced illumination. HE-MSR-COM overcomes the disadvantages of MSR-enhanced images, such as halo and poor visibility. It also overcomes the shortcomings of HE-enhanced images, such as blurred edges and high-frequency noise. The fusion of MSR and HE processes in the proposed HE-MSR-COM is defined in (10).

$$I_{out} = IDCT(\alpha(I) \times DCT(I_{MSR}) * mask_{MSR} + \beta(I_{HE}) \times DCT(I_{HE}) * mask_{HE}) \tag{10}$$

where I_{out} is the output enhanced image; I_{MSR} and I_{HE} denote the MSR- and HE-enhanced images, respectively; $DCT\ (\cdot)$ is the discrete cosine transform, and $IDCT\ (\cdot)$ is the inverse discrete cosine transform; $mask_{MSR}$ is the high-pass filter, and $mask_{HE}$ is the low-pass filter; $*$ represents the multiplication of the corresponding positions of two matrices of the same size; $\alpha(I)$ is an edge-adaptive coefficient, which is a function of the original input image I; $\beta(I_{HE})$ is an adaptive coefficient as a result of a function of the image illumination.

The frequency-domain diagram after DCT transformation is shown in Figure 4. The low-frequency information is concentrated near the origin of the coordinates, and the high-frequency information is distributed in other areas. The frequency-domain filter design is shown in Figure 5.

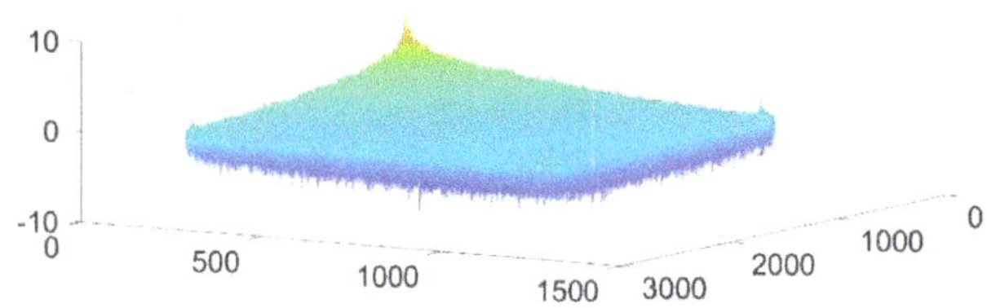

Figure 4. DCT transform spectrum diagram.

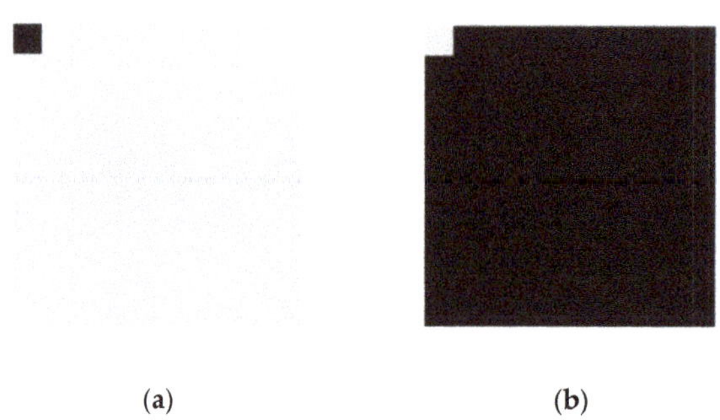

Figure 5. The filter is a logical matrix of the same size as the original image. The matrix element value of the gray part is 1, and the matrix element value of the black part is 0. (**a**) High−pass filter. (**b**) Low−pass filter.

The mean gradient is an evaluation of edge information, defined as follows:

$$g(I) = \frac{1}{(M-1) \times (N-1)} \times \sum_{i=1}^{M-1} \sum_{j=1}^{N-1} \sqrt{\frac{(I(i,j) - I(i+1,j))^2 + (I(i,j) - I(i,j+1))^2}{2}} \tag{11}$$

where M, N define the size of the image, I is the image, and i, j are the coordinates of the pixels.

$\alpha(I)$ is determined by the edge information of the original image, and it is defined as follows:

$$\alpha(I) = \alpha \times g(I)/mean(g) \quad (12)$$

where $mean(g)$ represents the mean gradient values of the selected images in the dataset, and $g(I)$ is the mean gradient of the current image; α is an adjustable parameter in a range of 0.8–1.2. If α is too small, the edge information will be lost. If α is too large, the edge of the object will be too bright, and the enhanced image will not be natural.

$\beta(I_{HE})$ is determined by the HE-enhanced image. It is used to adjust for excessive enhancement effects that HE may bring. It is defined as follows:

$$\beta(I_{HE}) = \beta \times mean\left(mid\left(I_{day}\right)\right)/mid(I_{HE}) \quad (13)$$

where $mid(\cdot)$ is the median brightness of an image, which is a statistical function to reasonably judge the brightness distribution of an image. I_{day} is a subset of the normal illuminated images in the dataset. The subset can be selected manually from daylight images or automatically selected according to the calculated brightness values of the images. β is an adjustable parameter in a range of 0.7–1.0. The image is over-dark if β is less than 0.7, and over-bright if β is larger than 1, which both degrade the image's visibility. HE tends to have excessive enhancement, so a value less than 1 is generally selected. γ is a mean memory parameter that is used to update $mean(g)$ and $mean\left(mid\left(I_{day}\right)\right)$ with an additive contribution rate of the current image. If γ is too small, the mean values change slowly and reduce the adaptability, and if γ is too large, the enhanced performance becomes unstable.

The filter parameters of $mask_{MSR}$ and $mask_{HE}$ are mainly determined by prior knowledge of the dataset that contains both day and night images.

Algorithm 1 HE-MSR-COM

Input: Low-light input image I;
Output: Enhanced image I_{out};
Initialization:
$mean(g)$ is the mean gradient obtained from the sampling data of the dataset;
I_{day} samples from selected normal lighting images;
Mean memory parameter γ;
Calculate MSR weight parameter α, HE weight parameter β;
Dataset sampling to obtain prior filter parameters $mask_{MSR}$, $mask_{HE}$.
1: **while (Input $\neq \varnothing$) do**
2: Update $mean(g)$ by $mean(g) = \gamma \times g(I) + (1 - \gamma) \times mean(g)$
3: **if** (I is normal illumination image) **then**
4: Update $mean\left(mid\left(I_{day}\right)\right)$ by
$mean\left(mid\left(I_{day}\right)\right) = \gamma \times mid(I) + (1 - \gamma) \times mean(mid(I_{day}))$;
5: **else**
6: Estimate initial illumination L via (5), (6);
7: Estimate reflectance R(I_{MSR}) via (4)
8: Obtain HE-enhanced image I_{HE} via (7), (8), (9);
9: Calculate weight parameters $\alpha(I)$ via (12);
10: Calculate weight parameters $\beta(I_{HE})$ via (13);
11: Fuse enhanced image via (10) to obtain I_{out};
12: **end if**
13: **end while**

4. Experiments

4.1. Datasets

From GTA5 [28] and Cityscapes [29], driving images with low light and normal lighting were selected as data sources for the experiment. The GTA5 dataset contains

24,966 high-resolution composite images and is a commonly used dataset for semantic segmentation training in the field of autonomous driving. The Cityscapes dataset consists of 25,000 street images from 50 different cities, collected using different devices under varying lighting conditions.

4.2. Evaluation Metrics

There are two main ways to evaluate the performance of enhanced images: subjective evaluation and objective evaluation. Subjective evaluation is based on human vision and involves human interaction. Objective evaluations are performed by different defined mathematical metrics based on image information. In this paper, entropy, mean gradient, PSNR (peak signal-to-noise ratio), and contrast ratio are used to evaluate the enhanced image.

Entropy is a common objective metric of image quality evaluation. It reflects the richness of an image. In general, the greater the entropy of the image, the richer the information, and the better the quality. It is defined as follows:

$$\mathrm{E}(I) = -\sum_{i=0}^{L-1} P(i) \times \log_2(P(i)) \tag{14}$$

where $P(i)$ is the probability of the pixels with gray level of i in the image, and L is the pixel's gray dispersion level—generally 256.

PSNR is used to measure the distortion degree of the enhanced image. The larger the PSNR, the more semantic information the enhanced image retains and the less noise it introduces. It is defined as follows:

$$\mathrm{MSE} = \frac{1}{M \times N} \sum_{i=0}^{M-1} \sum_{j=0}^{N-1} [I(i,j) - K(i,j)]^2 \tag{15}$$

$$\mathrm{PSNR} = 20 \times \log_{10}\left(\frac{MAX_I}{\sqrt{MSE}}\right) \tag{16}$$

where M, N represent the size of the image. I is the original image, and i, j are the coordinates of the pixels. K is the enhanced image. MAX_I is the maximum pixel value—for general RGB images, it is 255.

Contrast ratio usually shows the sharpness of an image. The higher the contrast, the higher the resolution of the image. It is defined as follows:

$$\mathrm{C(I)} = \sum_{\delta} \delta(i,j)^2 P_\delta(i,j) \tag{17}$$

where $\delta(i,j)$ is the gray difference between adjacent pixels, and $P_\delta(i,j)$ is the probability of pixels with a gray difference of δ.

CE (comprehensive evaluation): for the above four evaluation metrics, the maximum value is taken as 100%, and the CE of each algorithm is calculated. It is defined as follows:

$$\mathrm{CE}(I) = \left(\frac{E(I)}{MAX_E} + \frac{g(I)}{MAX_g} + \frac{PSNR(I)}{MAX_{PSNR}} + \frac{C(I)}{MAX_C}\right)/4 \tag{18}$$

4.3. Experimental Results and Analysis

Frequency components in different spectral ranges are separated from the dataset, and their corresponding mean gradients are calculated. The experimental results are shown in Table 1. Thus, the values of the filter parameters of $mask_{MSR}$ and $mask_{HE}$ are adjusted.

Table 1. Mean gradients of different frequency components.

Spectrum	Mean Gradient
[0, 4)	0.022
[2, 8)	0.062
[4, 16)	0.145
[8, 32)	0.292
[16, 64)	0.550
[32, 128)	0.982
[64, 256)	1.574
[128, 512)	2.151
[256, 1024)	2.322
[1024, 2048)	0.078

Filter parameters' selection ranges are determined based on the algorithm characteristics of HE and MSR. HE has advantages in low-frequency information enhancement, while MSR is better at high-frequency information enhancement. Therefore, the basic parameter selection ranges can be determined, and the violation of this range will lead to poor enhancement results.

It can be noted from Table 1 that the edge information is mainly concentrated in the frequency range (16, 1024), so $mask_{MSR}$ selects these image frequency components in this range to obtain the edge information of the image, while $mask_{HE}$ selects the low-frequency component of the image (0, 16) for generating the enhanced illumination information of the image.

The PSNR of the original HE-enhanced image was compared with that of the filtered HE-enhanced image. As shown in Figure 6, the ordinate is the filtered PSNR minus the original PSNR. The PSNR of the filtered image is larger than that of the original image on both datasets. This proves that a larger PSNR of the image can be obtained by using frequency filtering of the high-frequency noise introduced by the HE enhancement.

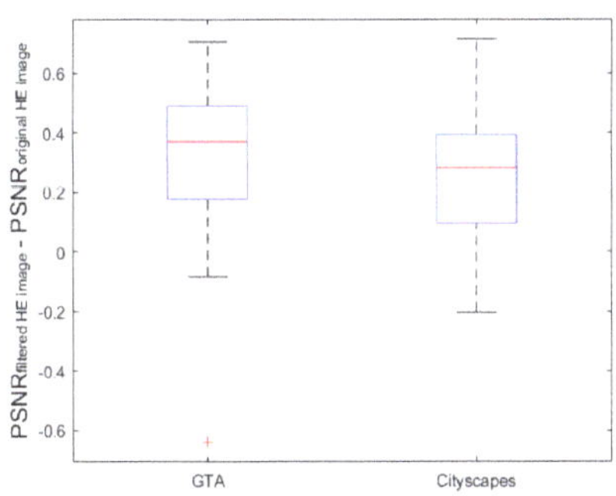

Figure 6. PSNR comparison between the filtered HE−enhanced image and the original HE−enhanced image.

The next step of the proposed method is combining the filtered HE-enhanced result with the MSR-enhanced result using a set of designed weights. Therefore, using a HE-enhanced image with a higher PSNR contributes to a better final result after fusing with the MSR result. The original HE-enhanced image is not used in the subsequent fusion process.

More experiments are underway to tackle more accurate selection of the filter parameters and the weight parameters (α and β) for a better performance by using optimization algorithms, such as genetic algorithms. These results will be presented in a subsequent paper.

α is an adjustable weight of edge information. The larger α is, the larger the weight of the edge information will be. β is used to adjust the enhanced brightness. β values are generally 0.7–1.0, because HE tends to produce over-bright enhanced results. Five criteria have been adopted to study the performance of different selections of α and β, as illustrated in Table 2.

Table 2. Enhancement performance of different parameter combinations.

	Entropy	Mean Gradient	PSNR	Contrast Ratio	CE
$\alpha = 0.8,\ \beta = 0.7$	7.30	3.59	**63.71**	65.25	77.53%
$\alpha = 1.0,\ \beta = 0.7$	7.35	4.44	63.48	100.06	87.89%
$\alpha = 1.2,\ \beta = 0.7$	7.38	**5.24**	63.17	**139.12**	**98.69%**
$\alpha = 0.8,\ \beta = 0.85$	7.45	3.60	61.82	64.95	77.27%
$\alpha = 1.0,\ \beta = 0.85$	7.50	4.43	61.68	98.92	87.42%
$\alpha = 1.2,\ \beta = 0.85$	7.52	5.21	61.50	136.43	97.86%
$\alpha = 0.8,\ \beta = 1.0$	7.70	3.59	58.83	63.17	76.51%
$\alpha = 1.0,\ \beta = 1.0$	7.72	4.36	58.78	94.31	85.82%
$\alpha = 1.2,\ \beta = 1.0$	**7.73**	5.07	58.73	127.94	95.24%

The experimental results show that when $\alpha = 1.2,\ \beta = 0.7$, it can obtain the optimal CE on the Cityscapes dataset. Thus, the two weights $\alpha = 1.2, \beta = 0.7$ were selected for the follow-up experiments. The mean memory parameter γ was selected as 0.02 for all of the above experiments. The definitions of α and β take into account the content differences between different images in the dataset. Using adaptive weights makes the enhancement results more stable across different images. Ordinary HE and MSR algorithms can be regarded as methods with weights of 0 or 1, so α and β values close to 1 were selected to prevent over-enhancement. Since HE tends to over-enhance, three discrete values of less than or equal to 1 were selected for β. The high-frequency information processed by MSR can allow for a larger range in selecting α values. Therefore, three groups of representative values of alpha and beta were selected in the experiment. Other parameter values may result in better performance, which will be further studied in our future work. This paper mainly shows that HE and MSR can obtain better enhancement results in frequency-domain combination. CE is a comprehensive consideration of a variety of indicators, so it was selected as the primary evaluation factor. Entropy sometimes cannot accurately represent the image quality. For example, HE usually equally distributes the pixel values, which can theoretically produce the maximum entropy, but there is still room for improvement in the HE enhancement results. The CE result of the selected combination of α and β was extremely close to the highest PSNR result, with a difference of only 0.85%.

The enhancement results of different algorithms—HE, CLAHE, MSR, GC, and HE-MSR-COM (our algorithm)—were compared. The enhanced performance is shown in Figure 7.

All of the methods are implemented via MATLAB programming. MSR and GC were implemented by code in MATLAB. HE and CLAHE were implemented by calling histeq () and adapthisteq (), respectively, using library functions provided by MATLAB.

The HE-enhanced image has blurred edges, and the image brightness is over-enhanced. CLAHE makes up for HE's over-enhancement issue, but there is still a loss of edge information. MSR enhancement produces sharp edges, but their visibility depends on time-consuming manual adjustment of parameters. The image enhancement results are not stable when using the same parameters for different images. GC directly maps the pixel values nonlinearly. It maps over-light or over-dark pixels to medium brightness. GC achieves good visibility and robust enhancement in brightness, but the image edge information has a great loss and suffers from a foggy effect. The proposed HE-MSR-COM retains the advantages of HE in better adaptive illumination enhancement and those of MSR in adaptive edge information enhancement. The highest visual enhancement performance was obtained in the above evaluations.

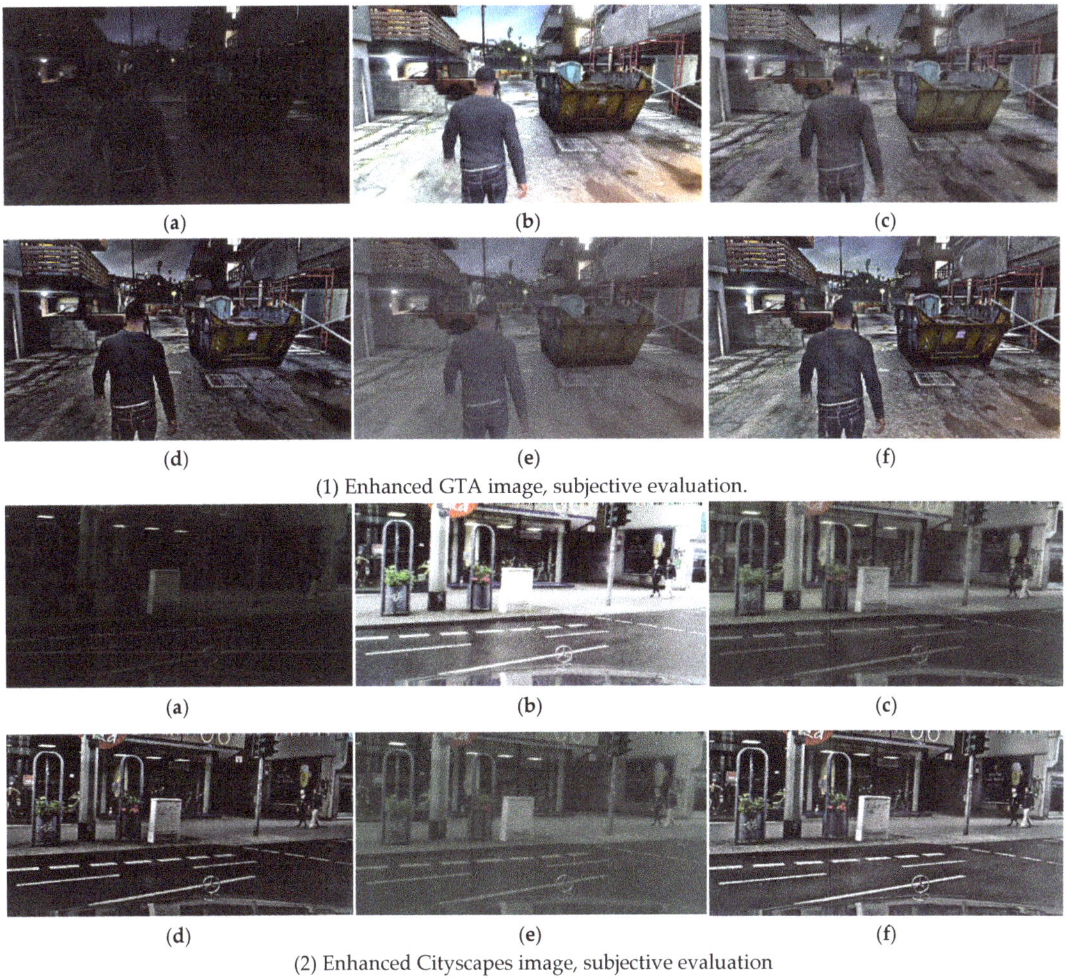

Figure 7. Results comparison: (**a**) original image; (**b**) HE; (**c**) CLAHE; (**d**) MSR; (**e**) GC; (**f**) HE-MSR-COM.

The performance comparisons of different algorithms on the GTA5 dataset are shown in Table 3 and Figure 8.

Table 3. Mean performance comparison in GTA5.

Method	Original	HE	CLAHE	MSR	GC	HE-MSR-COM
Entropy	6.52	**7.80**	7.24	3.47	6.81	7.48
Mean Gradient	2.25	5.85	5.34	4.12	2.40	**9.88**
PSNR	Inf [1]	57.60	**61.83**	61.70	60.14	61.45
Contrast Ratio	47.14	200.15	140.64	242.51	38.78	**424.19**
CE	54.40%	74.89%	70.02%	60.79%	54.51%	**98.84%**

[1] The PSNR score of the original image was deemed to be 100%.

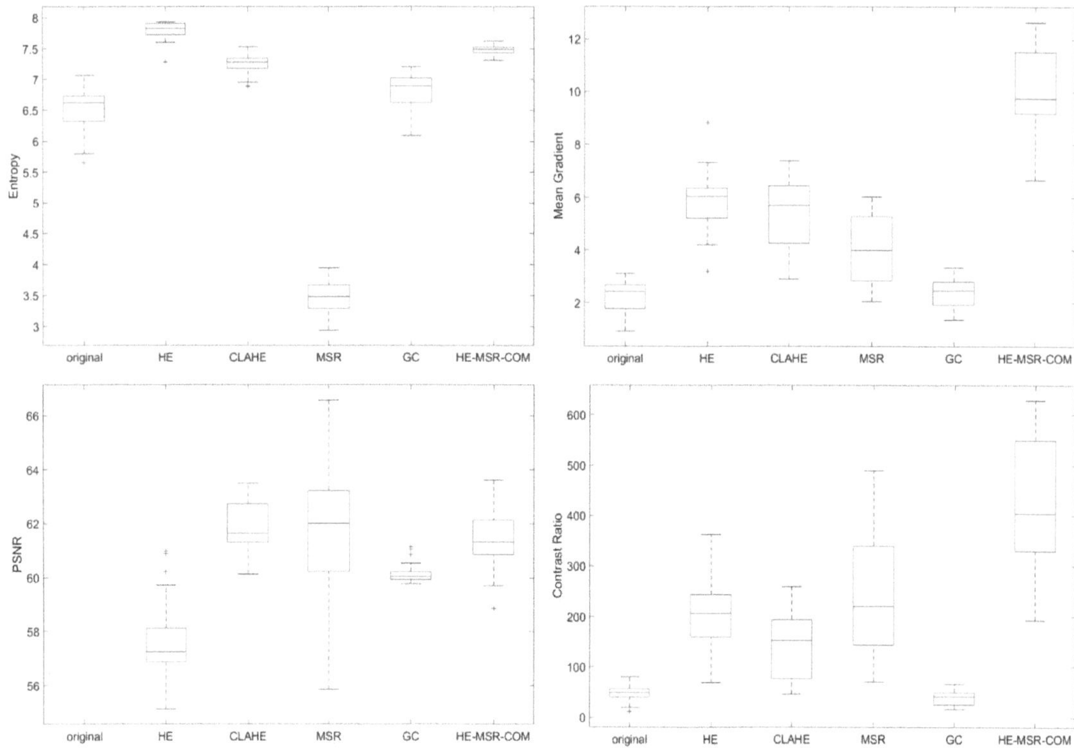

Figure 8. Evaluation distribution in GTA5.

Figure 8 presents the results of five different image enhancement methods under four evaluation metrics, including entropy, mean gradient, PSNR, and contrast ratio. The "+" symbols represent the outliers. The upper and lower boundaries of the dashed line are the maximum and minimum values, respectively. The upper and lower boundaries of the box are the upper and lower quartiles, respectively, and the inner red line is the median value. The boxplots can adequately represent the distribution of enhanced image performance. As shown in Figure 8, HE achieves the maximum entropy by evenly distributing the grayscale of the pixels. HE-MSR-COM retains the advantages of HE in low-frequency information with respect to brightness and color, so HE-MSR-COM has the second-highest score in entropy. HE-MSR-COM magnifies the advantages of MSR by adaptive weight and achieves the highest mean gradient. HE-MSR-COM overcomes the problem of high-frequency noise caused by HE and achieves the second-highest PSNR, which is almost equal to the highest PSNR. HE-MSR-COM has the highest average value in contrast ratio, but it also brings large variance. However, the subjective evaluation shows that the contrast ratio of HE-MSR-COM is significantly enhanced, and it receives the highest score in the CE, as indicated in Table 3. HE-MSR-COM shows optimized performance in many evaluation metrics on the GTA5 dataset.

The performance comparison of different algorithms on the Cityscapes dataset is shown in Table 4 and Figure 9.

Table 4. Mean performance comparison in Cityscapes.

Method	Original	HE	CLAHE	MSR	GC	HE-MSR-COM
Entropy	6.55	**7.85**	7.25	3.42	6.66	7.48
Mean Gradient	1.98	5.81	4.71	2.93	2.05	**6.88**
PSNR	Inf [1]	58.34	62.50	60.52	59.76	**63.74**
Contrast Ratio	20.68	135.90	66.42	85.75	16.88	**176.62**
CE	55.98%	88.23%	74.14%	57.42%	54.47%	**98.83%**

[1] The PSNR score of the original image was deemed to be 100%.

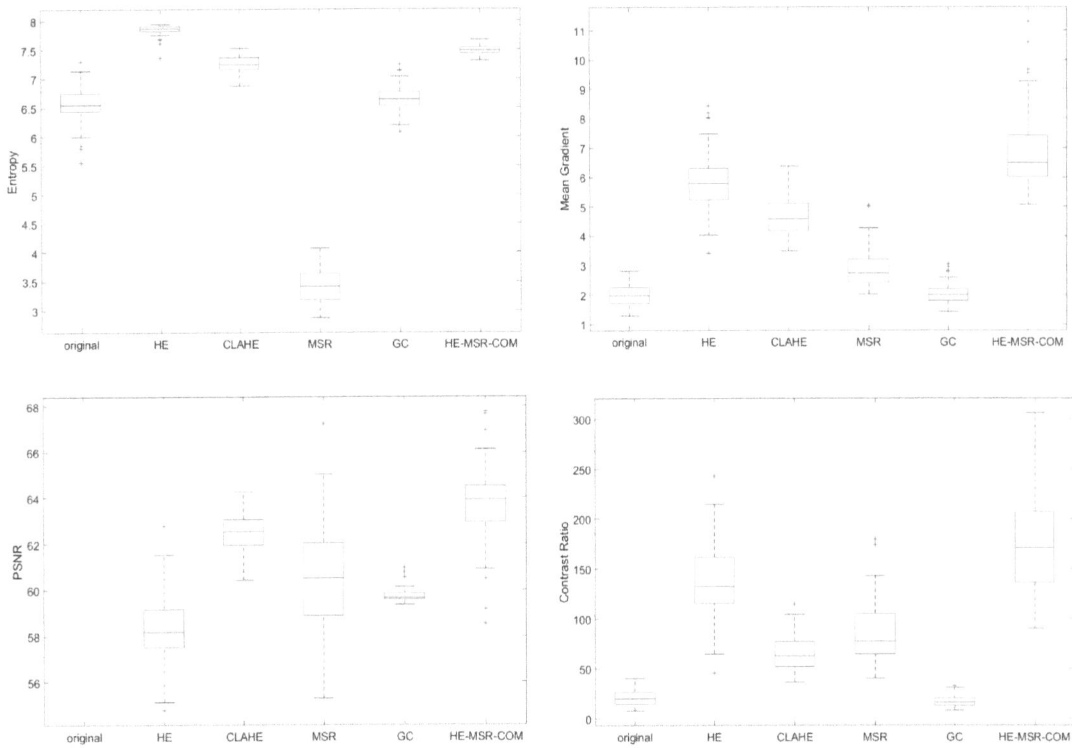

Figure 9. Evaluation distribution in Cityscapes.

As shown in Figure 9, HE and HE-MSR-COM have the highest and second-highest score in entropy, respectively. HE-MSR-COM has the highest mean gradient in all methods. HE-MSR-COM removes noise as much as possible through frequency-domain filtering and obtains the highest PSNR. HE-MSR-COM has the highest average value in contrast ratio, but it also brings large variance. Because the images of Cityscapes come from different devices, the contrast between images is different, and the high variance is consistent with the actual images. Our method also receives the highest score in the CE, as indicated in Table 4. HE-MSR-COM has advanced performance in entropy, mean gradient, PSNR, and contrast ratio in many real scenes of the Cityscapes dataset.

In the above GTA5 and Cityscapes datasets, HE achieved stable enhancement performance in different datasets, but its PSNR was low due to the introduction of high-frequency noise and over-enhancement. MSR requires manual adjustment of parameters to achieve optimal performance. Using the same parameters in different datasets brings performance instability. HE obtained the highest CE score in both datasets—23.95% and 10.6% higher than those of the second-highest method, respectively.

4.4. Discussion

The experiment compared subjective evaluation with objective evaluation. The results reveal that HE is simple and reliable, but this comes at the expense of losing edge information and excessively enhancing brightness. On the other hand, the stability of MSR is low. The optimal enhancements usually require manual adjustment of the parameters independently. It is difficult to adapt fixed parameters for complex realistic conditions. HE-MSR-COM uses HE-enhanced images to obtain good brightness enhancement with an adjustable weight and excellent subjective visual evaluation performance, and MSR is used in the proposed method to obtain the edge information, achieving good performance under the different metrics such as contrast ratio and mean gradient. Frequency-domain processing can effectively combine the advantages of HE and MSR, as well as avoiding the problems of high-frequency noise caused by HE and unstable brightness in the low-frequency domain of MSR. Based on the experimental performance, HE-MSR-COM showed stable and superior enhancement performance on different datasets. HE-MSR-COM is simple, reliable, and efficient, and it can be used as a preprocessing module for low-illumination images for most visual algorithms.

5. Conclusions

This paper proposes a method combining HE with MSR, called HE-MSR-COM. The image enhancement method proposed in this paper focuses on enhancing the low-illumination image, which is used as an image preprocessing module. When the image is collected by the autopilot system, it is first processed by the image enhancement before it is taken as an input for the subsequent autonomous driving visual tasks, such as semantic segmentation, target detection, etc. We aim to improve performance in the visual tasks of autonomous driving by providing a higher quality of the visual image. Our experiments showed that HE-MSR-COM has the advantages of both HE and MSR, enabling it to achieve higher performance and balance in the overall illumination and edge details. The HE-MSR-COM night image enhancement algorithm has two advantages: (1) The enhanced illumination component is obtained from HE-enhanced image. The low-pass filter in the frequency domain retains the advantage of enhanced illumination and removes the high-frequency noise. This successfully ensures good adaptive illumination. (2) The enhanced reflectance component is obtained from the MSR-enhanced image. The high-pass filter in the frequency domain retains the advantage of enhanced edge information. This successfully reserves more semantic information. HE-MSR-COM achieves excellent night image enhancement performance. It can be embedded into common visual algorithms for autonomous driving to improve their visual detection performance in night scenes.

In the future, HE-MSR-COM will be deployed in autonomous driving semantic segmentation networks. The night image enhancement can be further evaluated and optimized by combining it with practical autonomous driving visual algorithms in real night scenes.

Author Contributions: Conceptualization, X.C. and Y.H. (Yi Han); methodology, Y.H. (Yi Han), X.C., and Y.Z.; software, X.C., Y.H. (Yi Han), and Y.Z.; validation, X.C., P.H., Y.H. (Yanqing Huang), and Z.L.; formal analysis, X.C., Y.Z., Y.H. (Yanqing Huang), and Z.L.; investigation, X.C., Z.L., Z.Y., and Q.L.; resources, X.C., P.H., and Y.H. (Yi Han); data curation, X.C., Y.H. (Yi Han), and Z.L.; writing—original draft preparation, X.C. and Y.H. (Yi Han); writing—review and editing, Y.H. (Yi Han), Z.Y., and Q.L.; visualization, X.C. and Y.H. (Yi Han); supervision, Y.H. (Yi Han), P.H., and Z.Y.; project administration, Y.Z.; funding acquisition, Y.Z. All authors have read and agreed to the published version of the manuscript.

Funding: This work was supported by a grant from the National Natural Science Foundation of China (Grant No. 61801341). This work was also supported by the Research Project of Wuhan University of Technology Chongqing Research Institute (No. YF2021-06).

Institutional Review Board Statement: Not applicable.

Informed Consent Statement: Not applicable.

Data Availability Statement: The simulation data used to support the findings of this study are available from the corresponding author upon request.

Conflicts of Interest: The authors declare no conflict of interest.

References

1. Wang, W.; Wu, X.; Yuan, X.; Gao, Z. An experiment-based review of low-light image enhancement methods. *IEEE Access* **2020**, *8*, 87884–87917. [CrossRef]
2. Sobbahi, R.A.; Tekli, J. Comparing deep learning models for low-light natural scene image enhancement and their impact on object detection and classification: Overview, empirical evaluation, and challenges. *Signal Process. Image Commun.* **2022**, *109*, 116848. [CrossRef]
3. Land, E.H. The retinex theory of color vision. *Sci. Am.* **1977**, *237*, 108–129. [CrossRef] [PubMed]
4. Land, E.H.; McCann, J.J. Lightness and retinex theory. *JOSA* **1971**, *61*, 1–11. [CrossRef] [PubMed]
5. Wei, C.; Wang, W.; Yang, W.; Liu, J. Deep retinex decomposition for low-light enhancement. *arXiv* **2018**, *preprint*. arXiv:1808.04560.
6. Gonzalez, R.C.; Woods, R.E. *Digital Image Processing*, 2nd ed.; Prentice Hall: Upper Saddle River, NJ, USA, 2002; p. 793.
7. Dhal, K.G.; Das, A.; Ray, S.; Gálvez, J.; Das, S. Histogram equalization variants as optimization problems: A review. *Arch. Comput. Methods Eng.* **2021**, *28*, 1471–1496. [CrossRef]
8. Sande, K.V.D.; Gevers, T.; Snoek, C. Evaluating color descriptors for object and scene recognition. *IEEE Trans. Pattern Anal. Mach. Intell.* **2009**, *32*, 1582–1596. [CrossRef]
9. Jobson, D.J.; Rahman, Z.; Woodell, G.A. Properties and performance of a center/surround retinex. *IEEE Trans. Image Process.* **1997**, *6*, 451–462. [CrossRef]
10. Jobson, D.J.; Rahman, Z.; Woodell, G.A. A multiscale retinex for bridging the gap between color images and the human observation of scenes. *IEEE Trans. Image Process.* **1997**, *6*, 965–976. [CrossRef]
11. Wang, P.; Wang, Z.; Lv, D.; Zhang, C.; Wang, Y. Low illumination color image enhancement based on Gabor filtering and Retinex theory. *Multimed. Tools Appl.* **2021**, *80*, 17705–17719. [CrossRef]
12. Yang, X.; Jian, L.; Wu, W.; Liu, K.; Yan, B.; Zhou, Z.; Peng, J. Implementing real-time RCF-Retinex image enhancement method using CUDA. *J. Real-Time Image Process.* **2019**, *16*, 115–125. [CrossRef]
13. Tao, F.; Yang, X.; Wu, W.; Liu, K.; Zhou, Z.; Liu, Y. Retinex-based image enhancement framework by using region covariance filter. *Soft Comput.* **2018**, *22*, 1399–1420. [CrossRef]
14. Kim, Y.T. Contrast enhancement using brightness preserving bi-histogram equalization. *IEEE Trans. Consum. Electron.* **1997**, *43*, 1–8.
15. Chen, S.D.; Ramli, A.R. Minimum mean brightness error bi-histogram equalization in contrast enhancement. *IEEE Trans. Consum. Electron.* **2003**, *49*, 1310–1319. [CrossRef]
16. Ooi, C.H.; Kong, N.S.P.; Ibrahim, H. Bi-histogram equalization with a plateau limit for digital image enhancement. *IEEE Trans. Consum. Electron.* **2009**, *55*, 2072–2080. [CrossRef]
17. Ooi, C.H.; Isa, N.A.M. Quadrants dynamic histogram equalization for contrast enhancement. *IEEE Trans. Consum. Electron.* **2010**, *56*, 2552–2559. [CrossRef]
18. Salah-ELDin, A.; Nagaty, K.; ELArif, T. An enhanced histogram matching approach using the retinal filter's compression function for illumination normalization in face recognition. In Proceedings of the International Conference Image Analysis and Recognition, Póvoa de Varzim, Portugal, 25–27 June 2008; pp. 873–883.
19. Tan, T.L.; Sim, K.S.; Tso, C.P. Image enhancement using background brightness preserving histogram equalisation. *Electron. Lett.* **2012**, *48*, 155–157. [CrossRef]
20. Shome, S.K.; Vadali, S.R.K. Enhancement of diabetic retinopathy imagery using contrast limited adaptive histogram equalization. *Int. J. Comput. Sci. Inf. Technol.* **2011**, *2*, 2694–2699.
21. Lin, S.C.F.; Wonga, C.Y.; Rahman, M.A.; Jiang, G.; Liu, S.; Kwoka, N.; Shi, H.; Yu, Y.H.; Wu, T. Image enhancement using the averaging histogram equalization (AVHEQ) approach for contrast improvement and brightness preservation. *Comput. Electr. Eng.* **2015**, *46*, 356–370. [CrossRef]
22. Chen, Y.; Kang, J.U.; Zhang, G.; Zhang, G.; Cao, J.; Xie, Q.; Kwan, C. Real-time infrared image detail enhancement based on fast guided image filter and plateau equalization. *Appl. Opt.* **2020**, *59*, 6407–6416. [CrossRef]
23. Kwan, C.; Gribben, D. Target Detection and Classification Improvements using Contrast Enhanced 16-bit Infrared Videos. *Signal Image Process. Int. J. (SIPIJ)* **2021**, *12*. [CrossRef]
24. Liao, K.C.; Wu, H.Y.; Wen, H.T. Using Drones for Thermal Imaging Photography and Building 3D Images to Analyze the Defects of Solar Modules. *Inventions* **2022**, *7*, 67. [CrossRef]
25. Lore, K.G.; Akintayo, A.; Sarkar, S. LLNet: A deep autoencoder approach to natural low-light image enhancement. *Pattern Recognit.* **2017**, *61*, 650–662. [CrossRef]
26. Jiang, Y.; Gong, X.; Liu, D.; Cheng, Y.; Fang, C.; Shen, X.; Yang, J.; Zhou, P.; Wang, Z. Enlightengan: Deep light enhancement without paired supervision. *IEEE Trans. Image Process.* **2021**, *30*, 2340–2349. [CrossRef] [PubMed]

27. Qu, H.; Yuan, T.; Sheng, Z.; Zhang, Y. A pedestrian detection method based on yolov3 model and image enhanced by retinex. In Proceedings of the International Congress on Image and Signal Processing, BioMedical Engineering and Informatics (CISP-BMEI), Beijing, China, 13–15 October 2018; pp. 1–5.
28. Richter, S.R.; Vineet, V.; Roth, S.; Koltun, V. Playing for data: Ground truth from computer games. In Proceedings of the European Conference on Computer Vision (ECCV), Amsterdam, The Netherlands, 11–14 October 2016; pp. 102–118.
29. Cordts, M.; Omran, M.; Ramos, S.; Rehfeld, T.; Enzweiler, M.; Benenson, R.; Franke, U.; Roth, S.; Schiele, B. The cityscapes dataset for semantic urban scene understanding. In Proceedings of the Conference on Computer Vision and Pattern Recognition (CVPR), Las Vegas, NV, USA, 26–30 June 2016; pp. 3213–3223.

electronics

MDPI

Article

Closed-Loop Residual Attention Network for Single Image Super-Resolution

Meng Zhu [1] and Wenjie Luo [1,2,*]

[1] School of Cyber Security and Computer, Hebei University, Baoding 071002, China; lwj12111@hbu.edu.cn
[2] Hebei Machine Vision Engineering Research Center, Hebei University, Baoding 071002, China
* Correspondence: luowenjie@hbu.edu.cn;

Abstract: Recent research on single image super-resolution (SISR) using convolutional neural networks (CNNs) with the utilization of residual structures and attention mechanisms to utilize image features has demonstrated excellent performance. However, previous SISR techniques mainly integrated extracted image features within a deep or wide network architecture, ignoring the interaction between multiscale features and the diversity of features. At the same time, SISR is also a typical ill-posed problem in that it allows for several predictions for a given LR image. These problems limit the great learning ability of CNNs. To solve these problems, we propose a closed-loop residual attention network (CLRAN) to extract and interact with all the available diversity of features features efficiently and limit the space of possible function solutions. Specifically, we design an enhanced residual attention block (ERA) to extract features, and it dynamically assigns weight to the internal attention branches. The ERA combines multi-scale block (MSB) and enhanced attention mechanism (EAM) base on the residual module. The MSB adaptively detects multiscale image features of different scales by using different 3 × 3 convolution kernels. The EAM combines multi-spectral channel attention (MSCA) and spatial attention (SA). Therefore, the EAM extracts different frequency component information and spatial information to utilize the diversity features. Furthermore, we apply the progressive network architecture and learn an additional map for model monitoring, which forms a closed-loop with the mapping already learned by the LR to HR function. Extensive experiments demonstrate that our CLRAN outperforms the state-of-the-art SISR methods on public datasets for both ×4 and ×8, proving its accuracy and visual perception.

Keywords: image super-resolution; attention mechanism; convolutional neural networks; deep learning

Citation: Zhu, M.; Luo, W. Closed-Loop Residual Attention Network for Single Image Super-Resolution. *Electronics* **2022**, *11*, 1112. https://doi.org/10.3390/electronics11071112

Academic Editor: Gemma Piella

Received: 7 March 2022
Accepted: 30 March 2022
Published: 31 March 2022

1. Introduction

Single image super-resolution (SISR) refers to the technology of reconstructing an underlying high-resolution (HR) image from a single low-resolution (LR) image of the scene. It is known as a typical ill-posed problem, as several HR outputs may correspond to the input LR image. To tackle this inverse problem, numerous algorithms have been proposed. According to the three tier classification of [1], SISR algorithms can be divided into two types: learning methods [2–4] and reconstruction methods [5,6]. The SISR algorithms based on deep learning try to hallucinate the missing details of the super-resolution (SR) images. The methods based on the reconstruction requires the degradation model and explicit prior information to define constraints for the target HR image.

In recent years, numerous studies based on deep learning methods with utilization of residual structures and attention mechanisms have demonstrated outstanding performance in SISR challenges. Dong et al. [7] proposed a super-resolution convolutional neural network (SRCNN) in 2014, which is the first successful effort at introducing CNN with its three convolution layers into SISR. Subsequently, a number of CNN-based SISR models have been proposed to learn the mapping between LR and HR images. Ledig et al. [8] proposed SRResNet, which introducing residual learning to train deep network in SISR.

Kim et al. [9] proposed VDSR, which inspired by the proposal of the residual network [10] and extended the depth of CNN to twenty layers. Lim et al. [11] proposed a 69-layer model named EDSR to improve the high-frequency details, which was inspired by SRResNet and removed redundant modules and expanded model. The success of EDSR also illustrated the efficiency of network deepening. On this foundation, Zhang et al. [12] proposed a 400-layer model named RCAN, which combined the residual structure with the attention mechanism and achieved state-of-the-art performance. The success of RCAN also illustrated the efficiency of deep network combined residual structure and attention block.

However, there are still some limitations for CNN-based SISR models. First, very deep and very wide SISR networks lead to a huge computational cost, which is difficult to apply in real-world applications. Second, most of the deepened networks with stacked convolution operations neglect the full utilization of the feature information in the LR image.

To tackle these problems, Lai et al. [13] designed a pyramid network in a coarse-to-fine fashion to gradually predict sub-band residuals. Li et al. [14] proposed a multi-scale residual network (MSRN), which is not designed to be very deep and very wide, but employs different kernel sizes (3×3 and 5×5) in two-bypass convolution layers to exploit the multi-scale spatial features. Furthermore, MSRN employed the hierarchical feature fusion (HFF) technique to combine the outputs of all residual blocks, utilizing the intermediate features. MSRN obtained equivalent performance with a 7-times smaller model size than EDSR. Subsequently, Muqeet et al. [15] proposed HRAN, which employed dilated convolution layers with different dilation factors to attain a larger receptive field and exploited the channel and spatial dependencies. HRAN proposed the binarized feature fusion (BFF) structure, considering that the HFF is difficult to integrate the features extracted from the CNN smoothly. Behjati et al. [16] combined channel attention mechanisms with residual blocks following two independent but parallel computing paths to attend to relevant features and preserve higher frequency details. Dense connections were employed in prior work [17], which extended each feature to subsequent features through residual connections. Instead of the residual block, Wang et al. [18] proposed a residual in a residual dense block (RRDB), which combines a multi-layer residual network and a dense connection to improve the perceptual quality of the SR image in deep models. Musunuri et al. [19] employed RRDB to replace the residual block in EDSR, yielding better reconstruction results and achieving perceptual quality. The SISR models based on CNN, which combine multiscale feature extraction and attention mechanisms, have achieved excellent performance. However, most networks do not limit the function space when designing the network. The channel attention may discard relevant details contained in other frequency components, which ignores the diversity of features. Moreover, not all attention mechanisms improve network performance, and attention employed across all levels is inefficient, as also described in [16,20].

In this paper, we propose a novel closed-loop residual attention network (CLRAN) that combines residual structures and attention mechanisms to utilize the multiscale features and the diversity of features. The CLRAN also limits the space of possible functions while learning the mapping from LR to HR. We introduce a progressive framework for the reconstruction from LR to HR. The framework is based on the cascade of deep CNNs to gradually reconstruct the HR image and naturally apply deep supervision simultaneously at each level of CLRAN, and it is easily extended to other upscaling factors. Guo et al. [21] proposed that, ideally, the SR image can be downsampled to obtain the same LR image as the input LR image. With this limitation, it is possible to estimate the underlying downsample kernel and reduce the space of potential functions to learn a more effective map. Therefore, we employ an extra map that the SR image uses to reconstruct the input LR image to limit the potential space. The extra mapping utilizes the features from the process of gradually reconstructing the HR image, which plays a supervising function in our model. Specifically, the CLRAN is trained by the Charbornnier penalty loss function [13] to achieve a better visual SR result.

The framework employs basic architecture block (Basic-CLRAN) to gradually obtain the HR image. To achieve multi-scale SR, we only need to modify the number of Basic-CLRANs. In this way, the parameters are shared between different scales, and the network parameters are reduced in our model. Considering the structure of our model is simple, in Basic-CLRAN, we employ HFF technique rather than BFF technique [15] to combine local multi-scale features and global features.

In Basic-CLRAN, an enhanced residual attention block (ERA) is proposed as the basic building block to interact features between each other and extract the diversity features for more powerful feature representations. The ERA contains a multi-scale feature extraction part and an enhanced attention part. In the multi-scale feature extraction part, we propose the multi-scale block (MSB) to obtain the multiscale image features. Considering the stacking multiple dilated convolutions used in [15] to attain a larger receptive field that caused some pixel information not be utilized in the network, we adopt two 3 × 3 convolutions instead of 5 × 5 convolutions in multi-scale residual block (MSRB) [14] of MSRN and introduce two 1 × 1 convolutions, which not only obtain the same effect, but also reduce parameters and indirectly increase the depth of the network. In the enhanced attention part, motivated by the attention mechanism [22–24], we propose an enhanced attention mechanism (EAM) to improve the interactions of the deep multi-scale features and utilize the diversity features. The EAM mainly contains a multi-spectral channel attention (MSCA) block and spatial attention (SA) block. The MSCA block has the ability to capture other frequency component channel-wise information for more powerful feature representations. The SA block further extracts the spatial information and helps the network discriminate "where" to concentrate the features. Considering the drawback described in [16,20], we design a non-attention branch to concentrate on the information that is ignored by the enhanced attention branch. The weights of the two branches are automatically calculated by introducing an attention dropout module (ADM) [20].

In order to verify the effectiveness of the proposed methods, we propose a closed-loop residual attention network (CLRAN), combining the progressive framework with the Basic-CLRAN. In summary, the main contributions of this paper are threefold:

(1) We propose an extra mapping that limits the potential space with the progressive framework in our model, thus forming a closed loop to enhance the performance of the SR model.

(2) We propose an enhanced residual attention block (ERA). This block is based on the residual structure that fuses features at several scales by introducing the multi-scale block (MSB) and utilizes diversity of features by introducing the enhanced attention mechanism (EAM). The MSB and the EAM also can be employed for feature extraction in other computer vision tasks.

(3) We propose a closed-loop residual attention network (CLRAN). The network extracts diversity of features from the input LR image and integrates them with the features throughout the middle process to obtain high accuracy SR images. By employing a progressive framework, the CLRAN gradually obtains the SR result. At the same time, the network is easily extended to certain upscaling factors by modifying the number of Basic-CLRANs in the progressive framework.

The rest of this paper is organized as follows. In Section 2, related work on image-super resolution and attention mechanisms is introduced. In Section 3, the details of the proposed methods are presented. In Section 4, the experimental process, the results, and analysis of the proposed method on different benchmark datasets are presented. Additionally, the ablation study on the proposed network is presented. Model complexity comparisons are also included. In Section 5, the conclusions of the paper are presented.

2. Related Work

In recent years, with the development of neural networks, the image super-resolution algorithms have made remarkable progress. In order to address the ill-posed issue in SISR, researchers continuously widen and deepen the network. However, only broadening and deepening the network did not achieve the expected significant improvement. Therefore, researchers designed some network structures and learning strategies such as residual networks, recursive networks, dense connections, progressive structure designs, attention mechanisms, and GAN models. In this section, we first describe the related SISR algorithms based on CNNs. We then discuss the attention mechanism.

2.1. CNN-Based Networks

Dong et al. [7] first proposed a shallow three-layer convolutional neural network (SR-CNN) for learning a nonlinear mapping function from LR $\rightarrow$ HR. Subsequently, He et al. [10] proposed a residual learning technique. Ledig et al. [8] proposed SRResNet introducing residual learning to SISR. Kim et al. proposed VDSR [9] with the deep (20 layers) CNN and global residual connection and DRCN [25] with a recursive block to increase the depth without introducing new parameters. Based on DRCN, Tai et al. [26] proposed DRRN combined residual learning and recursive learning. These approaches extract features from an interpolated LR image, which takes much memory and computation time. To address this problem, Dong et al. [27] proposed the FSRCNN, which improves the training speed of SRCNN. Shi et al. [28] proposed ESPCN, designing a sub-pixel convolution layer. Subsequently, numerous networks were proposed to boost the reconstruction performance of HR images. Lim et al. [11] proposed EDSR with an extremely deep and broad network structure that was based on SRResNet and removed unnecessary modules in residual blocks, resulting in considerable promotion. SRDenseNet [17] introduced dense connections [29] in SISR. Tai et al. [30] proposed MemNet, adopting memory blocks consisting of recursive and gate units. RDN [31] employed the dense connections to utilize all the hierarchical features of the convolutional layers. Wang et al. [18] proposed ESRGAN, in which a residual in a residual dense block (RRDB) combined residual blocks, and a dense connection was proposed to improve the perceptual quality of the SR image. Subsequently, Musunuri et al. [19] employed to RRDB replace the residual block in EDSR, yielding better reconstruction results. Recently, some networks have focused on balancing the performance and memory consumption of SISR. Lai et al. [13] proposed LapSRN, which employs the Laplacian pyramid structure to progressively reconstruct the sub-band residuals of the HR image. Ahn et al. [32] proposed CARN, which employs group convolution and learns high-frequency details by locally and globally cascading connections. For multiscale feature extraction techniques, Li et al. [14] proposed MSRN, which employs different kernel size convolution to exploit multiscale spatial features. Muqeet et al. [15] proposed HRAN, which employs different dilation factors dilated convolution layers to exploit the multiscale features.

2.2. Attention-Based Networks

The attention mechanism in deep learning is comparable to the attention mechanism in human vision. It is viewed as a means of biasing the allocation of available computational resources towards the most informative components of a signal [22]. The attention mechanism has recently been widely applied in computer vision tasks such as image classification [33] and image captioning [22]. This mechanism aims to bias the allocation of available resources towards the most informative parts of an input signal [34]. Hu et al. [22] proposed the squeeze-and-excitation (SE) block, which is focused on the channel-to-channel relationship. Woo et al. [24] proposed convolutional block attention module (CBAM), in which channel attention mechanism and spatial attention mechanism are combined. Dai et al. [35] proposed second-order channel attention (SOCA) to adaptively rescale features by considering second-order statistics of features, so the network could focus on more informative features and enhance discriminative learning ability. Qin et al. [23] proposed multi-spectral channel attention by compressing channels in the channel attention mechanism by applying a discrete cosine transform (DCT).

Some researchers have successfully applied attention mechanisms to CNN-based image enhancement methods, especially to SISR. Liu et al. [36] originally proposed employing non-local operations in a recurrent neural network for image restoration. Zhang et al. [12] considered that if all channels of features were treated equally, the network would lack the ability to discriminate and learn, thus proposed a channel attention (CA) mechanism that employed the residual channel attention network (RCAN), in which the features of each channel were adaptively re-scaled by modeling the interdependence between feature channels. Subsequently, some models [15,34,37] that combined channel attention and spatial attention mechanisms were proposed to learn more discriminative features.

Recently, researchers have started to introduce more sophisticated attention mechanisms to further improve the performance of SISR. Liu et al. [38] proposed enhanced spatial attention (ESA), which reduces the number of channels and adopts a larger stride convolution to shrink spatial dimensions, effectively enlarging the receptive field. Inspired by ESA, Muqeet et al. [39] proposed a cost-efficient attention mechanism (CEA) with dilated convolutions to refine the features. Zhao et al. [40] designed PAN, introducing a pixel-wise channel attention to SISR. Mei et al. [41] designed PANet to capture multi-scale feature. Behjati et al. [16] combined channel attention mechanisms with residual blocks following two independent but parallel computational paths, in which features and attention are processed simultaneously.

3. Proposed Method

3.1. Network Architecture

The complete framework of the proposed network is shown in Figure 1. As we have discussed in Section 1, the CLRAN employs a progressive framework by Basic-CLRAN to reconstruct the HR image from the LR image step by step. For 4× SR task, we employ two Basic-CLRANs, in which we obtain 2× SR for each input image. The Basic-CLRAN in Figure 1 is composed of two parts: feature extraction and reconstruction. We set the original LR image (I_{LR}) as the input of the Basic-CLRAN; the shallow feature E_0 is obtained through initial feature extraction with a 3 × 3 convolutional layer

$$E_0 = H^3_{HF}(I_{LR}) \tag{1}$$

where $H^i_{HF}(\cdot)$ denotes the convolution operation and i denotes the size of convolution kernel.

The extracted feature E_0 is sent to the enhanced residual attention feature extraction part with several ERA modules. We denote the proposed the ERA module as $H_{ERA_i}(\cdot)$, given by

$$E_i = H_{ERA_i}(E_0) \tag{2}$$

where $E_i (i \neq 0)$ is the output feature map of the ith ERA module. After enhanced residual attention feature extraction, we introduce HFF structure expressed as follows:

$$F_{DFS} = \omega * [E_0, E_1, E_2, \ldots, E_n] + b \tag{3}$$

where $[E_0, E_1, E_2, \ldots, E_n]$ denotes the connection operation and denotes the input features of reconstruction part.

The extracted features F_{DFS} from the feature extraction are sent to the reconstruction part; the configuration information for the reconstruction module is shown in Table 1. We employ a PixelShuffle [28] layer upsampled to the same dimensions as HR. We use $I_{HR'}$ to denote the final output from the reconstruction module. Therefore, the final output SR image I_{SR} from Basic-CLRAN is expressed as follows:

$$I_{SR} = H_{UP}(I_{LR}) + I_{HR'} \tag{4}$$

where $H_{UP}(\cdot)$ and I_{SR} denote an upsampled module that contains a pixelshuffle layer.

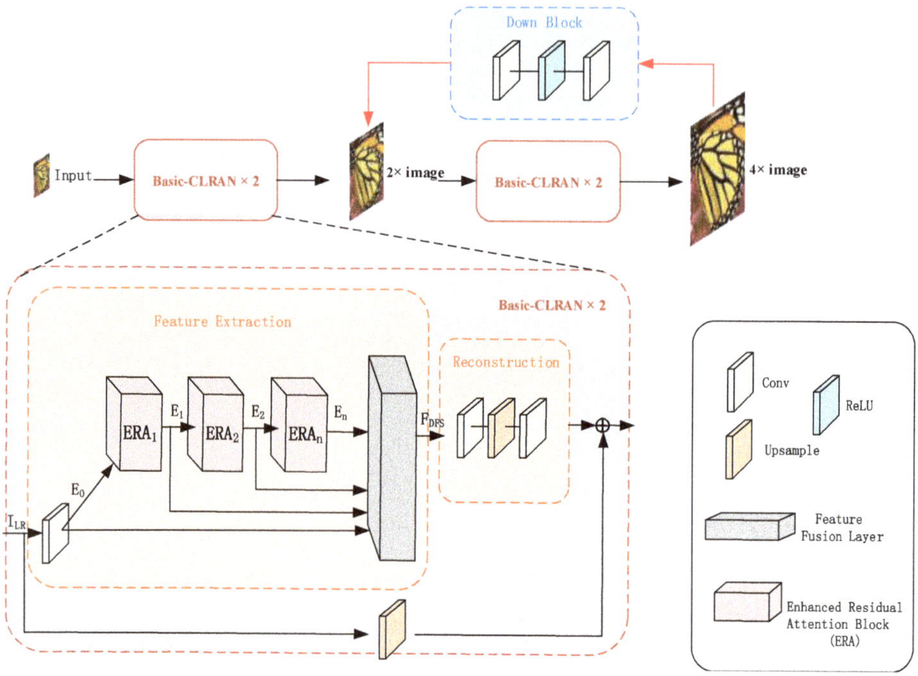

Figure 1. The complete architecture of the closed-loop residual attention network (CLRAN) for 4× SR. The CLRAN contains a primal network (marked with black lines) and a dual regression network (marked with red lines).

Table 1. Detailed configuration information for the reconstruction structure.

Layer	Input Channel	Output Channel	Kernel Size
Input conv	64	64 × 2 × 2	3 × 3
PixelShuffle (×2)	64 × 2 × 2	64	/
Input conv	64	1	3 × 3

In CLRAN, we incorporate progressive architecture into our network. Therefore, for different upscaling factors, we only need to change the number of Basic-CLRANs. The details of our network for different SR tasks are shown in Table 2.

Table 2. The design details for different upscaling factors in our network.

Upscaling Factor	Number of Basic-CLRANs	Upscaling Factor in PixelShuffle	Number of ERAs
×4	2	×2	2
×8	3	×2	2

Loss Function: Different from most networks that have used L1 loss function, we choose the Charbornnier penalty function [13] to train our model. Our ultimate goal is to learn an end-to-end mapping function f from LR $\rightarrow$ HR. However, the space of the possible mapping functions is extremely large, making the function training difficult. Guo et al. [21] provided the derivation of the generalization error bound for the dual regression scheme to prove that introducing dual regression mapping (DRM) to limit the space of the possible mapping functions is effective. Inspired by Guo et al. [21], we learn the primary mapping P for HR reconstruction and the dual regression mapping D for LR reconstruction simultaneously. Given a training dataset $\{I_{LR}^i, I_{HR}^i\}_{i=1}^N$, we address the following problem in our network:

$$\sum_{i=1}^{N} L_P(P(I_{LR}^i), I_{HR}^i) + \lambda L_D(D(P(I_{LR}^i), I_{LR}^i)) \tag{5}$$

where L_P and L_D denote the loss function for the primal mapping and DRM tasks, respectively. The weight of the DRM loss is controlled by λ. Guo et al. [21] discussed the sensitivity of λ; according to the analysis, we set $\lambda = 0.1$ during our training.

In CLRAN, we input an LR image, and then the SR image is progressively predicted at $\log_2 S$ levels, where S is the scale factor. The expression $I_{HR'}$ denotes the output SR image at level s. We denote the desired output SR image at level s by y_s. The overall loss function is defined as:

$$L_{tP} = \sum_S^{\log_2 S} L_P(y_s, I_{HR}^S) \tag{6}$$

$$L_{tD} = \sum_S^{\log_2 S-1} L_D(D(y_{s+1}), y_s) \tag{7}$$

$$L_T = L_{tP} + L_{tD} \tag{8}$$

where L_{tP} and L_{tD} denote the total loss for the primal mapping and DRM tasks in our network, respectively, and L_T represents the overall loss of our work.

3.2. Enhanced Residual Attention Block (ERA)

The enhanced residual attention block (ERA) of the proposed network, shown in Figure 2, is composed of two parts: the multi-scale part and the enhanced attention part. The multi-scale part contains the MSB, and the enhanced attention part consists of the enhanced attention branch and the non-attention branch.

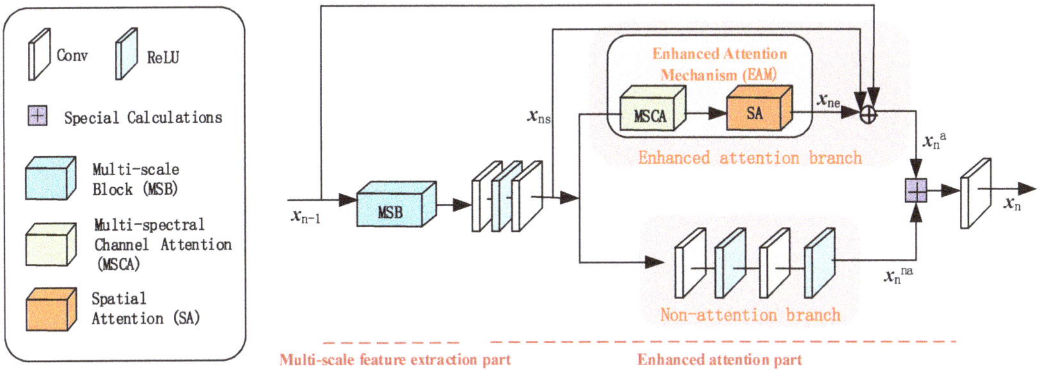

Figure 2. The structure of the enhanced residual attention block (ERA). The purple box denotes special calculations where each added component is multiplied by an automatically generated trainable scalar parameter by the ADM.

Inspired by [16,20], we design the non-attention branch to learn the information that is ignored by the enhanced attention branch. The two branches enable CNNs to make the best use of existing feature information and fully explore the correlation and dependence between the features.

We also introduce the ADM [20] into ERA to balance the enhanced attention branch and non-attention branch. Formally, we have:

$$x_n = f_{1\times1}(\pi_n^{na} \times x_n^{na} + \pi_n^a \times x_n^a) \tag{9}$$

where x_n^{na} is the output feature of the non-attention branch, and x_n^a is the output feature of the enhanced attention branch; π_{na} and π_a are weights of the non-attention branch and the enhanced attention branch, respectively. The dynamic weights are computed by the ADM block; $f_{1\times1}(\cdot)$ denotes the convolution function of 1×1 kernel convolution, and x_n is the output feature of the ERA.

Local Residual Learning Structure: The residual learning and shortcut connections alleviate the difficulty of learning between the LR and HR images. We adopt residual structure in the enhanced attention branch to maximize the utilization of the local residual features and enable the network to be more efficient. The utilization of local residual learning in our network significantly reduces the computational complexity, and the performance of the network is enhanced.

As shown in Figure 2, we use x_{n-1} to describe the input feature maps sent to the ERA, x_{ns} to describe the input feature maps sent to the enhanced attention part, and x_{ne} to describe the output feature maps from the EAB. Formally, we describe the output of the enhanced attention branch x_n^a as

$$x_n^a = x_{n-1} + x_{ns} + x_{ne} \tag{10}$$

where the operation $x_{n-1} + x_{ns} + x_{ne}$ is performed by a shortcut connection and element-wise addition.

3.3. *Multi-Scale Block (MSB)*

Several studies [14,15] have proposed a block to extract the multiscale spatial features. Although the dilated convolution used in [15] achieves much larger receptive fields, not all pixels are used for calculation, resulting in the loss of extracted information details. Therefore, we still use the conventional convolution layers to extract features. As shown in Figure 3a, the multi-scale residual block (MSRB) is used in MSRN [14] to extract the multiscale spatial features. Inspired by the successful application of MSRB, we propose the multi-scale block (MSB) to detect image features at different scales. As shown in Figure 3b, we adopt two 3×3 convolutions instead of 5×5 convolutions and introduce two 1×1 convolutions to reduce parameters and accelerate calculation. In addition, we remove the local shortcut connection (LSC) in MSB and directly follow the attention enhanced attention part to extract diversity of features. In this way, redundancy is reduced in feature utilization and the cost of computational complexity is reduced. The whole operation is defined as

$$S_1 = \upsilon_3^1(\upsilon_1^1(E_{n-1})) \tag{11}$$

$$P_1 = \sigma_3^3(\sigma_3^2(\sigma_1^2(E_{n-1}))) \tag{12}$$

$$S_2 = \sigma_3^4(\sigma_1^3([S_1, P_1])) \tag{13}$$

$$P_2 = \sigma_3^6(\sigma_3^5(\sigma_1^4([P_1, S_1]))) \tag{14}$$

$$E_n = \sigma_1^5([S_2, P_2]) \tag{15}$$

where E_{n-1} represents the feature maps sent to the MSB, and E_n represents the output feature maps of MSB; σ_i^j denotes a fusion function that combines the convolution function and the ReLU function, where i denotes the size of the convolution kernel and j denotes the number of σ_i; $[S_1, P_1]$, $[S_2, P_2]$, and $[P_1, S_1]$ denote the concatenation operation.

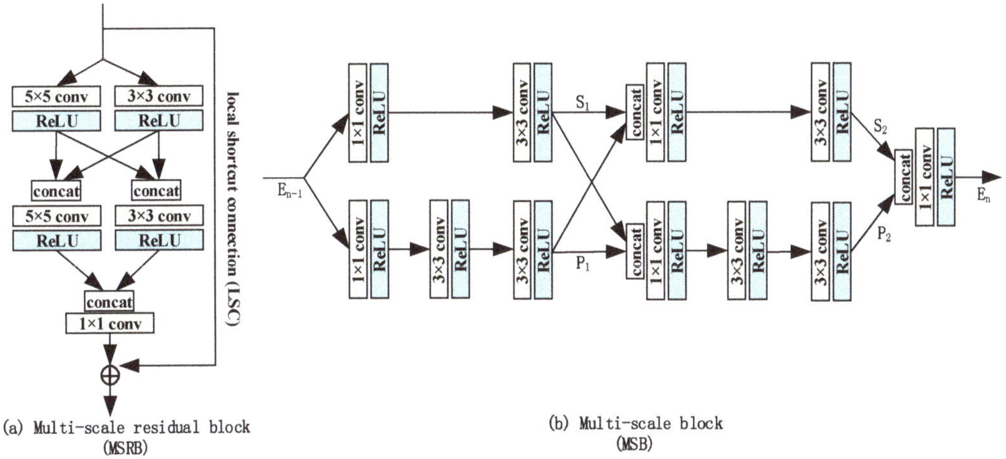

Figure 3. The structure of multi-scale residual block (MSRB) and multi-scale block (MSB), respectively.

3.4. Enhanced Attention Mechanism

Enhanced Attention Mechanism (EAM): In the enhanced attention part, we introduce multi-spectral channel attention (MSCA) and spatial attention (SA) mechanisms into our network. Convolution operations extract meaningful features by combining channel and spatial information together. However, the MSCA block depicted in Figure 4 only utilizes the inter-channel relationship, which neglects spatial information. SA is critical in determining "where" to concentrate. In our work, we propose the EAM that focuses on features in both channel and spatial dimensions. As shown in Figure 5, the EAM infers attention feature maps sequentially along two distinct dimensions, channel and spatial, and attention feature maps multiply with the input feature maps for adaptive feature refinement. Our module contributes significantly to the efficient flow of information within a network. The EAM is expressed as

$$F' = M_f(F) \otimes F \tag{16}$$

$$F'' = M_s(F') \otimes F' \tag{17}$$

where $F \in R^{C\times H\times W}$ denotes input feature maps, M_f denotes the MSCA block, M_s denotes the SA block, $\otimes$ denotes element-wise multiplication, and F'' is the final refined output features.

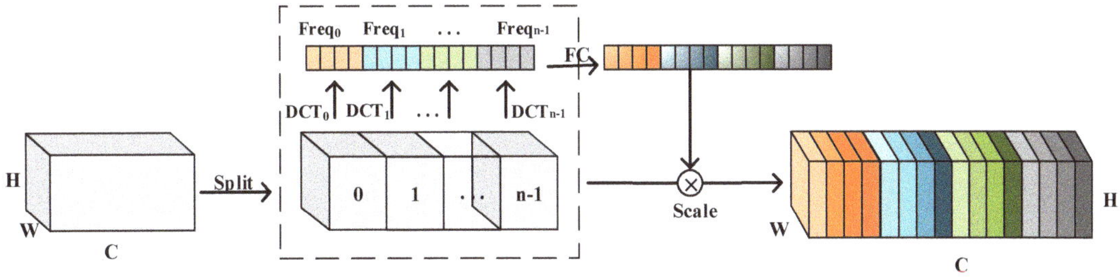

Figure 4. The structure of the multi-spectral channel attention (MSCA) block.

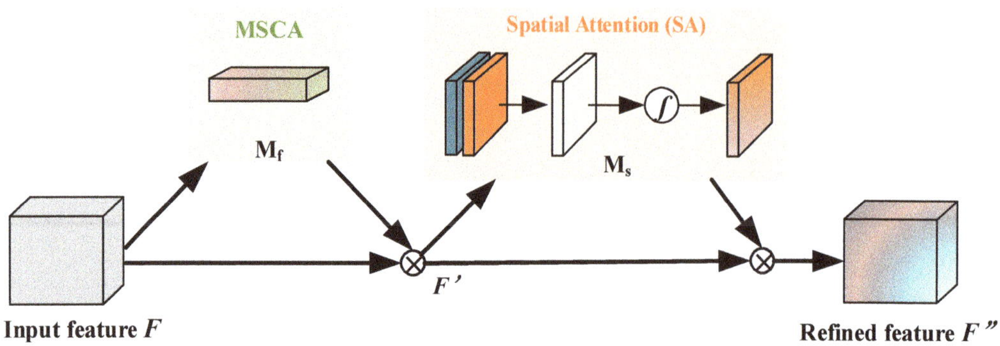

Figure 5. The structure of the enhanced attention mechanism (EAM).

Multi-spectral Channel Attention (MSCA) Module: The channel attention (CA) mechanism uses a scalar to represent and evaluate the importance of each channel and automatically distributes weights to different channels so as to extract critical and important information so that the model makes accurate judgments and will not incur greater overhead in the calculation and storage of the model.

Low-level and mid-level features, in addition to high-level features, are important for reconstructing an SR image. Due to massive information loss, the channel attention mechanism that uses a scalar to represent a channel is difficult. Qin et al. [23] proposed that using global average pooling (GAP) in the channel attention mechanism means only preserving the lowest frequency information and discarding the useful information in representing the channels from other frequencies. Their proposed MSCA mechanism generalizes GAP to more frequency components of 2D discrete cosine transform (DCT).

As shown in Figure 4, the input features $F \in R^{C\times H\times W}$ are split along the channel dimension into several parts. For each part, a corresponding 2D DCT frequency component $Freq^i$ is assigned by employing selection criterion. Finally, the multi-spectral vector $Freq \in R^C$ is obtained by concatenation:

$$Freq = cat([Freq^0, Freq^1, \ldots, Freq^{n-1}]) \tag{18}$$

The feature maps from MSCA module is then expressed as

$$M_f(F) = sigmoid(f_c(Freq)) \tag{19}$$

where *sigmoid* denotes the sigmoid function, and f_c represents fully connected layer.

Spatial Attention (SA) Module: SA tells the network on which informative part it should be focused. As shown in Figure 5, in the SA block, the input features $F^{'} \in R^{C\times H\times W}$ first apply average-pooling and max-pooling operations along the channel axis and then concatenate the outputs to generate an efficient feature map. The combined output is convolved with the convolution function of 7 × 7 kernel convolution, producing our 2D spatial attention map. In short, the spatial attention weight is expressed as follows:

$$M_s(F^{'}) = sigmoid(f^{7\times 7}([AvgPool(F^{'}), MaxPool(F^{'})])) \tag{20}$$

where *sigmoid* denotes the sigmoid function, and $f^{7\times 7}$ represents the convolutional layer with the filter size 7 × 7.

4. Experiments

In this section, we evaluate the performance of our model on several benchmark test datasets. The datasets used for training and testing are introduced first, and next the implementation details are discussed. Following that, we compare our model to several other methods. Finally, we conducted an ablation study to validate and evaluate the effectiveness of our proposed methods. Specially, we employed the PyTorch framework to all of the implementations.

4.1. Datasets and Metrics

We trained on the DIV2K dataset [42], which contains 800 training images. Bicubic downsampling is employed to obtain the LR images. We evaluated our model using the standard and publicly available benchmark datasets Set5 [43], Set14 [44], B100 [45], Urban100 [46], and Manga109 [47]. Set5 [43], Set14 [44], and B100 [45] contain animals, humans, and natural settings, whereas Urban100 [46] focuses only on urban settings. Urban100 contains rich structure contents. The PSNR and SSIM metrics are employed to evaluate the SR results on the Y channel of the transformed YCbCr color space.

4.2. Implementation Details

In this section, we specify the implementation details of our proposed model. We provided two models, namely a small model CLRAN-S and a large model CLRAN-L for 4× and 8× SR. In our model, we employed two enhanced residual attention blocks (ERA, N = 2) in each Basic-CLRAN, and the output from each ERA was 64 feature maps. We chose the Charbornnier penalty [37] function to train our model.

In each training batch, we randomly extracted 16 LR patches with a size of 128 × 128 and 1500 epochs. We trained our model with ADAM optimizer [48] with $\beta_1 = 0.9$, $\beta_2 = 0.999$, and $\epsilon = 10^{-8}$. The learning rate was initialized as 1×10^{-4}. We employed the PyTorch framework to implement our models with GeForce RTX 2080 GPU.

4.3. Results

We compared our model with several state-of-the-art methods in terms of quantitative results and visual results. For quantitative comparison, we compared the PSNR and SSIM values of different methods for 4× and 8× SR. The results of all comparison approaches were derived from their pre-trained models, publicly available code, or original papers.

The results of the PSNR and SSIM values are presented in Table 3. It was found that CLRAN yielded promising performance. CLRAN achieved comparable or superior results compared with all the other methods, including the extremely competitive MSRN. CLRAN-S has the best PSNR on Set5, Set14, B100 and best SSIM on Set5, Set14, B100, and Manga109 for scale ×4. Our CLRAN-L also has excellent SSIM performance on Set5, B100, and Manga109 for scale ×8. Compared with other methods, we found that CLRAN-S and CLRAN-L had achieved almost the best SSIM performance on all benchmark datasets. This confirms that CLRAN is able to gradually aggregate, select, and save relevant details throughout the network. That was mainly because we employed the Charbornnier penalty function [13], thus our model was capable of aggregating rich structured information to generate more representative features. Our model employed YCbCr color space.

For quality comparison, we provided visual comparisons between our method and the considered methods (see Figure 6). We observe that the majority of the approaches were unable to properly recover the tiniest details and so lost the structures, as well as a hazy effect in the majority of the methods. Our model was capable of reconstructing clear and natural images and outperformed other approaches evaluated.

In order to fully utilize the features from the input LR image, our network combined residual structures and attention mechanisms to extract multiscale and diversity of features. Inspired by [14], we proposed the MSB to extract multiscale features. Muqeet et al. [15] was also inspired by [14], which used different dilated convolution layers and channel and spatial attention mechanisms. However, dilated convolution is not friendly to pixel level prediction, and a network based on dilated convolution to design needs some skills, which makes it difficult to migrate directly to other tasks. Moreover, not all attention mechanisms improve network performance, and attention mechanisms may discard relevant details. Behjati et al. [16] designed the network to integrate channel attention mechanisms with residual blocks via two independent but parallel processing routes. However, in [16], the features of the attention branch and residual branch connect directly and neglect spatial information. The residual blocks are simple and neglect to extract the multiscale features. Wang et al. [18] proposed RRDB combined residual network and dense connection. Musunuri et al. [19] employed RRDB to replace the residual block in EDSR, improving the perceptual quality of the SR image. However, as EDSR is a deep and wide network, training this model will cost more memory, space, and datasets. In short, these models do not limit the space of the possible functions and neglect to extract the diversity of features. Moreover, our model employing the loss function is different from these models.

Figure 6. Visual comparison of different methods for 4× image SR.

Table 3. Quantitative results with the BI degradation model for all upscaling factors ×4 and ×8. The red number indicates the best result, and the blue number indicates the second best result. "-" denotes the results that are not reported.

Algorithms	Scale	Set5 PSNR/SSIM	Set14 PSNR/SSIM	B100 PSNR/SSIM	Urban100 PSNR/SSIM	Manga109 PSNR/SSIM
Bicubic	4	28.42/0.810	26.10/0.702	25.96/0.667	23.15/0.657	24.92/0.789
SRCNN [7]		30.48/0.863	27.50/0.751	26.90/0.710	24.52/0.722	27.58/0.856
FSRCNN [27]		30.72/0.866	27.61/0.775	26.98/0.715	24.62/0.728	27.90/0.861
VDSR [9]		31.35/0.883	28.02/0.768	27.29/0.726	25.18/0.754	28.83/0.887
SRDenseNet [29]		32.02/0.893	28.50/0.778	27.53/0.733	26.05/0.781	29.49/0.899
DRCN [25]		31.56/0.881	28.15/0.763	27.24/0.715	25.15/0.753	28.98/0.882
LapSRN [13]		31.54/0.881	28.19/0.772	27.32/0.728	25.21/0.756	29.09/0.890
DCSR [49]		31.58/0.887	28.21/0.772	27.32/0.726	27.24/0.831	-/-
MemNet [30]		31.74/0.889	28.26/0.772	27.40/0.728	25.50/0.763	29.42/0.894
SRMDNF [50]		31.96/0.893	28.35/0.779	27.49/0.734	25.68/0.773	30.09/0.902
MSRN [14]		32.07/0.890	28.60/0.775	27.52/0.727	26.04/0.790	30.17/0.903
CARN [32]		32.13/0.894	28.60/0.781	27.58//0.735	26.07/0.784	-/-
IMDN [51]		32.21/0.895	28.58/0.781	27.56/0.735	26.04/0.784	30.45/0.908
CLRAN-S(Ours)		32.24/0.898	28.65/0.781	27.59/0.735	26.05/0.785	30.37/0.908
Bicubic	8	24.39/0.657	23.19/0.568	23.67/0.547	20.74/0.515	21.47/0.649
SRCNN [7]		25.34/0.647	23.86/0.544	24.14/0.504	21.29/0.513	22.46/0.661
FSRCNN [27]		20.13/0.552	19.75/0.482	24.21/0.568	21.32/0.538	22.39/0.673
SCN [52]		25.59/0.707	24.02/0.603	24.30/0.570	21.22/0.557	22.68/0.696
VDSR [9]		25.73/0.674	23.20/0.511	24.34/0.517	21.48/0.529	22.73/0.669
SRDenseNet [29]		25.99/0.704	24.23/0.581	24.45/0.530	21.67/0.562	23.09/0.712
DRCN [25]		25.93/0.674	24.25/0.551	24.49/0.517	21.71/0.529	23.20/0.669
LapSRN [13]		26.14/0.737	24.35/0.620	24.54/0.585	21.81/0.580	23.39/0.734
MemNet [30]		26.16/0.741	24.38/0.620	24.58/0.584	21.89/0.583	23.56/0.739
MSLapSRN [53]		26.34/0.756	24.57/0.627	24.65/0.590	22.06/0.596	23.90/0.756
MSRN [14]		26.59/0.725	24.88/0.596	24.70/0.541	22.37/0.598	24.28/0.752
CLRAN-L(Ours)		26.97/0.776	24.85/0.637	24.76/0.593	22.35/0.610	24.35/0.773

4.4. Discussion

To validate the effectiveness of our work, we conduct a set of experiments to compare the performance of the MSB, DRM, ADM, and attention mechanisms [22–24], and the number of ERAs in SISR tasks. The results are displayed in Tables 4–6. In Table 4, we conduct the ablation study to validate the effectiveness of MSB, DRM, and ADM. All comparative experiments employ attention mechanisms with the MSCA and SA. In Table 5, we conduct the ablation study to validate the effectiveness of different attention mechanisms in the enhanced attention branch of ERA, and all comparative experiments employ ADM.

Effects of MSB: We propose MSB, which is an efficient multiscale feature extraction structure. This module adaptively detects image features at different scales and fully utilizes the potential features of images. To validate the effectiveness of MSB, we visualize the output feature maps of MSB. The result is shown in Figure 7. With the deepening of the number of network layers, the features extracted by the module become more and more abstract, which is not conducive to our observation. Therefore, we visualize the features extracted by the first application of MSB in the network. From Figure 7, we can observe that the output of MSB retains almost all the information of the original image.

When we employed MSB in our network, 32.47 dB PSNR was obtained with 3.33 M parameters; when we employed without MSB, and the performance of our network with 0.88 parameters decreased by 0.32 dB. Although employing our proposed module increases memory consumption, the effect on performance is obvious, so employing this MSB block in our network is necessary.

Effects of ADM and DRM: In order to evaluate the effects of ADM and DRM, we conducted the comparative experiments. As shown in Table 2, the experiments without

ADM and DRM have lower PSNR than the experiments that employed the ADM and DRM. Therefore, our modules are designed reasonably.

Effects of Different Attention Mechanisms: As shown in Figure 8, in the same way as the visualization of the application of MSB, we visualized the MSCA block heatmaps and the CA block heatmaps. As can be seen, for the MSCA employed in our model, the image structure is clear and the high-frequency and low-frequency regions of the feature map are correctly detected, but the CA employed is incapable of precisely locating them.

From Table 5, we also displayed the comparative experiment results to evaluate the performance of different attention mechanisms in our model. As can be seen, the combination of the MSCA and the SA in our model achieved the best performance. Therefore, we apply the MSCA block and the SA block in the enhanced attention branch of ERA. For case 1, our model did not have the attention mechanism, and the performance was much lower than those cases combined with the attention mechanism. Therefore, the attention mechanism applied to our model is necessary.

Effects of Increasing the Number of ERAs: It is well established that increasing the depth of the network may effectively increase network performance. In our work, increasing the number of ERAs is the easiest way to obtain better SR results. In order to verify the influence of the number of ERAs on the network, we conducted a series of experiments. As shown in Table 6, our network performance improved quickly with increasing ERAs.

In order to gain a more intuitive sense of the effect of the number of ERAs on our model, we plotted the changes in the model metrics during the first 50 epochs, with every 5 epochs as a sample. Given the parameter size of the ERA module itself, we increased the number of ERA from 1 to 5. As shown in Figure 9, the improvement of our model was obvious with the growing number of ERAs, although increasing the number of ERAs in our model will lead to a more complex network. Considering balancing network performance and complexity, we employed two ERAs (N = 2) in our network, which resulted in the optimal balance of performance and model parameters.

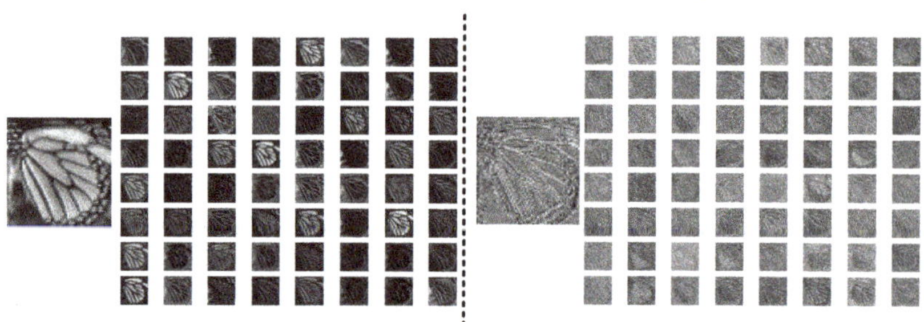

Figure 7. Feature map visualization. On the left: the input feature map of the MSB. On the right: the output feature map of MSB. The 64-channel summation feature map and each channel feature map are shown, respectively.

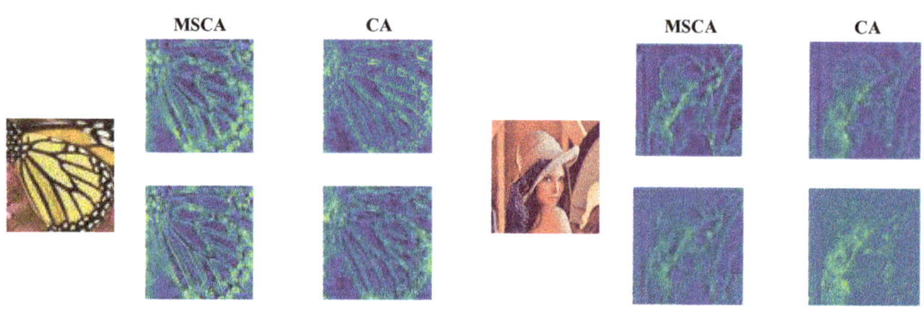

Figure 8. Attention block heatmaps for the MSCA block and the CA block. **The first row:** averaged input feature map of attention layers. **The second row:** averaged output feature map of attention layers.

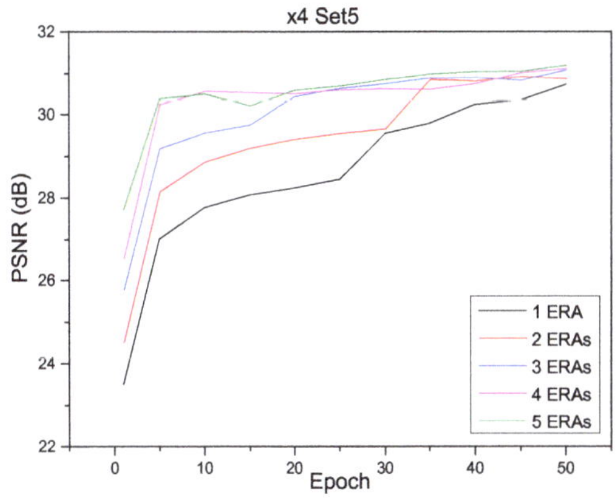

Figure 9. Performance comparison of CLRAN-S with a different number of ERAs.

Table 4. Ablation study: effect of different components of CLRAN-S. Test on Set5 (×4).

Case Index	1	2	3	4
MSB	×	√	√	√
DRM	√	×	√	√
ADM	√	√	×	√
Parameter (M)	0.88	3.32	3.32	3.33
PSNR (dB)	31.92	32.20	32.12	32.24

Table 5. Ablation study: effect of different attention mechanisms of CLRAN-S. Test on Set5 (×4).

Case Index	1	2	3	4	5
SA	×	√	×	×	×
CA+SA	×	×	×	√	×
MSCA	×	×	√	×	×
MSCA+SA	×	×	×	×	√
Parameter (M)	3.15	3.32	3.32	3.32	3.33
PSNR (dB)	32.04	32.14	32.15	32.17	32.24

Table 6. Effect of the number of ERAs on the performance of CLRAN-S (testing on Set5) for 4× SR.

N	1	2	3	4	5
PSNR	32.04	32.24	32.26	32.30	32.34

4.5. Model Complexity Analysis

As shown in Figure 10, we visualize a cost effectiveness analysis between PSNR and model size. CLRAN-S comparisons were done with seven state-of-the-art methods: SRCNN [7], VDSR [9], LapSRN [13], DRCN [25], SRDenseNet [29], MSRN [14], and CARN [32]. CLRAN-S with approximately 3.33M parameters obtained the best performance, which verifies the effectiveness of our model. CLRAN-L comparisons were made with four state-of-the-art methods: SRCNN [7], VDSR [9], LapSRN [13], and MSRN [14]. CLRAN-L with approximately 4.89M parameters obtains best performance, which verifies the effectiveness of our model. In comparison to these methods, CLRAN-S and CLRAN-L achieve higher PSNR with a slightly larger model, demonstrating that the trade-off between performance and model complexity is reasonable.

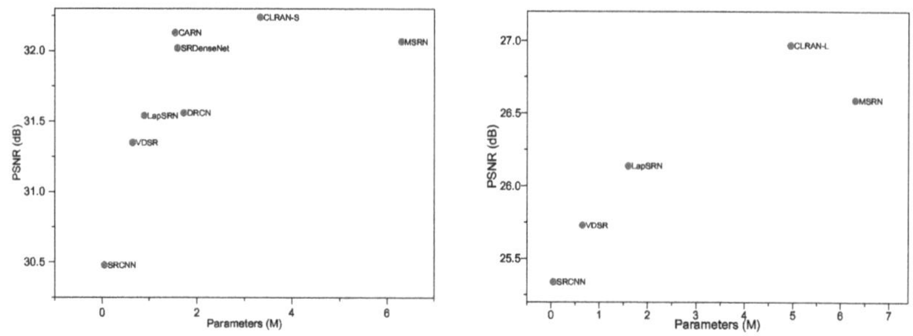

(a) Results on Set5 (4×) (b) Results on Set5 (8×)

Figure 10. PSNR vs. parameters on Set5.

5. Conclusions

In this paper, we proposed a new closed-loop residual attention network (CLRAN) for single image super-resolution. Specifically, the basic architecture block of closed-loop residual attention network (Basic-CLRAN) allowed CLRAN to fully utilize both local and hierarchical diversity of features and easily migrated to achieve other upscaling factor SR tasks. Additionally, the enhanced residual attention block (ERA) extracted the multiscale and diversity image features. The multi-scale block (MSB) was proposed to fuse features at several scales, and the enhanced attention mechanism (EAM) combined a multi-spectral channel mechanism and a spatial attention mechanism proposed to utilize different frequency components channel features and spatial information. Furthermore, we proposed additional mapping and a progressive framework in our model, restricting the space of possible functions and obtaining the SR result step-by-step, taking into account the ill-posed SR problem and limiting the generation of distinct SR images. Comprehensive experiments and ablation studies on benchmark datasets demonstrate the effectiveness of each proposed module, which suggests our model is reasonable.

Author Contributions: Funding acquisition, W.L.; Resources, W.L.; Supervision, W.L.; Writing—original draft, M.Z.; Writing—review & editing, W.L. All authors have read and agreed to the published version of the manuscript.

Funding: This research was funded by the Natural Science Foundation of Hebei Province (F2019201451).

Data Availability Statement: The datasets used in this paper are public datasets. The DIV2K could be found from https://data.vision.ee.ethz.ch/cvl/DIV2K/ (accessed on 1 March 2022).

Conflicts of Interest: The authors declare no conflict of interest.

References

1. Bevilacqua, M. Algorithms for Super-Resolution of Images and Videos Based On Learning Methods. Ph.D. Thesis, Université Rennes 1, Rennes, France, 2014.
2. Gao, X.; Zhang, K.; Tao, D.; Li, X. Image super-resolution with sparse neighbor embedding. *IEEE Trans. Image Process.* **2012**, *21*, 3194–3205. [PubMed]
3. Dong, C.; Loy, C.C.; He, K.; Tang, X. Image super-resolution using deep convolutional networks. *IEEE Trans. Pattern Anal. Mach. Intell.* **2015**, *38*, 295–307. [CrossRef] [PubMed]
4. Timofte, R.; De Smet, V.; Van Gool, L. A+: Adjusted anchored neighborhood regression for fast super-resolution. In *Computer Vision–ACCV 2014, Proceedings of the 12th Asian Conference on Computer Vision, Singapore, 1–5 November 2014*; Springer: Berlin/Heidelberg, Germany, 2014; pp. 111–126.
5. Dai, S.; Han, M.; Xu, W.; Wu, Y.; Gong, Y. Soft edge smoothness prior for alpha channel super resolution. In Proceedings of the 2007 IEEE Conference on Computer Vision and Pattern Recognition, Minneapolis, MN, USA, 17–22 June 2007; pp. 1–8.
6. Yang, X.; Zhang, Y.; Zhou, D.; Yang, R. An improved iterative back projection algorithm based on ringing artifacts suppression. *Neurocomputing* **2015**, *162*, 171–179. [CrossRef]

7. Dong, C.; Loy, C.C.; He, K.; Tang, X. Learning a deep convolutional network for image super-resolution. In *Computer Vision–ECCV 2014, Proceedings of the 13th European Conference, Zurich, Switzerland, 6–12 September 2014*; Springer: Berlin/Heidelberg, Germany, 2014, pp. 184–199.
8. Ledig, C.; Theis, L.; Huszár, F.; Caballero, J.; Cunningham, A.; Acosta, A.; Aitken, A.; Tejani, A.; Totz, J.; Wang, Z.; et al. Photo-realistic single image super-resolution using a generative adversarial network. In Proceedings of the IEEE Conference on Computer Vision and Pattern Recognition, Honolulu, HA, USA, 21–26 July 2017; pp. 4681–4690.
9. Kim, J.; Lee, J.K.; Lee, K.M. Deeply-recursive convolutional network for image super-resolution. In Proceedings of the IEEE Conference on Computer Vision and Pattern Recognition, Las Vegas, NV, USA, 26 June–1 July 2016; pp. 1637–1645.
10. He, K.; Zhang, X.; Ren, S.; Sun, J. Deep residual learning for image recognition. In Proceedings of the IEEE Conference on Computer Vision and Pattern Recognition, Las Vegas, NV, USA, 26 June–1 July 2016; pp. 770–778.
11. Lim, B.; Son, S.; Kim, H.; Nah, S.; Mu Lee, K. Enhanced deep residual networks for single image super-resolution. In Proceedings of the IEEE Conference on Computer Vision and Pattern Recognition Workshops, Honolulu, HA, USA, 21–26 July 2017; pp. 136–144.
12. Zhang, Y.; Li, K.; Li, K.; Wang, L.; Zhong, B.; Fu, Y. Image super-resolution using very deep residual channel attention networks. In Proceedings of the European Conference on Computer Vision (ECCV), Munich, Germany, 8–14 September 2018; pp. 286–301.
13. Lai, W.S.; Huang, J.B.; Ahuja, N.; Yang, M.H. Deep laplacian pyramid networks for fast and accurate super-resolution. In Proceedings of the IEEE Conference on Computer Vision and Pattern Recognition, Honolulu, HI, USA, 21–26 July 2017; pp. 624–632.
14. Li, J.; Fang, F.; Mei, K.; Zhang, G. Multi-scale residual network for image super-resolution. In Proceedings of the European Conference on Computer Vision (ECCV), Munich, Germany, 8–14 September 2018; pp. 517–532.
15. Muqeet, A.; Iqbal, M.T.B.; Bae, S.H. HRAN: Hybrid residual attention network for single image super-resolution. *IEEE Access* **2019**, *7*, 137020–137029. [CrossRef]
16. Behjati, P.; Rodriguez, P.; Mehri, A.; Hupont, I.; Tena, C.F.; Gonzalez, J. Hierarchical Residual Attention Network for Single Image Super-Resolution. *arXiv* **2020**, arXiv:2012.04578.
17. Tong, T.; Li, G.; Liu, X.; Gao, Q. Image super-resolution using dense skip connections. In Proceedings of the IEEE International Conference on Computer Vision, Venice, Italy, 22–29 October 2017; pp. 4799–4807.
18. Wang, X.; Yu, K.; Wu, S.; Gu, J.; Liu, Y.; Dong, C.; Qiao, Y.; Change Loy, C. Esrgan: Enhanced super-resolution generative adversarial networks. In Proceedings of the European Conference on Computer Vision (ECCV) Workshops,Munich, Germany, 8–14 September 2018 pp. 0–0.
19. Musunuri, Y.R.; Kwon, O.S. Deep residual dense network for single image super-resolution. *Electronics* **2021**, *10*, 555. [CrossRef]
20. Chen, H.; Gu, J.; Zhang, Z. Attention in Attention Network for Image Super-Resolution. *arXiv* **2021** arXiv:2104.09497.
21. Guo, Y.; Chen, J.; Wang, J.; Chen, Q.; Cao, J.; Deng, Z.; Xu, Y.; Tan, M. Closed-loop matters: Dual regression networks for single image super-resolution. In Proceedings of the IEEE/CVF Conference on Computer Vision and Pattern Recognition, Seattle, WA, USA, 14–19 June 2020; pp. 5407–5416.
22. Hu, J.; Shen, L.; Sun, G. Squeeze-and-excitation networks. In Proceedings of the IEEE Conference on Computer Vision and Pattern Recognition, Salt Lake City, UT, USA, 18–22 June 2018; pp. 7132–7141.
23. Qin, Z.; Zhang, P.; Wu, F.; Li, X. Fcanet: Frequency channel attention networks. In Proceedings of the IEEE/CVF International Conference on Computer Vision, Montreal, QC, Canada, 10–17 October 2021; pp. 783–792.
24. Woo, S.; Park, J.; Lee, J.Y.; Kweon, I.S. Cbam: Convolutional block attention module. In Proceedings of the European Conference on Computer Vision (ECCV), Munich, Germany, 8–14 September 2018; pp. 3–19.
25. Kim, J.; Lee, J.K.; Lee, K.M. Accurate image super-resolution using very deep convolutional networks. In Proceedings of the IEEE Conference on Computer Vision and Pattern Recognition, Honolulu, HA, USA, 21-26 July 2016; pp. 1646–1654.
26. Tai, Y.; Yang, J.; Liu, X. Image super-resolution via deep recursive residual network. In Proceedings of the IEEE Conference on Computer Vision and Pattern Recognition, Honolulu, HA, USA, 21–26 July 2017; pp. 3147–3155.
27. Dong, C.; Loy, C.C.; Tang, X. Accelerating the super-resolution convolutional neural network. In *Computer Vision—ECCV 2016, Proceedings of the 14th European Conference, Amsterdam, The Netherlands, 11–14 October 2016*; Springer: Berlin/Heidelberg, Germany, 2016; pp. 391–407.
28. Shi, W.; Caballero, J.; Huszár, F.; Totz, J.; Aitken, A.P.; Bishop, R.; Rueckert, D.; Wang, Z. Real-time single image and video super-resolution using an efficient sub-pixel convolutional neural network. In Proceedings of the IEEE Conference on Computer Vision and Pattern Recognition, Honolulu, HA, USA, 21–26 July 2016; pp. 1874–1883.
29. Huang, G.; Liu, Z.; Van Der Maaten, L.; Weinberger, K.Q. Densely connected convolutional networks. In Proceedings of the IEEE Conference on Computer Vision and Pattern Recognition, Honolulu, HA, USA, 21–26 July 2017; pp. 4700–4708.
30. Tai, Y.; Yang, J.; Liu, X.; Xu, C. Memnet: A persistent memory network for image restoration. In Proceedings of the IEEE International Conference on Computer Vision, Venice, Italy, 22–29 October 2017; pp. 4539–4547.
31. Zhang, Y.; Tian, Y.; Kong, Y.; Zhong, B.; Fu, Y. Residual dense network for image super-resolution. In Proceedings of the IEEE Conference on Computer Vision and Pattern Recognition, Salt Lake City, UT, USA, 18–22 June 2018; pp. 2472–2481.
32. Ahn, N.; Kang, B.; Sohn, K.A. Fast, accurate, and lightweight super-resolution with cascading residual network. In Proceedings of the European Conference on Computer Vision (ECCV), Munich, Germany, 8–14 September 2018; pp. 252–268.

33. Show, A. Tell: Neural Image Caption Generation with Visual Attention Kelvin Xu. Available online: https://kelvinxu.github.io/projects/capgen.html (accessed on 1 March 2022).
34. Zamir, S.W.; Arora, A.; Khan, S.; Hayat, M.; Khan, F.S.; Yang, M.H.; Shao, L. Learning enriched features for real image restoration and enhancement. In *Computer Vision—ECCV 2020, Proceedings of the European Conference on Computer Vision, Glasgow, UK, 23–28 August 2020*; Springer: Berlin/Heidelberg, Germany, 2020; pp. 492–511.
35. Dai, T.; Cai, J.; Zhang, Y.; Xia, S.T.; Zhang, L. Second-order attention network for single image super-resolution. In Proceedings of the IEEE/CVF Conference on Computer Vision and Pattern Recognition, Long Beach, CA, USA, 15–20 June 2019; pp. 11065–11074.
36. Liu, D.; Wen, B.; Fan, Y.; Loy, C.C.; Huang, T.S. Non-local recurrent network for image restoration. *arXiv* **2018**, arXiv:1806.02919.
37. Hu, Y.; Li, J.; Huang, Y.; Gao, X. Channel-wise and spatial feature modulation network for single image super-resolution. *IEEE Trans. Circuits Syst. Video Technol.* **2019**, *30*, 3911–3927. [CrossRef]
38. Liu, J.; Zhang, W.; Tang, Y.; Tang, J.; Wu, G. Residual feature aggregation network for image super-resolution. In Proceedings of the IEEE/CVF Conference on Computer Vision and Pattern Recognition, Seattle, WA, USA, 13–19 June 2020; pp. 2359–2368.
39. Muqeet, A.; Hwang, J.; Yang, S.; Kang, J.; Kim, Y.; Bae, S.H. Multi-attention based ultra lightweight image super-resolution. In *Computer Vision—ECCV 2020 Workshops, Proceedings of the European Conference on Computer Vision, Glasgow, UK, 23–28 August 2020*; Springer: Berlin/Heidelberg, Germany, 2020; pp. 103–118.
40. Zhao, H.; Kong, X.; He, J.; Qiao, Y.; Dong, C. Efficient image super-resolution using pixel attention. In *Computer Vision—ECCV 2020 Workshops, Proceedings of the European Conference on Computer Vision, Glasgow, UK, 23–28 August 2020*; Springer: Berlin/Heidelberg, Germany, 2020; pp. 56–72.
41. Mei, Y.; Fan, Y.; Zhang, Y.; Yu, J.; Zhou, Y.; Liu, D.; Fu, Y.; Huang, T.S.; Shi, H. Pyramid attention networks for image restoration. *arXiv* **2020**, arXiv:2004.13824.
42. Agustsson, E.; Timofte, R. Ntire 2017 challenge on single image super-resolution: Dataset and study. In Proceedings of the IEEE Conference on Computer Vision and Pattern Recognition Workshops, Honolulu, HA, USA, 21–26 July 2016; pp. 126–135.
43. Bevilacqua, M.; Roumy, A.; Guillemot, C.; Alberi-Morel, M.L. *Low-Complexity Single-Image Super-Resolution Based On Nonnegative Neighbor Embedding*; BMVA Press: Swansea, UK, 2012.
44. Yang, J.; Wright, J.; Huang, T.S.; Ma, Y. Image super-resolution via sparse representation. *IEEE Trans. Image Process.* **2010**, *19*, 2861–2873. [CrossRef] [PubMed]
45. Martin, D.; Fowlkes, C.; Tal, D.; Malik, J. A database of human segmented natural images and its application to evaluating segmentation algorithms and measuring ecological statistics. In Proceedings Eighth IEEE International Conference on Computer Vision. ICCV 2001, Vancouver, BC, Canada, 7–14 July 2001; Volume 2, pp. 416–423.
46. Huang, J.B.; Singh, A.; Ahuja, N. Single image super-resolution from transformed self-exemplars. In Proceedings of the IEEE Conference on Computer Vision and Pattern Recognition (CVPR), Boston, MA, USA, 7–12 June 2015.
47. Matsui, Y.; Ito, K.; Aramaki, Y.; Fujimoto, A.; Ogawa, T.; Yamasaki, T.; Aizawa, K. Sketch-based manga retrieval using manga109 dataset. *Multimed. Tools Appl.* **2017**, *76*, 21811–21838. [CrossRef]
48. Kingma, D.P.; Ba, J. Adam: A method for stochastic optimization. *arXiv* **2014** arXiv:1412.6980.
49. Zhang, Z.; Wang, X.; Jung, C. DCSR: Dilated convolutions for single image super-resolution. *IEEE Trans. Image Process.* **2018**, *28*, 1625–1635. [CrossRef]
50. Zhang, K.; Zuo, W.; Zhang, L. Learning a single convolutional super-resolution network for multiple degradations. In Proceedings of the IEEE Conference on Computer Vision and Pattern Recognition, Salt Lake City, UT, USA, 18–23 June 2018; pp. 3262–3271.
51. Hui, Z.; Gao, X.; Yang, Y.; Wang, X. Lightweight image super-resolution with information multi-distillation network. In Proceedings of the 27th ACM International Conference on Multimedia, Nice, France, 21–25 October 2019; pp. 2024–2032.
52. Wang, Z.; Liu, D.; Yang, J.; Han, W.; Huang, T. Deep networks for image super-resolution with sparse prior. In Proceedings of the IEEE International Conference on Computer Vision, Santiago, Chile, 7–13 December 2015; pp. 370–378.
53. Lai, W.S.; Huang, J.B.; Ahuja, N.; Yang, M.H. Fast and accurate image super-resolution with deep laplacian pyramid networks. *IEEE Trans. Pattern Anal. Mach. Intell.* **2018**, *41*, 2599–2613. [CrossRef] [PubMed]

electronics

MDPI

Article

A Dual CNN for Image Super-Resolution

Jiagang Song [1,†], Jingyu Xiao [2,*,†], Chunwei Tian [3,4,5], Yuxuan Hu [2], Lei You [6] and Shichao Zhang [2]

1 School of Computer Science and Engineering, Guangxi Normal University, Guilin 541004, China; songjg@stu.gxnu.edu.cn
2 School of Computer Science, Central South University, Changsha 410083, China; hyxx_08@163.com (Y.H.); zhangsc@csu.edu.cn (S.Z.)
3 Research & Development Institute, Northwestern Polytechnical University, Shenzhen 518057, China; chunweitian@nwpu.edu.cn
4 School of Software, Northwestern Polytechnical University, Xi'an 710129, China
5 Yangtze River Delta Research Institute, Northwestern Polytechnical University, Taicang 215400, China
6 School of Biomedical Informatics, University of Texas Houston Science Center at Houston, Houston, TX 77030, USA; Lei.You@uth.tmc.edu
* Correspondence: jyxiao@csu.edu.cn
† These authors contributed equally to this work.

Abstract: High-quality images have an important effect on high-level tasks. However, due to human factors and camera hardware, digital devices collect low-resolution images. Deep networks can effectively restore these damaged images via their strong learning abilities. However, most of these networks depended on deeper architectures to enhance clarities of predicted images, where single features cannot deal well with complex screens. In this paper, we propose a dual super-resolution CNN (DSRCNN) to obtain high-quality images. DSRCNN relies on two sub-networks to extract complementary low-frequency features to enhance the learning ability of the SR network. To prevent a long-term dependency problem, a combination of convolutions and residual learning operation is embedded into dual sub-networks. To prevent information loss of an original image, an enhanced block is used to gather original information and obtained high-frequency information of a deeper layer via sub-pixel convolutions. To obtain more high-frequency features, a feature learning block is used to learn more details of high-frequency information. The proposed method is very suitable for complex scenes for image resolution. Experimental results show that the proposed DSRCNN is superior to other popular in SR networks. For instance, our DSRCNN has obtained improvement of 0.08 dB than that of MemNet on Set5 for ×3.

Keywords: dual networks; enhanced CNN; fine learning block; image super-resolution

Citation: Song, J.; Xiao, J.; Tian, C.; Hu, Y.; You, L.; Zhang, S. A Dual CNN for Image Super-Resolution. *Electronics* **2022**, *11*, 757. https://doi.org/10.3390/electronics11050757

Academic Editor: Soon Ki Jung

Received: 22 January 2022
Accepted: 22 February 2022
Published: 1 March 2022

1. Introduction

Due to effects of human factor and camera hardware, captured images often are not clear. To overcome these challenges, single super-resolution (SISR) techniques are presented [1]. For instance, priori knowledge is used to guide the SR model [2] Zha et al. [3] embedded a sparse idea into dictionary learning to repair high-quality images. To obtain richer information, Zhang et al. [4] combined non-local and local priors to achieve a non-local mean SR model with steering kernel regression. Zhang et al. [5] used a Monte Carlo-based Markov chain to train an SR model for improving visual effects. There are other popular SR methods, i.e., random forest [6], gradient profile [2], and regression [7]. Although these methods can repair low-resolution images well, they are faced with two challenges as follows:

(1) They referred to complex optimization methods to mine more detailed information for promoting super-resolutions of repaired images.
(2) They reply on manually chosen parameters to promote visual effects of predicted images.

To handle these problems, convolutional neural networks (CNNs) with strong self-learning abilities composed of common components are developed [8,9]. For instance Dong et al. [10] designed a shallow architecture via pixel mapping operations to automatically obtaining clearer images rather than manual setting parameters. To pursue more perfect restoration effects, residual learning techniques and concatenation operations are applied in image restoration [11]. Tai et al. [12] used two different residual operations to fuse obtained local features to improve the learning ability of CNN in SISR. Kim et al. [13] used residual learning operations to gather hierarchical features for the final layer in order to enhance the robustness of obtained features in SISR. Although these methods have obtained remarkable results in SISR, they upsampled given low-resolution images as inputs of CNNs, which can increase computational cost [14]. To solve this problem, an upsampling operation set in the final layer of CNN is developed [15]. For instance, Dong et al. [14] set a deconvolution operation as the final layer to reduce the complexity of the whole SR network. To improve the SR performance, Zhang et al. [16] enlarged the depth of SR network and repeatedly used concatenation operations to facilitate obtained features in SISR Although these methods perform well in SR, they may depend on deeper architectures to extract more accurate features in promoting SR performance, which may have higher requirements on hardware devices. In addition, obtained features from single architecture may not fully deal with complex screens.

In this paper, we propose a dual super-resolution CNN (DSRCNN) via three blocks (i.e., two sub-network enhanced block (TSEB), enhanced block (EB), and feature learning block (FLB) to obtain high-quality images. TSEB used two sub-networks to extract complementary low-frequency features to enhance the learning ability of SR networks. To prevent a long-term dependency problem, a combination of convolutions and residual learning operation is embedded into dual sub-networks. To prevent information loss of an original image, an enhanced block is used to gather original information and obtain high-frequency information of deeper layers via sub-pixel convolutions. To obtain more high-frequency features, a feature learning block is used to learn more details of high-frequency information.

The main contributions of the proposed DSRCNN are as follows:

(1) DSRCNN uses two sub-networks to extract complementary features to enhance the learning ability of an SR model, which is very suitable to complex screens. The multiple combinations of residual learning operation, convolutional layer, and ReLU are embedded into two sub-networks to enhance the memory abilities of shallow layers to deep layers and extract more accurate information as well as a large amount of information of dual networks in SISR.

(2) Combining low-frequency and high-frequency features to train a robust SR model.

The remainder of this paper is as follows: Section 2 gives the related work; Section 3 describes the proposed method; Section 4 shows experiments; and Section 5 presents the conclusions.

2. Related Work

2.1. Deep CNNs for Image Super-Resolution

Big data and strong hardware devices, i.e., a graphic processing unit (GPU), contribute the success of CNNs in image applications, i.e., image super-resolution [17]. These SR methods can be summarized as two kinds: upsampling low-resolution image-based CNNs and upsampling obtained low-frequency feature-based CNNs. The first method requires that input and output images have the same sizes. That is, they used upsampling operations to amplify low-resolution images as inputs of CNN to predict super-resolution images. Inspired by that, a deep network with a sparse coding algorithm was used to improve the SR execution speed and performance [18]. Kim et al. [19] combined a deeper architecture and residual learning operation to extract more accurate structure information in SR. Alternatively, Mao et al. [20] utilized skip connections to construct a symmetrical architecture to enhance the learning ability of designed CNN in SR. Although these methods were very effective in image super-resolution, they are still faced with big complexity. To address this issue, the second method is developed. The second method directly puts the given low-resolution images as inputs of CNNs and uses upsampling operations at deep layers to amplify obtained low-frequency features to obtain high-frequency features for constructing high-quality images. Motived by that, scholars conducted a lot of SR methods [21]. For instance, Tian et al. [22], respectively, enhanced low-frequency features and high-frequency features to achieve a robust SR model. Ahn et al. [22] combined residual blocks and smaller kernels to make a trade-off between SR performance and execution speed. In addition, Tian et al. [15] found local key features to obtain richer features in a horizontal and vertical way for promoting visual effects. Chen et al. [23] conducted an efficient network via multi-scale ideas. Geng et al. [24] used the combination of Shearlet and residual network to extract more accurate information in SISR. Nathan et al. [25] used the combination of attention ideas and multi-scale to improve the performance of SISR. According to mentioned illustrations, we can see that upsampling obtained low-frequency feature-based CNNs are popular in SISR. Thus, we use this idea in this paper.

2.2. Fusion of Multiple CNNs for Image Restoration

Some SR methods used a single network architecture to extract representative information to construct high-quality images. However, obtained features may be affected by different screens, which are not beneficial to complex screens. To address this issue, fusion multiple CNNs are employed in image restoration [26]. Tian et al. [27] utilized two different sub-networks to extract different features to enlarge differences of the CNN for promoting denoising effects. Pan et al. [28] used dual CNN to extract a different structure and detailed information in image restoration. Tian et al. [29] fused a signal processing idea, a sparse method into dual CNNs to remove the noise. Xin et al. [30] proposed a dual recursive network with a wavelet idea to predict high-quality images. Inspired by that, we also choose the fusion of multiple CNNs in SISR.

2.3. Peak Signal-to-Noise Ratio (PSNR)

Given a clean image I and noise image K with size MXN, MSE is defined as [27]:

$$MSE = \frac{1}{mn}\sum_{i=0}^{m-1}\sum_{j=0}^{n-1}[I(i,j) - K(i,j)]^2 \tag{1}$$

Then, PSNR(dB) is defined as [27]:

$$PSNR = 10 \cdot \log_{10}\left(\frac{MAX_I^2}{MSE}\right) \tag{2}$$

2.4. Structural SIMilarity (SSIM)

The SSIM formula is based on three comparative measures between samples X and Y: luminance, contrast, and structure [27].

$$l(x,y) = \frac{2\mu_x\mu_y + c_1}{\mu_x^2 + \mu_y^2 + c_1} \tag{3}$$

$$l(x,y) = \frac{2\mu_x\mu_y + c_1}{\mu_x^2 + \mu_y^2 + c_1} \tag{4}$$

$$c(x,y) = \frac{2\sigma_x\sigma_y + c_2}{\sigma_x^2 + \sigma_y^2 + c_2} \tag{5}$$

Usually, take $c_3 = c_2/2$. $\mu_x$is the mean of x. μ_y is the mean of y. σ_x^2 is the variance of x. σ_y^2 is the variance of y. σ_x^2 is the covariance of x and y. $c_1 = (k_1L)^2, c_2 = (k_2L)^2$

are two constants, avoiding division by zero. L is the range of pixel values, $2^B - 1$. $c_1 = (k_1 L)^2, c_2 = (k_2 L)^2$ default value [27]:

$$\text{SSIM}(x,y) = \left[l(x,y)^\alpha \cdot c(x,y)^\beta \cdot s(x,y)^\gamma\right] \tag{6}$$

Setting $\alpha, \beta, \gamma$to 1, you can obtain [27]:

$$\text{SSIM}(x,y) = \frac{(2\mu_x\mu_y + c_1)(2\sigma_{xy} + c_2)}{\left(\mu_x^2 + \mu_y^2 + c_1\right)\left(\sigma_x^2 + \sigma_y^2 + c_2\right)} \tag{7}$$

In each calculation, a window of N×N is taken from the picture, and then the window is continuously sliding for calculation, and finally the average value is taken as the global SSIM [27].

3. The Proposed Method

The proposed DSRCNN is given in Figure 1. DSRCNN contains TSEB, EB, and FLB. TSEB depended on two sub-networks to extract complementary low-frequency features to improve the SR performance of DSRCNN. Specifically, the simultaneous use of convolutions and residual learning operation are used to enhance the effects of hierarchical features to prevent a long-term dependency problem. To prevent information loss of the given low-resolution image, EB employs a residual operation and sub-pixel convolutions to gather obtained different high-frequency features. FLB used several stacked convolutions to refine high-frequency features for obtaining more accurate high-frequency features in SISR. More information can be shown as follows.

3.1. Network Architecture

The proposed 23-layer DSRCNN consists of TSEB, EB, and FLB. The 17-layer TSEB utilizes dual CNNs to obtain complementary low-frequency information to promote learning ability of an SR model. In addition, using residual learning technique to fuse local hierarchical features in the TSEB can maintain memory ability of shallow layers for SISR. Then, EB fuses features of two different paths via a residual operation and sub-pixel convolutions to prevent information loss of given low-resolution image. Finally, FLB is used to refine high-frequency information to better represent predicted high-quality images. To clearer express the process above, the following symbols are defined. Let I_{LR} and I_{SR} be defined as the given LR image and predicted SR image. f_{TSEB} , f_{EB} and f_{FLB} are functions of TSEB, EB, and FLB, respectively. The execution process of the DSRCNN can be expressed as follows:

$$\begin{aligned} I_{SR} &= f_{FLB}(f_{EB}(f_{TSEB}(I_{LR}))) \\ &= f_{DSRCNN}(I_{LR}) \end{aligned} \tag{8}$$

where f_{DSRCNN} stands for the function of DSRCNN. In addition, DSRCNN relies on the following loss function to find optimal parameters.

3.2. Loss Function

The mean squared error (MSE) [31] is used to test the difference between a real high-quality image and predicted SR image for finding optimal parameters. The MSE value is computed via a training pair of $\{I_{LR}^k, I_{HR}^k\}_{(k=1)}^N$, where I_{LR^k} and I_{HR^k} express the k-th given low-resolution image and high-resolution image, respectively. In addition, N is the total of training samples. In addition, we minimize the loss function to train DSRCNN as follows:

$$l(p) = \frac{1}{2N}\sum_{(k=1)}^{N} \|f_{DSRCNN}(I_{LR}^k - I_{HR}^k)\|^2 \tag{9}$$

where l stands for loss function, and p is used to represent the parameter set of training a DSRCNN.

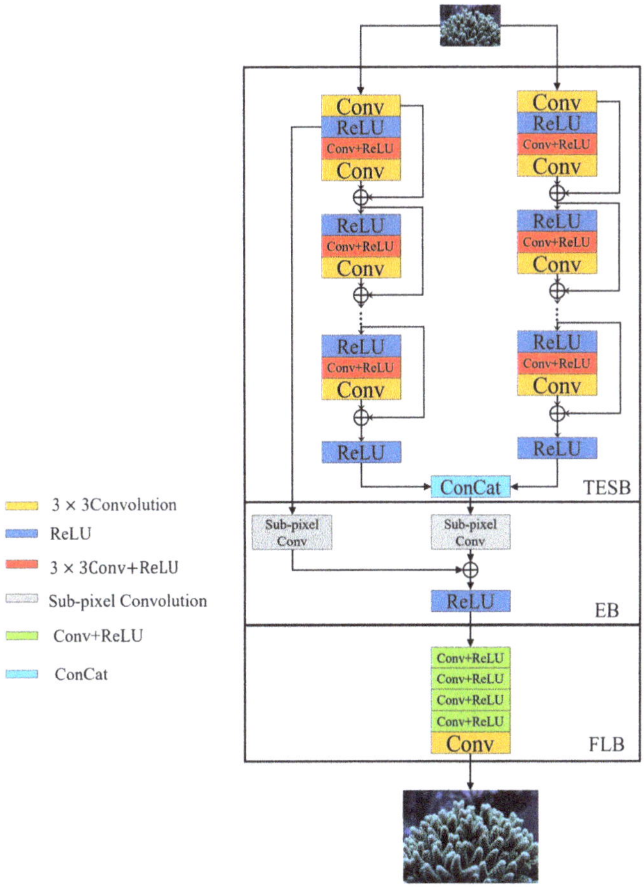

Figure 1. Network architecture of the proposed DSRCNN.

3.3. Two Sub-Network Enhanced Block

According to previous illustrations, it is known that obtained features of single architecture cannot fully deal with complex screens. In this paper, a two sub-network enhanced block is used to overcome this phenomenon. The two sub-network enhanced block consists of two phases. The first phase fuses two sub-networks via a concatenation operation to extract robust low-frequency features in SISR, where a concatenation operation is shown as Figure 2. The second phase is used to gather local hierarchical information to enhance the memory ability of shallow layers for improving the SR effect. Each sub-network from two phases is composed of 17 combinations of convolution and Rectified Linear Unit (ReLU) [32], and a single convolution layer. In addition, input and output channels of the first convolutional layer in each sub-network are 3 and 64. Input and output channels from 2nd to 16th convolutional layers are 64. Two sub-networks are fused via a concatenation operation at the end of the 16th convolutional layer. Thus, the input channel of the 17th convolutional layer is 128. To flexibility operate the convolution layer, the output channel of the 17th convolutional layer is 64. The input and output channels of the 18th convolutional layer are 128 and 64. In addition, kernel sizes of all convolutions are 3×3. The mentioned process can be explained via the following Equation (10).

$$\begin{aligned} O_{TSEB} &= f_{TSEB}(I_{LR}) \\ &= (Cat(O_{(TSEB_1)}, O_{(TSEB_2))}) \end{aligned} \tag{10}$$

where O_{TSEB_1} and O_{TSEB_2} express the outputs of two sub-networks. *Cat* denotes a concatenation operation as well as operation in Figures 1 and 2. O_{TSEB} stands for output of TSEB. More detailed information of two sub-networks can be shown in the second phase of TSEB. The second phase repeatedly uses a combination of residual operation and convolutions to enhance the effect of local hierarchical information for improving SR effects. That is, obtained features of odd convolutional layers from the 1st layer to the 17th layer can be gathered via residual operations to enhance robustness of obtained features for SISR. According to Figure 1 and illustrations above, the second phase in two sub-networks can be shown as follows:

$$O_{(TSEB_k)} = R(O_{L1} + O_{L3} + O_{L5} + O_{L7} + O_{Li} \ldots + O_{17}) \tag{11}$$

where $O_{(TSEB_k)}$ denotes the output of the kth each sub-network, and O_{Li} is the output of the ith convolution layer, where $i = 2, \ldots, 17$. $O_{L1} = R(C(I_{LR}))$ and $O_{Li} = C(O_{(Li-1)})$, where R denotes the function of ReLU. In addition, $O_j = R(C(O_{(j-1)}))$, where $j = 2, 4, 6, 8, \ldots, 16$. In addition, the output TSEB acts an enhanced block as follows.

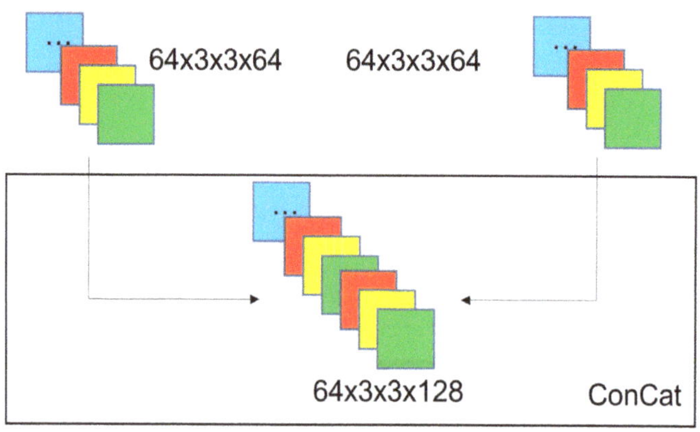

Figure 2. Sketch figure of concatenation operation.

3.4. Enhanced Block

A one-layer enhanced block is used to fuse information of an original image and obtained information from a deeper layer. This is implemented via the following steps. The first step obtains the information of the original image and a deeper layer, respectively. That is, we use sub-pixel operation to respectively amplify obtained features of the TSEB and the first layer of the below sub-network in Figure 1 as follows:

$$O_{(EB_s)} = Subp(R(O_{L1})) \tag{12}$$

$$O_{(EB_d)} = Subp(O_{TSEB}) \tag{13}$$

where $O_{(EB_s)}$ and $O_{(EB_d)}$ are obtained high-frequency features of shallow and deep layers, respectively. *Subp* denotes sub-pixel convolutional techniques, which is expressed as Subpixel Conv in Figure 1. In addition, the sub-pixel convolution consists of convolution with a kernel of 3×3, and the input and output channels in Equation (12) total 64. Input and output channels in Equation (13) are 128 and 64. The second step fuses the obtained features above via a residual operation to obtain more commentary information as follows:

$$\begin{aligned} O_{EB} &= f_{EB}(O_{TSEB}) \\ &= R(O_{(EB_s)} + O_{(EB_d)}) \end{aligned} \tag{14}$$

where O_{EB} is the output of EB. In addition, $+$ denotes a residual operation, which is expressed as $\oplus$ in Figure 1. In addition, O_{EB} acts as a feature learning block.

3.5. Feature Learning Block

To further learn high-frequency features, a 5-layer feature learning block is presented. It includes four Conv+ReLU and a Conv. Conv+ReLU denotes the combination of convolution and ReLU, where their input and output channels total 64, and a convolutional kernel is 3×3 . In addition, a Conv denotes a convolution, where its input and output channels are 64 and 3, and a convolutional kernel is 3×3. This is used to construct predicted high-quality images. The mentioned descriptions are visualized as Equation (15):

$$\begin{aligned} I_{SR} &= f_{FLB}(O_{EB}) \\ &= C(R(C(R(C(R(C(R(C(O_{EB}))))))))) \end{aligned} \tag{15}$$

where C is convolutional operation as well as Conv in Figure 1. Finally, we give a pseudo-code to show implementations of the proposed method as shown in Algorithm 1:

Algorithm 1 The process of converting an LR image into an SR image.

```
Input: Put an LR image I_LR into the DSRCNN model. Enter the scale factor
for Patch is 64 do
    The residual network is used to retain low-level features and fuse high-level features.
    The feature outputs of the two models are merged through ConCat.
    Updated I_LR feature (the first model O_TSEB_1 & the second model O_TSEB_2 )through a two sub-network
    enhanced block by Equation (10);
    The enhanced block contains two upsampling layers.The first upsampling roughly extracts the spliced
    low-frequency features and converts them into high-frequency features. The second upsampling extracts the
    features of the first layer and stacks them with the first upsampling.
    Updated O_EB_s and O_EB_d through an enhanced block by Equation (12) and (13);
    After five layers of convolutional layers (feature learning block ), the high-frequency features are further
    learned and extracted by Equation (15).
end for
Output: Obtain a super-resolution image I_SR with an inpainting scale as the input scale.
```

4. Experimental Results

4.1. Training Dataset

All DIV2K images are saved in PNG format and DIV2K dataset of RGB images with a large diversity of contents. To conduct fair experiments, the DIV2K dataset [33] is chosen to train a DSRCNN model. DIV2K consists of three scales, i.e., $\times 2$, $\times 3$ and $\times 4$. Each scale includes 800 training images. In addition, test images and validation images are 100 natural images. To enlarge differences of training images, we gather the given training dataset of DIV2K and the validation dataset as a new training dataset under the same scale. In addition, some data augmentation operations, i.e., random horizontal flips and 90° rotation operations, are used to enhance training data. To improve the speed of training DSRCNN, given LR images are cropped as an image patch with 64×64.

4.2. Test Dataset

To fairly and effectively test the SR performance, Set5 [34], Set14 [34], BSD100 (B100) [35] and Urban 100 (B100) [36] are chosen to conduct comparative experiments for $\times 2$, $\times 3$, and $\times 4$. The Set5 and Set14 are captured under the same conditions, which have five and fourteen natural images. B100 and U100 respectively contain 100 natural color images. These datasets can be further introduced as follows:

- The Set5 dataset is a dataset consisting of five images, i.e., baby, bird, butterfly, head, and woman [34].
- The Set14 dataset is a dataset consisting of 14 images, and it is commonly used for testing performance of Image Super-Resolution models [34].
- BSD is a dataset used frequently for image denoising and super-resolution. BSD100 have 100 images, which was conducted by Martin et al. The dataset is composed of a large variety of images ranging from natural images to object-specific such as plants, people, food, etc.
- The Urban100 dataset contains 100 images of urban scenes. It is commonly used as a test set to evaluate the performance of super-resolution models [36].

Because most of the SR methods use the Y channel to test the SR performance of their proposed methods, we also choose a Y channel to test the effect of our method on SR. That is, obtained RGB images of the DSRCNN are converted to the Y channel to verify the SR performance.

4.3. Implementation Details

This paper has the following initial parameters. We set a batch size as 64. In addition, the initial learning rate is 1×10^3, Beta_1 is 0.9, and Beta_2 is 0.999. In addition, the training steps are 6×10^5, for which the learning rate will be divided in half every 4×10^5. That is, or $1 \sim 4 \times 10^5$, the learning rate is 0.0001 for training a DSRCNN model. For $(4 \times 10^5)+1$ to 6×10^5, the learning rate is 0.00005. Epsilon size is 1×10^8. Additionally, other initial parameters can refer to Ref. [8]. In addition, we use an Adam optimizer [37] to update parameters. Codes of LSRCNN are programmed via Python of 0.41 and Python of 2.7. In addition, it runs on Ubuntu of 16.04, CPU of Inter Xeon 8163 and two NVIDIA Tesla P100. The Nvidia CUDA is 9.0 and CuDNN is 7.6.4.

4.4. Ablation Study

The proposed uses of TSEB, EB, and FLB to implement a robust SR model. In addition, TSEB is composed of two sub-networks and an enhanced technique. The enhanced technique uses local residual learning operation to enhance effects of local hierarchical layers to promote SR performance. These are verified in Table 1. DSRCNN is obtained higher PSNR and SSIM values than that of DSRCNN with residual learning operations, which shows the effectiveness of local residual learning operations. DSRCNN outperforms DSRCNN without RLO, one sub-network in Table 1, which shows the effectiveness of two sub-networks and RLO. It is known that enlarging the depth of a network is very useful to extract complementary information [38,39]. It is known that increasing the width of network can improve the performance of image tasks, according to GoogLeNet [40]. Although wider networks perform well in image applications, they will increase the complexity. In addition, two sub-networks can effectively address this question in image restoration. Thus, taking into account performance and complexity, we also choose two sub-networks to design network architecture in this paper. According to mentioned illustrations, we design two sub-networks for SISR [41–43]. In addition, it is proved that DSRCNN without RLO, EB_S outperforms improvement of 0.511dB compared to that of DSRCNN without RLO, one sub-network and EB_S in PSNR in Table 1, which shows the effectiveness of dual networks for SR. DSRCNN without RLO, one sub-network exceeds DSRCNN without RLO, one sub-network and EB_S in both PSNR and SSIM on U100 in Table 1, which tests the effectiveness of EB. Additionally, DSRCNN without RLO, one sub-network, and EB_S have obtained improvement of 0.16 dB in PSNR and 0.001 in SSIM than that of DSRCNN without RLO, one sub-network EB_S and RO in both PSNR and SSIM on U100 in Table 1, which tests the effectiveness of FLB. Specifically, EB_S and RO denote EB without enhancement from the shallow layer and FLB without four Conv+ReLU. In addition, several visual figures from an HR image, Bicubic, single branch model, and a two-branch model are conducted to test excellent performance of two-branch architecture, which can show the complementary of two branches as Figure 3 on page 9. In addition, these visual figures are obtained via amplifying one area of predicted high super-resolution images as observation areas. According to mentioned illustrations, we can see the effectiveness of key techniques for SISR.

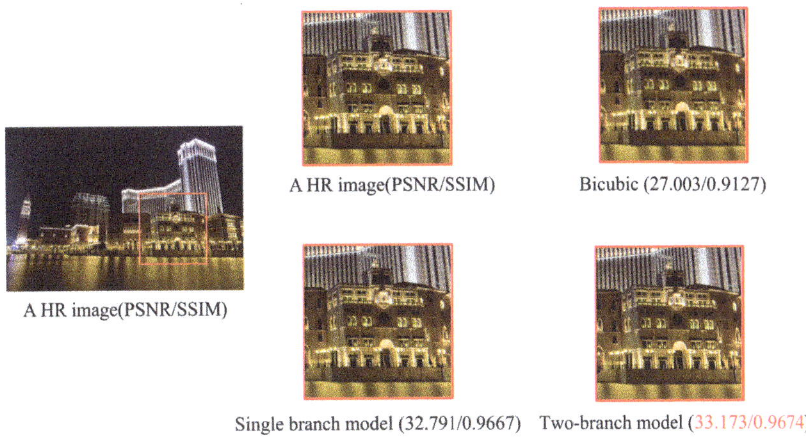

Figure 3. Visual figures of different methods.

Table 1. PSNR and SSIM of different methods on U100 for ×2.

	Methods	U100 PSNR/SSIM
Scale	DSRCNN	31.833/0.9252
	DSRCNN without RLO, one sub-network	31.676/0.9237
	DSRCNN without RLO, one sub-network and EB_S	31.220/0.9181
	DSRCNN without RLO, one sub-network EB_S and RO	31.060/0.9171
	DSRCNN without RLO, EB_S	31.649/0.9238
	DSRCNN without RLO	31.701/0.9241

4.5. Experiment Results

To fairly evaluate the SR performance of DSRCNN, quantitative and qualitative analysis are used to conduct experiments. The quantitative analysis includes PSNR [44] and SSIM [44] of popular methods, i.e., Bicubic, A+ [7], jointly optimized regressors (JOR) [45], RFL [6], self-exemplars super-resolution (SelfEx) [36], CSCN [18], RED [19], a denoising convolutional neural network (DnCNN) [46], trainable nonlinear reaction diffusion (TNRD) [47], fast dilated residual SR convolutional network (FDSR) [48], SRCNN [10], fast SR CNN (FSRCNN) [14], very deep SR network (VDSR) [19], deeply-recursive convolutional network (DRCN) [13], context wise network fusion (CNF) [49], Laplacian SR network (LapSRN) [50], deep persistent memory network (MemNet) [11], CARN-M [22], wavelet domain residual network (WaveResNet) [51], convolutional principal component (CPCA) [52], new architecture of deep recursive convolution networks for SR (NDRCN) [53], LESRCNN [8], LESRCNN-S [8], and DSRCNN on four public datasets, i.e., Set5, Set14, B100, and U100. In terms of quantitative analysis, our proposed DSRCNN has obtained the best SR results in most circumstances as shown in Tables 2–5. For example, our method has obtained gain PSNR of 0.08 dB than that of LESRCNN on Set 5 for ×2 in Table 2 In addition, DSRCNN has achieved gain PSNR of 0.16 dB and SSIM of 0.0035 than that of CARN-M on Set14 for ×3 in Table 3. In addition, our method has obtained an excellent SR performance on B100 in Table 4 and on U100 in Table 5, respectively. Our method is very competitive in complexity in Table 6. In terms of qualitative analysis, we choose Bicubic, SelfEx, SRCNN, and CARN-M as comparative methods to test the visual effects of DSRCNN. Amplify a chosen area from predicted high-resolution images from these methods as an observation area, where observation area is clearer and corresponding SR methods are better in SISR. As shown in Figures 4–6, we can see that the observation areas of our method are clearer than other SR methods. This shows that our method is more effective in SISR. According to quantitative analysis and qualitative analysis, our method is robust in different screens.

According to descriptions, we can see that the proposed method can reply with two sub-networks, the combination of residual learning, convolutional layer, and ReLU to obtain excellent SR performance. However, it has slower execution speed in SR than that of the single network with the same parameters. Thus, how to develop an efficient and robust SR network is very important for us in our work in the future.

Table 2. PSNR and SSIM of different techniques with scale factors of ×2, ×3 and ×4 on Set5.

Dataset	Model	×2	×3	×4
		PSNR/SSIM	**PSNR/SSIM**	**PSNR/SSIM**
Set5	Bicubic	33.66/0.9299	30.39/0.8682	28.42/0.8104
	A+ [7]	36.54/0.9544	32.58/0.9088	30.28/0.8603
	JOR [45]	36.58/0.9543	32.55/0.9067	30.19/0.8563
	RFL [6]	36.54/0.9537	32.43/0.9057	30.14/0.8548
	SelfEx [36]	36.49/0.9537	32.58/0.9093	30.31/0.8619
	CSCN [18]	36.93/0.9552	33.10/0.9144	30.86/0.8732
	RED [19]	37.56/0.9595	33.70/0.9222	31.33/0.8847
	DnCNN [46]	37.58/0.9590	33.75/0.9222	31.40/0.8845
	TNRD [47]	36.86/0.9556	33.18/0.9152	30.85/0.8732
	FDSR [48]	37.40/0.9513	33.68/0.9096	31.28/0.8658
	SRCNN [10]	36.66/0.9542	32.75/0.9090	30.48/0.8628
	FSRCNN [14]	37.00/0.9558	33.16/0.9140	30.71/0.8657
	RCN [54]	37.17/0.9583	33.45/0.9175	31.11/0.8736
	VDSR [19]	37.53/0.9587	33.66/0.9213	31.35/0.8838
	DRCN [13]	37.63/0.9588	33.82/0.9226	31.53/0.8854
	CNF [49]	37.66/0.9590	33.74/0.9226	31.55/0.8856
	LapSRN [50]	37.52/0.9590	-	31.54/0.8850
	MemNet [11]	37.78/0.9597	34.09/0.9248	31.74/0.8893
	CARN-M [22]	37.53/0.9583	33.99/0.9236	31.92/0.8903
	WaveResNet [51]	37.57/0.9586	33.86/0.9228	31.52/0.8864
	CPCA [52]	34.99/0.9469	31.09/0.8975	28.67/0.8434
	NDRCN [53]	37.73/0.9596	33.90/0.9235	31.50/0.8859
	LESRCNN [8]	37.65/0.9586	33.93/0.9231	31.88/0.8903
	LESRCNN-S [8]	37.57/0.9582	34.05/0.9238	31.88/0.8907
	DSRCNN(Ours)	37.73/0.9588	34.17/0.9247	31.89/0.8909

Table 3. PSNR and SSIM of different techniques with scale factors of ×2, ×3, and ×4 on Set14.

Dataset	Model	×2	×3	×4
		PSNR/SSIM	PSNR/SSIM	PSNR/SSIM
Set14	Bicubic	30.24/0.8688	27.55/0.7742	26.00/0.7027
	A+ [7]	32.28/0.9056	29.13/0.8188	27.32/0.7491
	JOR [45]	32.38/0.9063	29.19/0.8204	27.27/0.7479
	RFL [6]	32.26/0.9040	29.05/0.8164	27.24/0.7451
	SelfEx [36]	32.22/0.9034	29.16/0.8196	27.40/0.7518
	CSCN [18]	32.56/0.9074	29.41/0.8238	27.64/0.7578
	RED [19]	32.81/0.9135	29.50/0.8334	27.72/0.7698
	DnCNN [46]	33.03/0.9128	29.81/0.8321	28.04/0.7672
	TNRD [47]	32.51/0.9069	29.43/0.8232	27.66/0.7563
	FDSR [48]	33.00/0.9042	29.61/0.8179	27.86/0.7500
	SRCNN [10]	32.42/0.9063	29.28/0.8209	27.49/0.7503
	FSRCNN [14]	32.63/0.9088	29.43/0.8242	27.59/0.7535
	RCN [54]	32.77/0.9109	29.63/0.8269	27.79/0.7594
	VDSR [19]	33.03/0.9124	29.77/0.8314	28.01/0.7674
	DRCN [13]	33.04/0.9118	29.76/0.8311	28.02/0.7670
	CNF [49]	33.38/0.9136	29.90/0.8322	28.15/0.7680
	LapSRN [50]	33.08/0.9130	29.63/0.8269	28.19/0.7720
	MemNet [11]	33.28/0.9142	30.00/0.8350	28.26/0.7723
	CARN-M [22]	33.26/0.9141	30.08/0.8367	28.42/0.7762
	WaveResNet [51]	33.09/0.9129	29.88/0.8331	28.11/0.7699
	CPCA [52]	31.04/0.8951	27.89/0.8038	26.10/0.7296
	NDRCN [53]	33.20/0.9141	29.88/0.8333	28.10/0.7697
	LESRCNN [8]	33.320.9148	30.12/0.8380	28.44/0.7772
	LESRCNN-S [8]	33.30/0.9145	30.16/0.8384	28.43/0.7776
	DSRCNN(Ours)	33.43/0.9157	30.24/0.8402	28.46/0.7796

Table 4. PSNR and SSIM of different techniques with scale factors of ×2, ×3, and ×4 on B100.

Dataset	Model	×2	×3	×4
		PSNR/SSIM	PSNR/SSIM	PSNR/SSIM
B100	Bicubic	29.56/0.8431	27.21/0.7385	25.96/0.6675
	A+ [7]	31.21/0.8863	28.29/0.7835	26.82/0.7087
	JOR [45]	31.22/0.8867	28.27/0.7837	26.79/0.7083
	RFL [6]	31.16/0.8840	28.22/0.7806	26.75/0.7054
	SelfEx [36]	31.18/0.8855	28.29/0.7840	26.84/0.7106
	CSCN [18]	31.40/0.8884	28.50/0.7885	27.03/0.7161
	RED [19]	31.96/0.8972	28.88/0.7993	27.35/0.7276
	DnCNN [46]	31.90/0.8961	28.85/0.7981	27.29/0.7253
	TNRD [47]	31.40/0.8878	28.50/0.7881	27.00/0.7140

Table 4. *Cont.*

Dataset	Model	×2	×3	×4
		PSNR/SSIM	**PSNR/SSIM**	**PSNR/SSIM**
	FDSR [48]	31.87/0.8847	28.82/0.7797	27.31/0.7031
	SRCNN [10]	31.36/0.8879	28.41/0.7863	26.90/0.7101
	FSRCNN [14]	31.53/0.8920	28.53/0.7910	26.98/0.7150
	VDSR [19]	31.90/0.8960	28.82/0.7976	27.29/0.7251
	DRCN [13]	31.85/0.8942	28.80/0.7963	27.23/0.7233
	CNF [49]	31.91/0.8962	28.82/0.7980	27.32/0.7253
	LapSRN [50]	31.80/0.8950	-	27.32/0.7280
	MemNet [11]	32.08/0.8978	28.96/0.8001	27.40/0.7281
	CARN-M [22]	31.92/0.8960	28.91/0.8000	27.44/0.7304
	WaveResNet [51]	32.15/ 0.8995	28.86/0.7987	27.32/0.7266
	NDRCN [53]	32.00/0.8975	28.86/0.7991	27.30/0.7263
	LESRCNN [8]	31.95/0.8964	28.91/0.8005	27.45/0.7313
	LESRCNN-S [8]	31.95/0.8965	28.94/ 0.8012	27.47/0.7321
	DSRCNN(Ours)	32.05/0.8978	29.01/0.802927.50/0.7341	

Table 5. PSNR and SSIM of different techniques with scale factors of ×2, ×3, and ×4 on U100

Dataset	Model	×2	×3	×4
		PSNR/SSIM	**PSNR/SSIM**	**PSNR/SSIM**
	Bicubic	26.88/0.8403	24.46/0.7349	23.14/0.6577
	A+ [7]	29.20/0.8938	26.03/0.7973	24.32/0.7183
	JOR [45]	29.25/0.8951	25.97/0.7972	24.29/0.7181
	RFL [6]	29.11/0.8904	25.86/0.7900	24.19/0.7096
	SelfEx [36]	29.54/0.8967	26.44/0.8088	24.79/0.7374
	DnCNN [46]	30.74/0.9139	27.15/0.8276	25.20/0.7521
	TNRD [47]	29.70/0.8994	26.42/0.8076	24.61/0.7291
	FDSR [48]	30.91/0.9088	27.23/0.8190	25.27/0.7417
	SRCNN [10]	29.50/0.8946	26.24/0.7989	24.52/0.7221
	FSRCNN [14]	29.88/0.9020	26.43/0.8080	24.62/0.7280
U100	VDSR [19]	30.76/0.9140	27.14/0.8279	25.18/0.7524
	DRCN [13]	30.75/0.9133	27.15/0.8276	25.14/0.7510
	LapSRN [50]	30.41/0.9100	-	25.21/0.7560
	MemNet [11]	31.31/0.9195	27.56/0.8376	25.50/0.7630
	CARN-M [22]	31.23/0.9193	27.55/0.8385	25.62/0.7694
	WaveResNet [51]	30.96/0.9169	27.28/0.8334	25.36/0.7614
	CPCA [52]	28.17/0.8990	25.61/0.8123	23.62/0.7257
	NDRCN [53]	31.06/0.9175	27.23/0.8312	25.16/0.7546
	LESRCNN [8]	31.45/0.9206	27.70/0.8415	25.77/0.7732
	LESRCNN-S [8]	31.45/0.9207	27.76/0.8424	25.78/0.7739
	DSRCNN(Ours)	31.83/0.9252	27.99/0.8483	25.94/0.7815

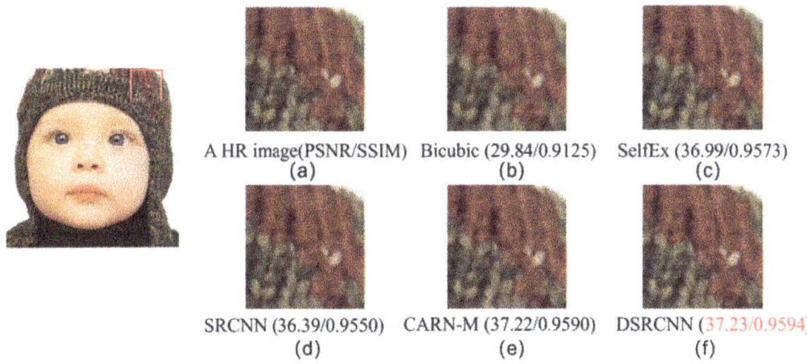

Figure 4. Visual effects of the different SR method on Set5 for ×2 scale: (**a**) A HR image (PSNR/SSIM), (**b**) Bicubic (29.84/0.9125), (**c**) SelfEx (36.99/0.9573), (**d**) SRCNN (36.39/0.9550), (**e**) CARN-M (37.22/0.9590), and (**f**) DSRCNN (37.23/0.9594).

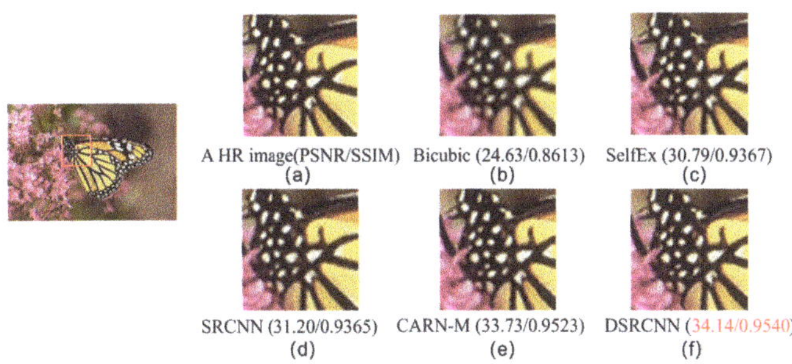

Figure 5. Visual effects of the different SR method on Set14 for ×4 scale: (**a**) A HR image (PSNR/SSIM), (**b**) Bicubic (24.63/0.8613), (**c**) SelfEx (30.79/0.9367), (**d**) SRCNN (31.20/0.9365), (**e**) CARN-M (33.73/0.9523), and (**f**) DSRCNN (34.14/0.9540).

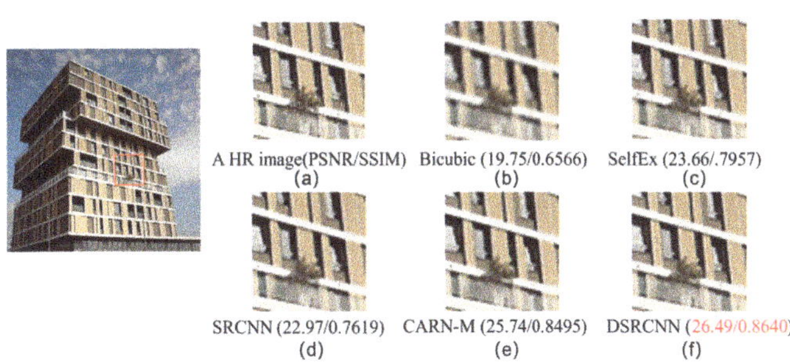

Figure 6. Visual effects of different SR method on U100 for ×4 scale: (**a**) A HR image (PSNR/SSIM), (**b**) Bicubic (19.75/0.6566), (**c**) SelfEx (23.66/.7957), (**d**) SRCNN (22.97/0.7619), (**e**) CARN-M(25.74/0.8495), and (**f**) DSRCNN (26.49/0.8640).

Table 6. Complexity of six networks for SISR.

Methods	Parameters	Flops
VDSR	665K	10.90G
DnCNN	556K	9.11G
DRCN	1,774K	29.07G
MemNet	677K	11.09G
LESRCNN	598K	3.56G
DSRCNN(our)	798 k	4.76G

5. Conclusions

In this paper, we propose a dual super-resolution CNN (DSRCNN) to obtain clear images. DSRCNN uses a two sub-network enhanced block (TSEB) to extract complementary low-frequency features to improve learning ability in SR. Combinations of convolutions and residual learning operation in TSEB are used to facilitate memory abilities of shallow layers, which can prevent a long-term dependency problem. To prevent information loss of an original image, an enhanced block is used to gather original information and obtain high-frequency information from a deeper layer via sub-pixel convolutions. To obtain more high-frequency features, a feature learning block is used to learn more details of high-frequency information. The proposed method can be applied to portable devices. We will use an attention mechanism to obtain more robust SR models in the future.

Author Contributions: Conceptualization, J.S. and J.X.; methodology, J.X.; software, J.S.; validation, C.T., Y.H. and L.Y.; formal analysis, J.S.; investigation, J.X.; resources, J.S.; data curation, J.S.; writing—original draft preparation, J.S.; writing—review and editing, C.T.; visualization, S.Z.; supervision, C.T.; project administration, S.Z.; funding acquisition, S.Z. All authors have read and agreed to the published version of the manuscript.

Funding: This paper is supported in part by the Shenzhen Science and Technology Program under Grant 2021A1515110079, in part by the Fundamental Research Funds for the Central Universities under Grant D5000210966, in part by the Basic Research Plan in Taicang under Grant TC2021JC23, and in part by the in part by the Key Project of NSFC under Grant 61836016.

Institutional Review Board Statement: Not applicable.

Informed Consent Statement: Not applicable.

Data Availability Statement: Not applicable.

Conflicts of Interest: There are no conflicts of interest.

References

1. Van Ouwerkerk, J. Image super-resolution survey. *Image Vis. Comput.* **2006**, *24*, 1039–1052. [CrossRef]
2. Sun, J.; Xu, Z.; Shum, H.-Y. Image super-resolution using gradient profile prior. In Proceedings of the IEEE Conference on Computer Vision and Pattern Recognition, Anchorage, AK, USA, 23–28 June 2008; pp. 1–8.
3. Zha, Z.; Yuan, X.; Wen, B.; Zhou, J.; Zhang, J.; Zhu, C. A benchmark for sparse coding: When group sparsity meets rank minimization. *IEEE Trans. Image Process.* **2020**, *29*, 5094–5109. [CrossRef]
4. Zhang, K.; Gao, X.; Tao, D.; Li, X. Single image super-resolution with non-local means and steering kernel regression. *IEEE Trans. Image Process.* **2012**, *21*, 4544–4556. [CrossRef]
5. Zhang, H.; Zhang, Y.; Li, H.; Huang, T.S. Generative bayesian image super resolution with natural image prior. *IEEE Trans. Image Process.* **2012**, *21*, 4054–4067. [CrossRef] [PubMed]
6. Schulter, S.; Leistner, C.; Bischof, H. Fast and accurate image upscaling with super-resolution forests. In Proceedings of the IEEE Conference on Computer Vision and Pattern Recognition, Boston, MA, USA, 7–12 June 2015; pp. 3791–3799.
7. Timofte, R.; Smet, V.D.; Van Gool, L. A+: Adjusted anchored neighborhood regression for fast super-resolution. In Proceedings of the Asian Conference on Computer Vision, Singapore, 1–5 November 2014; pp. 111–126.
8. Tian, C.; Zhuge, R.; Wu, Z.; Xu, Y.; Zuo, W.; Chen, C.; Lin, C.-W. Lightweight image super-resolution with enhanced cnn. *Knowl.-Based Syst.* **2020**, *205*, 106235. [CrossRef]

9. Lan, R.; Sun, L.; Liu, Z.; Lu, H.; Pang, C.; Luo, X. Madnet: A fast and lightweight network for single-image super resolution. *IEEE Trans. Cybern.* **2020**, *51*, 1443–1453. [CrossRef] [PubMed]
10. Dong, C.; Loy, C.C.; He, K.; Tang, X. Image super-resolution using deep convolutional networks. *IEEE Trans. Pattern Anal. Mach. Intell.* **2015**, *38*, 295–307. [CrossRef]
11. Tai, Y.; Yang, J.; Liu, X.; Xu, C. Memnet: A persistent memory network for image restoration. In Proceedings of the IEEE International Conference on Computer Vision, Venice, Italy, 22–29 October 2017; pp. 4539–4547.
12. Tai, Y.; Yang, J.; Liu, X. Image super-resolution via deep recursive residual network. In Proceedings of the IEEE Conference on Computer Vision and Pattern Recognition, Honolulu, HI, USA, 21–26 July 2017; pp. 3147–3155.
13. Kim, J.; Lee, J.K.; Lee, K.M. Deeply-recursive convolutional network for image super-resolution. In Proceedings of the IEEE Conference on Computer Vision and Pattern Recognition, Las Vegas, NV, USA, 27–30 June 2016; pp. 1637–1645.
14. Dong, C.; Loy, C.C.; Tang, X. Accelerating the super-resolution convolutional neural network. In Proceedings of the European Conference on Computer Vision, Amsterdam, The Netherlands, 11–14 October 2016; pp. 391–407.
15. Tian, C.; Xu, Y.; Zuo, W.; Lin, C.-W.; Zhang, D. Asymmetric CNN for image superresolution. *IEEE Trans. Syst. Man Cybern. Syst.* **2021**. [CrossRef]
16. Zhang, Y.; Tian, Y.; Kong, Y.; Zhong, B.; Fu, Y. Residual dense network for image super-resolution. In Proceedings of the IEEE Conference on Computer Vision and Pattern Recognition, Salt Lake City, UT, USA, 18–23 June 2018; pp. 2472–2481.
17. Tian, C.; Fei, L.; Zheng, W.; Xu, Y.; Zuo, W.; Lin, C.-W. Deep learning on image denoising: An overview. *Neural Netw.* **2020**, *131*, 251–275. [CrossRef]
18. Wang, Z.; Liu, D.; Yang, J.; Han, W.; Huang, T. Deep networks for image super-resolution with sparse prior. In Proceedings of the IEEE International Conference on Computer Vision, Santiago, Chile, 7–13 December 2015; pp. 370–378.
19. Kim, J.; Lee, J.K.; Lee, K.M. Accurate image super-resolution using very deep convolutional networks. In Proceedings of the IEEE Conference on Computer Vision and Pattern Recognition, Las Vegas, NV, USA, 26th June 2016; pp. 1646–1654.
20. Mao, X.; Shen, C.; Yang, Y.-B. Image restoration using very deep convolutional encoder-decoder networks with symmetric skip connections. *Adv. Neural Inf. Process. Syst.* **2016**, *29*, 2802–2810.
21. Wang, Z.; Chen, J.; Hoi, S.C. Deep learning for image super-resolution: A survey. *IEEE Trans. Pattern Anal. Mach. Intell.* **2020**, *43*, 3365–3387. [CrossRef] [PubMed]
22. Tian, C.; Xu, Y.; Zuo, W.; Zhang, B.; Fei, L.; Lin, C.-W. Coarse-to-fine cnn for image super-resolution. *IEEE Trans. Multimed.* **2020**, *23*, 1489–1502. [CrossRef]
23. Chen, W.; Yao, P.; Gai, S.; Da, F. Multi-scale feature aggregation network for image super-resolution. *Appl. Intell.* **2022**, *52*, 3577–3586. [CrossRef]
24. Geng, T.; Liu, X.-Y.; Wang, X.; Sun, G. Deep shearlet residual learning network for single image super-resolution. *IEEE Trans. Image Process.* **2021**, *30*, 4129–4142. [CrossRef] [PubMed]
25. Nathan, S.; Kansal, P. Leveraging multi scale backbone with multilevel supervision for thermal image super resolution. In Proceedings of the IEEE/CVF Conference on Computer Vision and Pattern Recognition, place virtually, 19–25 June 2021; pp. 4332–4338.
26. Ahn, N.; Kang, B.; Sohn, K.-A. Fast, accurate, and lightweight super-resolution with cascading residual network. In Proceedings of the European Conference on Computer Vision (ECCV), Munich, Germany, 6 August 2018; pp. 252–268.
27. Tian, C.; Xu, Y.; Zuo, W. Image denoising using deep cnn with batch renormalization. *Neural Netw.* **2020**, *121*, 461–473. [CrossRef]
28. Pan, J.; Liu, S.; Sun, D.; Zhang, J.; Liu, Y.; Ren, J.; Li, Z.; Tang, J.; Lu, H.; Tai, Y.-W.; et al. Learning dual convolutional neural networks for low-level vision. In Proceedings of the IEEE Conference on Computer Vision and Pattern Recognition, Salt Lake City, Utah, 18–22 June 2018; pp. 3070–3079.
29. Tian, C.; Xu, Y.; Zuo, W.; Du, B.; Lin, C.-W.; Zhang, D. Designing and training of a dual cnn for image denoising. *Knowl.-Based Syst.* **2021**, *226*, 106949. [CrossRef]
30. Xin, J.; Li, J.; Jiang, X.; Wang, N.; Huang, H.; Gao, X. Wavelet-based dual recursive network for image super-resolution. *IEEE Trans. Neural Netw. Learn. Syst.* **2020**, *33*, 707–720. [CrossRef]
31. Douillard, C.; Jézéquel, M.; Berrou, C.; Electronique, D.; Picart, A.; Didier, P.; Glavieux, A. Iterative correction of intersymbol interference: Turbo-equalization. *Eur. Trans. Telecommun.* **1995**, *6*, 507–511. [CrossRef]
32. Krizhevsky, A.; Sutskever, I.; Hinton, G.E. Imagenet classification with deep convolutional neural networks. *Adv. Neural Inf. Process. Syst.* **2012**, *25*, 1097–1105. [CrossRef]
33. Agustsson, E.; Timofte, R. Ntire 2017 challenge on single image super-resolution: Dataset and study. In Proceedings of the IEEE Conference on Computer Vision and Pattern Recognition Workshops, Honolulu, HI, USA, 21–26 July 2017; pp. 126–135.
34. Bevilacqua, M.; Roumy, A.; Guillemot, C.; Alberi-Morel, M.L. Low-complexity single-image super-resolution based on non-negative neighbor embedding. In Proceedings of the British Machine Vision Conference, Surrey, UK, 3–7 September 2012; pp. 135.1–135.10.
35. Martin, D.; Fowlkes, C.; Tal, D.; Malik, J. A database of human segmented natural images and its application to evaluating segmentation algorithms and measuring ecological statistics. In Proceedings Eighth IEEE International Conference on Computer Vision, ICCV 2001, Vancouver, BC, Canada, 7–14 July 2001; pp. 416–423.
36. Huang, J.-B.; Singh, A.; Ahuja, N. Single image super-resolution from transformed self-exemplars. In Proceedings of the IEEE Conference on Computer Vision and Pattern Recognition, Boston, MA, USA, 8–10 June 2015; pp. 5197–5206.

37. Kingma, D.P.; Ba, J. Adam: A method for stochastic optimization. *arXiv* **2014**, arXiv:1412.6980.
38. Shen, Z.; Wang, W.; Lu, X.; Shen, J.; Ling, H.; Xu, T.; Shao, L. Human-aware motion deblurring. In Proceedings of the IEEE/CVF International Conference on Computer Vision, Seoul, Korea, 27 October–2 November 2019; pp. 5572–5581.
39. Zhou, T.; Qi, S.; Wang, W.; Shen, J.; Zhu, S.-C. Cascaded parsing of human-object interaction recognition. *IEEE Trans. Pattern Anal. Mach. Intell.* **2021**. [CrossRef] [PubMed]
40. Szegedy, C.; Liu, W.; Jia, Y.; Sermanet, P.; Reed, S.E.; Anguelov, D.; Erhan, D.; Vanhoucke, V.; Rabinovich, A. Going deeper with convolutions. *CoRR* 2014. Available online: http://arxiv.org/abs/1409.4842 (accessed on 20 January 2022).
41. Wang, Y.; Gong, D.; Yang, J.; Shi, Q.; Hengel, A.v.d.; Xie, D.; Zeng, B. An effective two-branch model-based deep network for single image deraining. *arXiv* **2019**, arXiv:1905.05404.
42. Wang, L.; Li, Y.; Huang, J.; Lazebnik, S. Learning two-branch neural networks for image-text matching tasks. *IEEE Trans. Pattern Anal. Mach. Intell.* **2018**, *41*, 394–407. [CrossRef]
43. Faranda, R. A new parameters identification procedure for simplified double layer capacitor two-branch model. *Electr. Power Syst. Res.* **2010**, *80*, 363–371. [CrossRef]
44. Hore, A.; Ziou, D. Image quality metrics: Psnr vs. ssim. In Proceedings of the 2010 20th International Conference on Pattern Recognition, Istanbul, Turkey, 23–26 August 2010; pp. 2366–2369.
45. Dai, D.; Timofte, R.; Van Gool, L. Jointly optimized regressors for image super-resolution. In *Computer Graphics Forum*; Wiley Online Library: Hoboken, NJ, USA, 2015; Volume 34, No. 2, pp. 95–104.
46. Zhang, K.; Zuo, W.; Chen, Y.; Meng, D.; Zhang, L. Beyond a gaussian denoiser: Residual learning of deep cnn for image denoising. *IEEE Trans. Image Process.* **2017**, *26*, 3142–3155. [CrossRef] [PubMed]
47. Chen, Y.; Pock, T. Trainable nonlinear reaction diffusion: A flexible framework for fast and effective image restoration. *IEEE Trans. Pattern Anal. Mach. Intell.* **2016**, *39*, 1256–1272. [CrossRef]
48. Lu, Z.; Yu, Z.; Yali, P.; Shigang, L.; Xiaojun, W.; Gang, L.; Yuan, R. Fast single image super-resolution via dilated residual networks. *IEEE Access* **2018**, *7*, 109729–109738. [CrossRef]
49. Ren, H.; El-Khamy, M.; Lee, J. Image super resolution based on fusing multiple convolution neural networks. In Proceedings of the IEEE Conference on Computer Vision and Pattern Recognition Workshops, Honolulu, HI, USA, 21–26 July 2017; pp. 54–61.
50. Lai, W.-S.; Huang, J.-B.; Ahuja, N.; Yang, M.-H. Deep laplacian pyramid networks for fast and accurate super-resolution. In Proceedings of the IEEE Conference on Computer Vision and Pattern Recognition, Honolulu, HI, USA, 21–26 July 2017; pp. 624–632.
51. Bae, W.; Yoo, J.; Ye, J.C. Beyond deep residual learning for image restoration: Persistent homology-guided manifold simplification. In Proceedings of the IEEE Conference on Computer Vision and Pattern Recognition Workshops, Honolulu, HI, USA, 21–26 July 2017; pp. 145–153.
52. Xu, J.; Li, M.; Fan, J.; Zhao, X.; Chang, Z. Self-learning super-resolution using convolutional principal component analysis and random matching. *IEEE Trans. Multimed.* **2018**, *21*, 1108–1121. [CrossRef]
53. Cao, F.; Chen, B. New architecture of deep recursive convolution networks for super-resolution. *Knowl.-Based Syst.* **2019**, *178*, 98–110. [CrossRef]
54. Shi, Y.; Wang, K.; Chen, C.; Xu, L.; Lin, L. Structure-preserving image super-resolution via contextualized multitask learning. *IEEE Trans. Multimed.* **2017**, *19*, 2804–2815. [CrossRef]

electronics

MDPI

Article

A Novel Denoising Algorithm Based on Wavelet and Non-Local Moment Mean Filtering

Caixia Liu [1,2] and Li Zhang [3,*]

[1] College of Intelligent Education, Jiangsu Normal University, Xuzhou 221116, China
[2] Jiangsu Engineering Research Center of Educational Informationization, Xuzhou 221116, China
[3] Department of Information Science and Engineering, Zaozhuang University, Zaozhuang 277160, China
* Correspondence: chianleezl@sina.com

Abstract: Denoising is the basis and premise of image processing and an important part of image preprocessing. Denoising can effectively improve image quality, which contributes to subsequent image processing such as image segmentation, feature extraction, and so on. In this paper, we propose a novel image denoising method based on wavelet transform and nonlocal moment mean filtering approach (NMM). The noisy image is firstly denoised by a wavelet-based soft-thresholding denoising technique and NMM is then utilized to further eliminate the rest noises. Meanwhile, the fusion of moment invariants increases the robustness of our denoising algorithm due to the invariance of image scaling, translation, and rotation of color moments. Experiments show that our algorithm achieves a better denoising effect compared with some other denoising approaches.

Keywords: image denoising; wavelet transform; color moments; non-local mean filter

Citation: Liu, C.; Zhang, L. A Novel Denoising Algorithm Based on Wavelet and Non-Local Moment Mean Filtering. *Electronics* **2023**, *12*, 1461. https://doi.org/10.3390/electronics12061461

Academic Editor: Sergio Carrato

Received: 24 February 2023
Revised: 13 March 2023
Accepted: 16 March 2023
Published: 20 March 2023

1. Introduction

Image is often disturbed by random noise signals in the process of acquisition or transmission. Common image noises include salt and pepper noise, Gauss noise, Poisson noise, and so on. These noises reduce the quality of the images, which seriously hinders the subsequent image processing such as edge extraction, image segmentation, feature extraction, and so on. For example, Gaussian noise is a kind of noise whose probability density function obeys Gaussian distribution (i.e., normal distribution). If the amplitude distribution of noise is Gaussian, and its power spectral density is uniformly distributed, it is called Gaussian white noise. The effect of Gaussian noise on the image is random, which is a common noise in the image. The causes of this kind of noise mainly include: the light not being bright enough or uniform enough when the images are taken; the noise and interaction of circuit components; the temperature being too high because the sensor works for a long time. In the image, Gaussian noise is represented by the random change of pixel value, making the image become blurred or dotted with noise, which will lead to blurred or distorted details in the image, thus affecting the quality and subsequent image processing.

In order to obtain high-quality digital images, it is necessary to carry on the image noise reduction processing. Image denoising is a technology that uses context information of image sequence to remove noise and restore a clear image. It is one of the important research contents in the field of computer vision, that is, to maintain as much as possible the integrity of the original information (e.g., the main features) through a certain algorithm, but also to remove the useless information in the signal, so that the processed image is clearer. The quality of the image denoising algorithm is directly related to the effect of subsequent image processing.

Wavelet transform is a local transform of time and frequency domain, so it can extract information from signal effectively, and carry out a multi-scale detailed analysis of function or signal through operation functions such as scaling and shifting. It is widely used in image denoising. Meanwhile, the non-local Means (NLM) algorithm is one of the most popular image-denoising algorithms. It uses the redundancy of the natural image itself to restore the image polluted by noises and takes into account as much similarity-structure information as possible. At the same time, the denoising method based on the whole block information can better preserve the image edge and texture features.

In order to make full use of the advantages of the wavelet denoising and NLM method and improve the denoising effect, a novel hybrid filtering algorithm combining wavelet-based denoising technique (W) with a nonlocal mean moment filtering approach (NMM) is proposed (named W-NMM) in this paper. The technique can effectively remove noises while still retaining enough detailed information.

The remainder of this article is organized as follows: the relevant work is described in Section 2. Section 3 presents a detailed description of our W-NMM model. The experiments are displayed in Section 4 and the key issues of this paper are discussed in Section 5. In Section 6, we make a brief conclusion. A group of abbreviations and the corresponding nomenclature is shown in Table A1.

2. Related Work

The existence of noise reduces the image quality and hinders the subsequent processing of the image. In order to remove noise and improve image quality, many scholars have proposed a variety of image denoising methods including traditional techniques and neural network-based techniques as shown in Table 1.

The traditional image denoising methods can be divided into two categories: spatial denoising and frequency denoising. The former includes morphology filtering, mean filtering, Gauss filtering, morphological filtering, local filtering, non-local filtering, and so on [1]. The latter includes Wiener filtering, wavelet threshold denoising [2], and so on. For example, Chen et al. [3] proposed a multi-structural element auto-adapted determination weight algorithm combining morphology filter of opening and closing operations. According to the different characteristics of images contaminated by different kinds of noises, a hybrid denoising method was proposed by Guan et al. [4]. Firstly, the local threshold was used to classify the pixels as those polluted by Gauss noise and salt and pepper noise. Mean and median filtering approaches were used to denoise them. Hu et al. [5] analyzed mean filtering, median filtering, and wavelet transform, which are three conventional methods for image denoising processing. Because median filtering usually results in image blur, Zhao et al. [6] improved median filtering and put forward a weighted fast median filtering algorithm and a weighted adaptive median filtering algorithm. Aiming at the shortcomings of classical soft and hard thresholding methods in denoising, Yin et al. [7] presented an improved new threshold function, which could satisfy the continuous input-output curve while the decomposed wavelet coefficients were kept unchanged. Traditional soft and hard thresholding methods cannot effectively express energy distribution, so it is necessary to find a balance between denoising and edge information preserving. Zhang et al. [8] presented an improved threshold function integrating the advantages of the classical wavelet threshold function and other improved methods. Wang et al. [9] used a wavelet thresholding method to denoise the COVID-19 CT image, where the threshold function was obtained by the improved particle swarm optimization. Kazuaki et al. [10] removed quantum noise from the STEM image with a total variation denoising algorithm, where they defined an entropy of the STEM image that corresponds to the image contrast and then determined a hyperparameter to maximize the entropy. Guo et al. [11] presented a median filtering algorithm based on an adaptive two-stage threshold to improve the accuracy of CT image noise detection. In the method, an adaptive weighted median filter image denoising method was put forward based on a hybrid genetic algorithm. Yuan et al. [12] put forward an edge-preserving median filter and weighted coding with sparse nonlocal regularization for low-dose CT image denoising. In addition, the classical filtering algorithm also includes anisotropic diffusion [13], bilateral filtering [14], kernel singular value decomposition (K-SVD) [15], sparse 3-D transform-domain collaborative filtering [16], and so on.

Deep learning, especially convolutional neural network (CNN), has achieved good results in image recognition and other fields. In recent years, image denoising methods based on deep learning have also been developed. Wang et al. [17] proposed a multi-scale feature-extraction-based normalized attention neural network for image denoising. In the model, they employed a multi-scale feature extraction block to extract and combine features at distinct scales of the noisy image, and a normalized attention network was applied to learn the relationships between channels. Ahmed et al. [18] proposed a medical image denoising system based on the stacked convolutional autoencoder technique. Huang et al. [19] presented an unsupervised learning approach incorporating a pseudo-siamese network for image processing, where two independent branches of the network utilize different filling strategies, namely zero filling and adjacent pixel filling. Wang et al. [20] used an optimized denoising convolutional neural networks method based on sub-region processing and transfer learning to denoise the images. Usui et al. [21] compared the dose-dependent properties of a CNN-based denoising method for low-dose CT with those of other noise reduction methods on unique CT noise simulation images. They observed that the CNN model can eliminate noise and maintain image sharpness at these dose levels. Rajesh et al. [22] developed a differential evolution-based automatic network evolution model by exploring the fittest parameters. Furthermore, they adopted a transfer learning technique to accelerate the training process.

Table 1. A list of the literature on denoising methods.

Classification	Year	Author	Methods
Traditional image denoising methods	2003	Chen et al. [3]	Mathematics morphology
	2005	Guan et al. [4]	Mean and median filtering approaches
	2007	Hu et al. [5]	Mean filtering, median filtering, and wavelet transform
	2011	Zhao et al. [6]	Improved median filtering
	2018	Yin et al. [7]	Improved wavelet threshold
	2017	Zhang et al. [8]	Threshold with wavelet transform
	2022	Wang et al. [9]	Wavelet transform combined with improved PSO
	2022	Kazuaki et al. [10]	Total variation regularization
	2022	Guo et al. [11]	Adaptive threshold and optimized weighted median filter
	2021	Yuan et al. [12]	Edge-Preserving Median Filter and Weighted Coding with Sparse Nonlocal Regularization
	1990	Perona et al. [13]	Anisotropic diffusion
	1998	Tomasi [14]	Bilateral filtering
	2005	Aharon et al. [15]	K-SVD
	2007	Kostadin et al. [16]	Sparse 3-D transform-domain collaborative filtering
Deep learning approaches	2021	Wang et al. [17]	Attention neural network
	2021	Ahmed et al. [18]	Stacked convolutional autoencoder
	2021	Huang et al. [19]	Unsupervised pseudo-siamese network
	2021	Wang et al. [20]	Convolutional neural network
	2021	Usui et al. [21]	Convolutional neural network
	2022	Rajesh et al. [22]	An evolutionary block-based network

Although the traditional denoising method is simple, it also has many limitations. For example, the morphology method, neighborhood average method, and median filtering can suppress the noise, but also easily cause the image blur phenomenon, which is not suitable for the image with more details of points, lines, and peaks. The neural network–based techniques require a large number of training samples, and it is difficult to obtain all kinds of natural noise samples for training. Based on this, this paper aimed to propose a novel filtering algorithm based on wavelet and non-local moment mean filtering.

3. Methodology

The algorithm in this paper mainly includes two steps: multi-scale decomposition denoising and non-local moment mean filtering.

3.1. Multi-Scale Decomposition Denoising

Wavelet transform (see Figure 1 for a reference) [23] is a time-frequency localization analysis method in which the size of the window is fixed but its time window and frequency window can be changed. That is, the low-frequency part has a low time resolution and a high-frequency resolution, and the high-frequency part has a high time resolution and a low-frequency resolution, which is suitable for the analysis of non-stationary signals such as images and the extraction of local features from such signals.

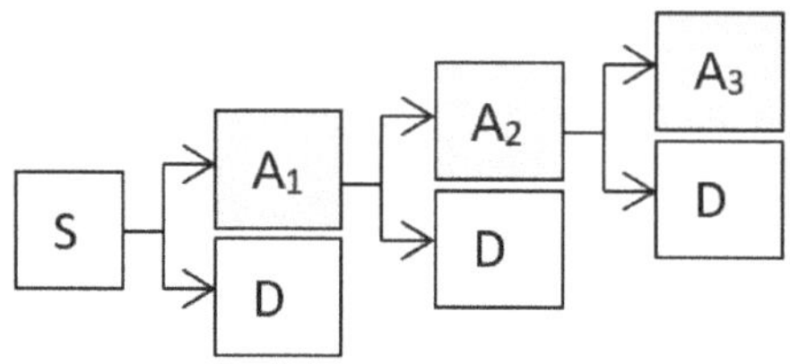

Figure 1. Image decomposition with wavelet transform.

Wavelet transform is to move a mother wavelet with a displacement τ, and then do the inner product with the analytic signal $x(t)$ at different scales a.

$$W'T_x(a,\tau) = \frac{\sqrt{a}}{2\pi}\int_{-\infty}^{\infty} X(\omega)\varphi^*(a\omega)e^{+j\omega\pi}d\omega \tag{1}$$

a is a scale factor and $a > 0$. τ reflects displacement. In the frequency domain, it is expressed as

$$WT_x(a,\tau) = \frac{1}{\sqrt{a}}\int_{-\infty}^{\infty} x(t)\psi^*\left(\frac{t-\tau}{a}\right)dt \tag{2}$$

Discrete Wavelet Transform (*DWT*) discretizes scale parameters according to power series, which is often used in multi-resolution analysis and signal decomposition and reconstruction.

$$DWTx(m,n) \leq x(t), \psi_{m,n}(t) \geq 2^{-\frac{m}{2}}\int_R x(t)\psi(2^{-m}t-n)dt \tag{3}$$

where the wavelet function is

$$\psi_{jk}(t) = 2^{-\frac{j}{2}}\psi\left(2^{-j}t-k\right) \tag{4}$$

In multi-resolution analysis, for example, orthogonal wavelet transform can be equivalent to a set of mirror filtering processes, i.e., signal S is decomposed through a high-pass filter and a low-pass filter. The high-frequency component, D_i, of the corresponding signal is called the detail component. The output of the low-pass filter corresponds to the relative signal A_i, which is called the approximate component, see Figure 1 for a reference.

In the wavelet domain, coefficients corresponding to the effective signal are usually very large, while those corresponding to noises are very small. At present, the commonly used threshold-based methods include hard threshold, soft threshold, and so on. The wavelet coefficients obtained by the soft threshold method have good continuity and no discontinuity. Here, we adopted the soft-threshold-based method to remove Gaussian noises.

When the absolute value of the wavelet coefficients is less than a given threshold value, it is zero; when the wavelet coefficients are larger than the threshold value, the threshold value is subtracted from the wavelet coefficients.

$$w_\lambda = \begin{cases} [\text{sgn}(w)](|w|-\lambda) & |w| \geq \lambda \\ 0 & |w| < \lambda \end{cases} \tag{5}$$

where sgn(x) returns "+1" if x is a positive value and "−1" otherwise cases. λ is calculated with [24].

$$\lambda = \sigma\sqrt{2\ln N} \tag{6}$$

here, σ = M/0.6745, and M is the median absolute deviation of detail coefficients at high-frequency sub-images. N is the length of the signal.

Figure 2 shows several denoising results with the soft threshold-based wavelet denoising method. From Figure 2, we can find that the noisy image was denoised well.

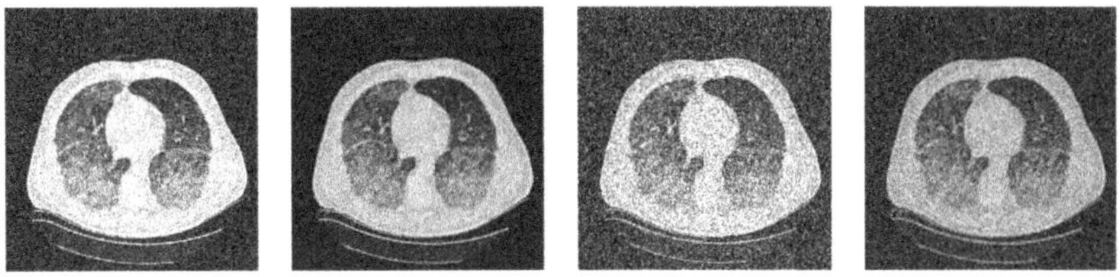

(a) **(b)** **(c)** **(d)**

Figure 2. Image denoising with wavelet transform. (**a**,**c**) are noisy images with Gaussian noise variance = 0.02 and variance = 0.1, respectively. (**b**,**d**) are the denoising results of (**a**,**c**), respectively.

3.2. Non-Local Moment Mean Filtering

Non-local mean filtering uses all the pixels in the image, and these pixels are weighted according to some kind of similarity. After filtering, the image clarity is high, and the details are not lost, so the structural information of the image is better protected [25]. If we take the noise image $v(i)$ as the sum of the image $u(i)$ and the noise $n(i)$ whose mean value is 0 without noise contamination, $v(i)$ can be expressed as

$$v(i) = u(i) + n(i) \tag{7}$$

For a given pixel i in an image v, the image block $N(i)$ sized $n \times n$ is an image block with i as the block center and $N(j)$ is an image block in the neighborhood of $N(i)$. The similarity between i and j is measured by Gaussian weighted Euclidean distance between the image blocks $N(i)$ and $N(j)$. The smaller the distance between $N(j)$ and $N(i)$ is, the more similar the pixel j is to the pixel i, and the greater the weight given by the pixel j in cumulative restoration.

Assuming that the denoised image is $I(i)$, for a pixel i, the NLM calculation is as follows

$$I(i) = \frac{\sum\limits_{j\in v} W(i,j)v(j)}{\sum\limits_{j\in v} w(i,j)} \tag{8}$$

We define $v(N_i)$ as a rectangular neighborhood centered on i, and the similarity coefficient $w(i,j)$ of the pixels i and j in the image v is as follows:

$$w(i,j) = \exp\left(-\frac{\|v(N_i) - v(N_j)\|^2_{2,\alpha}}{h^2}\right) \tag{9}$$

where α is the standard deviation of the Gaussian kernel function, $\|v(N_i) - v(N_j)\|^2_{2,\alpha}$ represents the weighted Euclidean distance between two image blocks; h is a filtering parameter to control the smoothness

$$\|V(N_i) - V(N_j)\|^2 = \frac{1}{d^2}\sum_{i+z\in N_i, j+z\in N_j} \|v(i+z) - v(j+z)\|^2 \tag{10}$$

The non-local mean filtering algorithm makes full use of the block information of the image and can keep the texture and edge of the image well. The filtering effect is better [26]. However, similarity measurement lacks robustness. In this paper, we replace the gray difference with the moment the difference in the weighted Euclidean distance between two image blocks and produce a novel denoising method, called the non-local moment mean denoising method, abbreviated as NMM.

Moments and the related invariants have been extensively analyzed to characterize the patterns of images in a variety of applications [27].

$$m_{pq} = \int_{-\infty}^{\infty}\int_{-\infty}^{\infty} x^p f(x,y)dxdy, \quad p,q = 0,1,2,\cdots \tag{11}$$

Hu [24] introduced seven-moment invariants $M = \{\phi_1, \phi_2, \phi_3, \phi_4, \phi_5, \phi_6, \phi_7\}$

$$\phi_1 = \eta_{20} + \eta_{02} \tag{12}$$

$$\phi_2 = (\eta_{20} - \eta_{02})^2 + 4\eta_{11}^2 \tag{13}$$

$$\phi_3 = (\eta_{30} - 3\eta_{12})^2 + (3\eta_{21} - \eta_{03})^2 \tag{14}$$

$$\phi_4 = (\eta_{30} + \eta_{12})^2 + (\eta_{21} + \eta_{03})^2 \tag{15}$$

$$\begin{aligned}\phi_5 = {} & (\eta_{30} - 3\eta_{12})(\eta_{30} + \eta_{12})\left[(\eta_{30} + \eta_{12})^2 - 3(\eta_{21} + \eta_{03})^2\right] \\ & +(3\eta_{21} - \eta_{03})(\eta_{21} + \eta_{03})\left[3(\eta_{30} + \eta_{12})^2 - (\eta_{21} + \eta_{03})^2\right]\end{aligned} \tag{16}$$

$$\phi_6 = (\eta_{20} - \eta_{02})\left[(\eta_{30} + \eta_{12})^2 - (\eta_{21} + \eta_{03})^2\right] + 4\eta_{11}(\eta_{30} + \eta_{12}) + (\eta_{21} + \eta_{03}) \tag{17}$$

$$\begin{aligned}\phi_7 = {} & (3\eta_{21} - \eta_{03})(\eta_{30} + \eta_{12})\left[(\eta_{30} + \eta_{12})^2 - 3(\eta_{21} + \eta_{03})^2\right] \\ & -(\eta_{30} - 3\eta_{12})(\eta_{21} + \eta_{03})\left[3(\eta_{30} + \eta_{12})^2 - (\eta_{21} + \eta_{03})^2\right]\end{aligned} \tag{18}$$

Moment invariants are useful properties of being unchanged under image scaling, translation, and rotation. In the end, the weighted Euclidean distance between two image blocks is

$$\|V(N_i) - V(N_j)\|^2 = \frac{1}{d^2}\sum \|v(N_{i,M}) - v(N_{j,M})\|^2 \tag{19}$$

where $N_{i,M}$ is the moment value of the image block N_i and $N_{j,M}$ is the moment value of the image block N_j.

Combined with wavelet-based denoising (W) and non-local moment mean filtering (NMM), the algorithm W-NMM is described as Algorithm 1.

Algorithm 1: W-NMM filtering

Input: image I to be filtered
t: radio of search window
f: radio of similarity window
h: degree of filtering
1. Take sym8 as the wavelet basis function to decompose the image in two layers.
2. Calculate the soft threshold according to Equation (6) on the high-frequency domains.
3. Denoise image I according to Equation (5) and obtained I′.
4. Symmetric padding I′;
5. For each pixel in I′(i,j) (i = f:M − f, j = f:N − f):
i1 ⇐ i + f; j1 ⇐ j + f;
Create objective window: W1= I′(i1 − f:i1 + f, j1 − f:j1 + f);
6. Set the borders of the neighboring window:
rmin⇐max(i1 − t, f + 1); rmax⇐min(i1 + t, m + f);
smin⇐max(j1 − t, f + 1); smax⇐min(j1 + t, n + f);

Algorithm 1: *Cont.*

```
7. For each pixel in W2(r,s):
                Set neighboring window: W2 = input2(r − f:r + f, s − f:s + f);
8.    Calculate the moments of W1 and W2:
n1 = hu_moments(W1);
                n2 = hu_moments(W2);
9. Calculate the similarity of W1 and W2 according to n1 and n2.
10. Calculate the Gaussian weight: w⇐exp − d/h);
11. Find the maximum of w: wmax.
                sweight⇐sweight + w;
                average⇐average + w × I'(r,s);
        end
12. Calculate the accumulation of
            average = average + wmax × I'(i,j);
            sweight = sweight + wmax;
13. Calculate denoised image Iout:
            if sweight > 0
                Iout'(i,j) = average/sweight;
            else
                Iout'(i,j) = I'(i,j);
            end
14. end
15. Extract the image Iout with the size same to I from Iout'.
```

4. Experiment

In order to evaluate the performance of our algorithm, we test it on a set of noisy images and several examples. The following metrics are utilized for evaluating the performance of image processing approaches Peak Signal to Noise Ratio (PSNR) [28] and Structural Similarity Index (SSIM) [29].

PSNR is widely used to evaluate image quality and is defined as

$$\text{PSNR} = 10 \cdot \log_{10}\left(\frac{MAX_I^2}{\text{MSE}}\right) = 20 \cdot \log_{10}\left(\frac{MAX_I}{\sqrt{\text{MSE}}}\right) \tag{20}$$

where MAX_I is the maximum value of an image I, MSE is Mean Square Error, and expressed as

$$\text{MSE} = \frac{1}{mn}\sum_{i=0}^{m-1}\sum_{j=0}^{n-1}\|I(i,j) - K(i,j)\|^2 \tag{21}$$

Where I and K can be taken as the denoised image and original image, respectively. The smaller the MSE and the bigger the PSNR, the better the image quality.

SSIM is one of the indicators to measure image quality. Given two images I and K, their SSIM can be defined as:

$$\text{SSIM} = \frac{(2\mu_I\mu_K + c_1)(2\sigma_{IK} + c_2)}{(\mu_I^2 + \mu_K^2 + c_1)(\sigma_I^2 + \sigma_K^2 + c_2)} \tag{22}$$

where, μ_I and μ_K are the means of I and K, respectively. σ_I^2 and σ_K^2 are variances of I and K, respectively. σ_{IK} is the covariance of I and K. $c_1 = (k_1L)^2$ and $c_2 = (k_2L)^2$ are constants keeping things stable. L is the dynamic range of the image pixel value, $k_1 = 0.01$, $k_2 = 0.03$.

The range of SSIM is [0, 1]. The larger the SSIM is, the better the image quality is. Figure 3 shows the denoising result with our W-NMM algorithm. In Figure 3a–d there are images with Gaussian white noise with variances 0.01, 0.02, 0.04, and 0.06, respectively. The first two lines are the corresponding noisy images with their partial histograms, and the last two lines are the denoised images with their partial histograms. It can be seen the algorithm removed the noises well. Figures 4 and 5 show the denoising results on the original noisy image and the corresponding ones after rotation, scaling, and translation. It can be seen that the results were similar, which shows the robustness of the method.

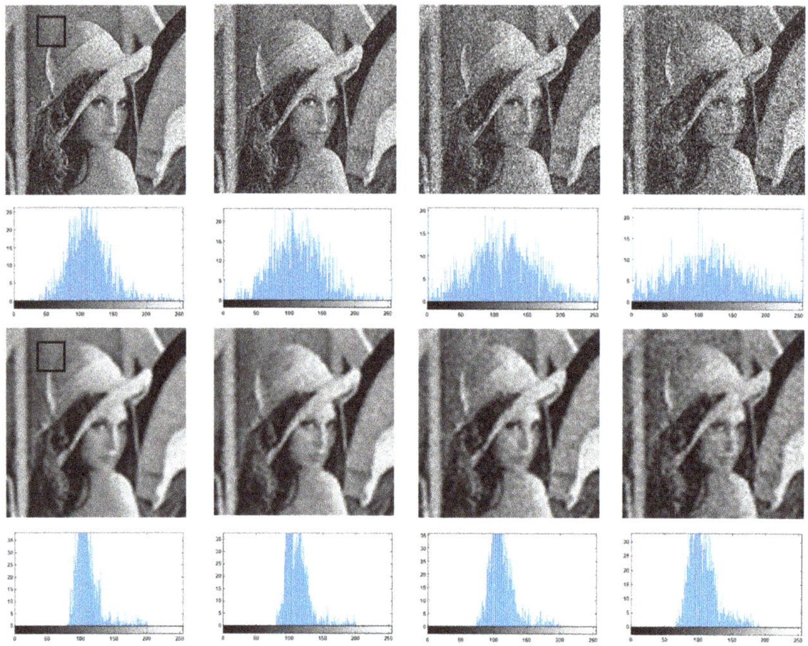

(a) (b) (c) (d)

Figure 3. The denoising result with W-NMM algorithm. (**a**–**d**) are images added Gaussian white noise of variance 0.01, 0.02, 0.04, and 0.06, respectively. The first two lines are the corresponding noisy images with their partial histograms, and the last two lines are the denoised images with their partial histograms.

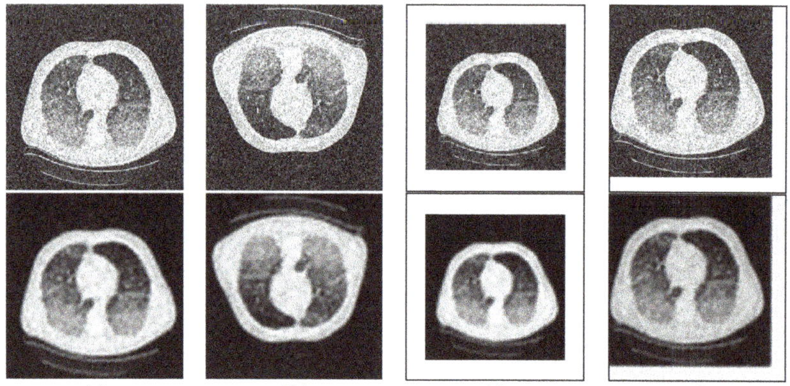

(a) (b) (c) (d)

Figure 4. The denoising results of the (**a**) original noisy image, (**b**) rotated image, (**c**) scaled image, and (**d**) translated image.

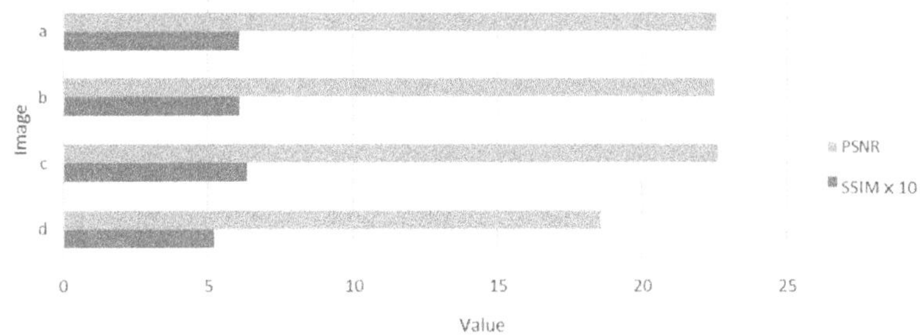

Figure 5. The evaluation of the denoising results of Figure 4 in terms of PSNR and SSIM. a~d refter to the images a–d in Figure 4.

We compare our W-NMM algorithm with the anisotropic diffusion filter (AD) [13], bilateral filter (BF) [14], Kernel Singular Value Decomposition (KSVD) [15], and block Matching and 3D collaborative filtering (BM3D) [16] on a group of CT images [30]. The visual results are shown in Figure 6, where the noisy CT images are added Gaussian white noise of variance = 0.02 as shown in Figure 6a–f) are the corresponding denoised results with AD, BF, KSVD, BM3D, and W-NMM. From Figure 6, we can find that the W-NMM algorithm has a better effect on Gaussian noise denoising. Compared with the other denoising methods, our algorithm can produce better results on noisy image denoising. The three-dimensional (3D) visualizations of the denoising effectiveness are exhibited in Figure 6 where two images are randomly selected from Figure 7 and their 3D visualizations are depicted before and after denoising with the proposed method. (a) are the noisy images and (b) are the corresponding denoised images with W-NMM. It can be observed that the proposed method removed the sharp noises, and the image regions become smooth while keeping the edges.

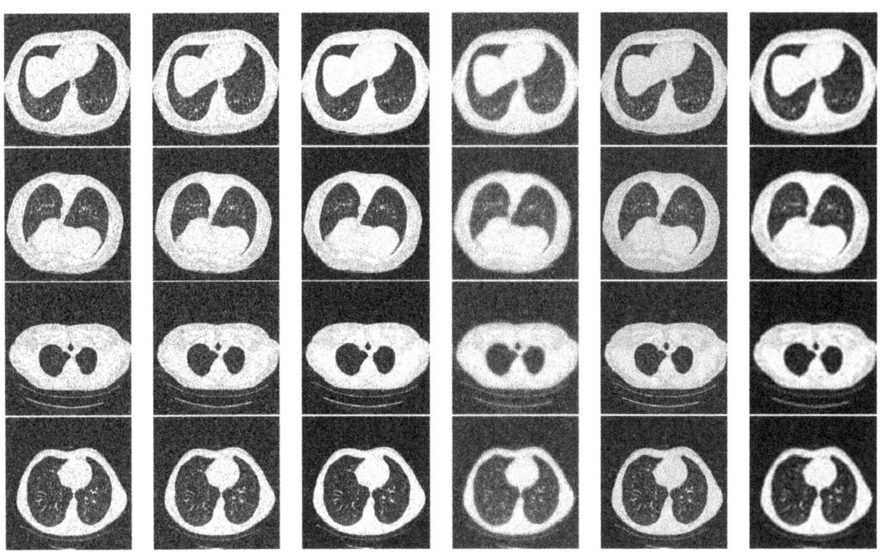

(a) (b) (c) (d) (e) (f)

Figure 6. Comparison of different denoising methods on four images. (**a**) are noisy images. (**b–f**) are the denoised images with AD, BF, KSVD, BM3D, and our W-NMM method, respectively.

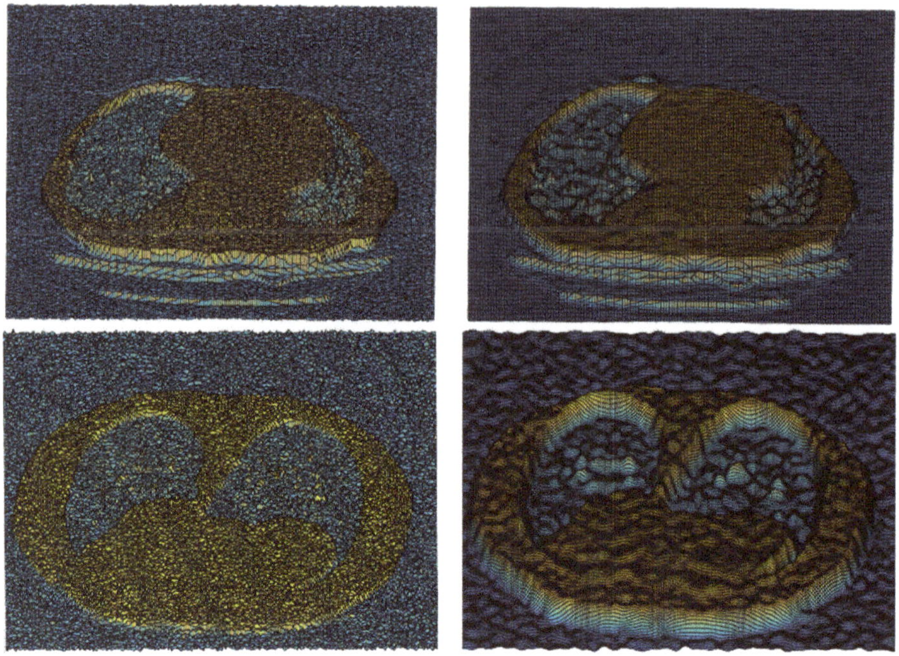

Figure 7. The 3D visualization of the denoising results of two images in Figure 6 with the WM-NLM method. (**a**) show the noisy images with Gaussian white noise, and (**b**) are the corresponding denoised results.

We evaluated the denoising methods (AD, BF, NLM, BM3D, and W-NMM) on the images in Figure 6 in terms of PSNR and SSIM. The results are shown in Table 2. We can find that our method achieved higher PSNR and SSIM than other methods. We test the methods on a group of medical images and compare their denoising effect, and the average results in terms of PSNR and SSIM are displayed in Figure 8. It can be observed that the W-NMM method is superior to the compared methods.

Table 2. Comparison of different denoising methods evaluated with PSNR and SSIM (The best results are shown in bold).

Image	I1		I2		I3		I4	
Metrics	PSNR	SSIM	PSNR	SSIM	PSNR	SSIM	PSNR	SSIM
AD	18.48	0.2964	18.38	0.2407	18.42	0.2352	18.27	0.3298
BF	20.37	0.3593	19.95	0.3004	20.21	0.2965	19.71	0.3731
KSVD	22.61	0.5578	22.49	0.3787	23.34	0.5512	**22.67**	0.5660
BM3D	20.66	0.5479	23.63	0.5736	22.80	0.4603	21.15	0.4830
W-NMM	**22.67**	**0.6219**	**24.20**	**0.6798**	**23.62**	**0.6557**	22.43	**0.6714**

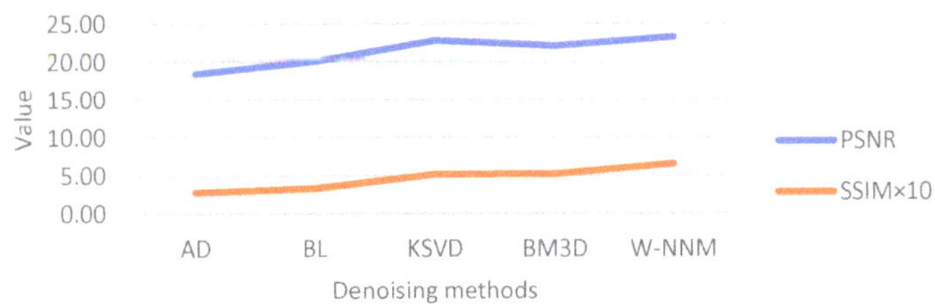

Figure 8. Comparison of the denoising performance with different denoising methods in terms of PSNR, and SSIM.

In order to test the validity of the two stages of filtering, we made an ablation experiment and the result is shown in Figure 9. S1 represents the result in the first stage, that is the corresponding image is denoised with wavelet filtering. S2 is the result in the second stage, that is the corresponding image was denoised with the NMM filter. It can be found that the SSIM was improved after the NMM filtering in the S2 stage.

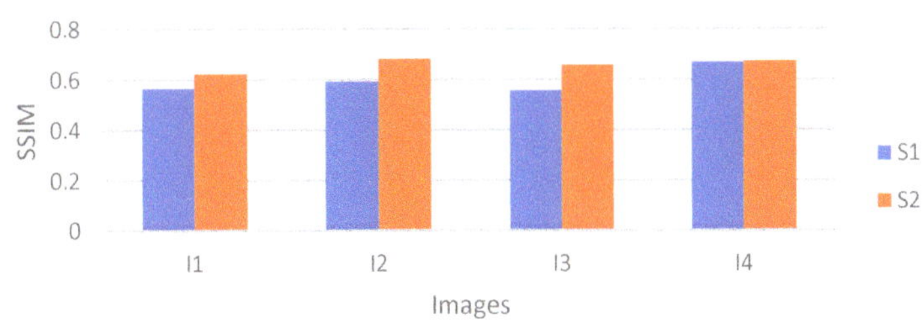

Figure 9. Ablation experiment of the W-NMM algorithm.

5. Discussion

In the process of digital image digitization and transmission, it is often affected by the noise of imaging equipment and the external environment, so that the image quality will be degraded. Image denoising is the process of reducing the noise in the digital image. The commonly used image denoising methods are suitable for processing images with low requirements on image details, that is to say, the loss of tiny details has little impact on the subsequent processing of image denoising. However, when dealing with medical images, such small mistakes are not allowed, because every small mistake in medical diagnosis or treatment can affect the doctor's treatment and even threaten the patient's life. So, we need good denoising techniques that can effectively remove noise while still preserving enough detail. It can be seen from the experiment that the algorithm proposed in this paper can effectively smooth the noise information in the image and keep the details of the image. Effective image denoising can not only help doctors diagnose the condition, but also be very conducive to the subsequent image segmentation, e.g., lung segmentation, providing help for computer-aided diagnosis.

Our algorithm achieved good performance on image denoising. However, the cost time is sometimes high, and the time efficiency is low due to the fusion of moments and the NLM approach. Our algorithm can generate the highest denoising effect with low time efficiency while the anisotropic diffusion filter has the highest time efficiency with the lowest PNSR. Accordingly, we can select suitable methods for different applications.

6. Conclusions

In this paper, we denoised the images with a wavelet-based non-local moment mean denoising algorithm. The proposed W-NMM algorithm combined frequency domain denoising with spatial domain denoising, and the introduction of moments increased the robustness of the denoising algorithm. The average of PSNR and SSIM achieved 23.3 and

0.66, respectively. In addition, it showed a better-denoised effect compared with several classical image denoising methods. It contributes to the subsequent image processing, such as image segmentation, 3D reconstruction, and so on. Nevertheless, the time cost was high because of the NLM operation. In the future, we will improve the time efficiency of our algorithm.

Author Contributions: Conceptualization, methodology, software, C.L.; validation, L.Z. All authors have read and agreed to the published version of the manuscript.

Funding: This research was supported by the National Natural Science Foundation of China (Grant No. 62007028), and the Doctoral Research Foundation of Jiangsu Normal University (NO.21XSRX005).

Data Availability Statement: The data used to support the findings of this study are available from the corresponding author upon request.

Conflicts of Interest: The authors declare no conflict of interest.

Appendix A

Table A1. List of abbreviations and the corresponding nomenclature.

Abbreviations	Nomenclature
W	Wavelet
NLM	Non-local mean filter
NMM	Non-local moment mean filter
AD	Anisotropic diffusion filter
BF	Bilateral filter
BM3D	Block matching and 3D collaborative filtering
KSVD	Kernel singular value decomposition
PSNR	Peak signal to noise ratio
SSIM	Structural similarity index

References

1. Xu, M.; Xie, X. An efficient feature-preserving PDE algorithm for image denoising based on a spatial-fractional anisotropic diffusion equation. *East Asian J. Appl. Math.* **2021**, *11*, 788–807.
2. Vaiyapuri, T.; Alaskar, H.; Sbai, Z.; Devi, S. GA-based multi-objective optimization technique for medical image denoising in wavelet domain. *J. Intell. Fuzzy Syst.* **2021**, *41*, 1575–1588. [CrossRef]
3. Chen, H.; Zhou, C.H.; Wang, S.Z. Research based on mathematics morphology image chirp method. *J. Eng. Graph.* **2003**, *2*, 116–119.
4. Guan, X.P.; Zhao, L.X.; Tang, Y.G. Mixed filter for image denoising. *J. Image Graph.* **2005**, *10*, 332–337.
5. Hu, L.; Zhang, W.; Tan, Y.Q. Application and analysis about some arithmetics for image denoising. *Inf. Technol.* **2007**, *7*, 81–83.
6. Zhao, G.C.; Zhang, L.; Wu, F.B. Application of improved median filtering algorithm to image de-noising. *J. Appl. Opt.* **2011**, *32*, 678–682.
7. Yin, Q.S.; Dai, S.G. Research on image denoising algorithm based on improved wavelet threshold. *Softw. Guide* **2018**, *17*, 89–91.
8. Zhang, X.; Turghunjan, A.T. Improvement of threshold image denoising algorithm with wavelet transform. *Comput. Technol. Dev.* **2017**, *27*, 81–84.
9. Wang, G.; Guo, S.; Han, L.; Cekderi, A.B.; Song, X.; Zhao, Z. Asymptomatic COVID-19 CT image denoising method based on wavelet transform combined with improved PSO. *Biomed. Signal Process. Control* **2022**, *76*, 103707. [CrossRef]
10. Kawahara, K.; Ishikawa, R.; Sasano, S.; Shibata, N.; Ikuhara, Y. Atomic-resolution STEM image denoising by total variation regularization. *Microscopy* **2022**, *5*, 302–310. [CrossRef]
11. Guo, S.; Wang, G.; Han, L.; Song, X.; Yang, W. COVID-19 CT image denoising algorithm based on adaptive threshold and optimized weighted median filter. *Biomed. Signal Process. Control* **2022**, *75*, 103552. [CrossRef] [PubMed]
12. Yuan, Q.; Peng, Z.; Chen, Z.; Guo, Y.; Yang, B.; Zeng, X. Edge-Preserving Median Filter and Weighted Coding with Sparse Nonlocal Regularization for Low-Dose CT Image Denoising Algorithm. *J. Healthc. Eng.* **2021**, *2021*, 6095676. [CrossRef] [PubMed]
13. Perona, P.; Malik, J. Scale-space and edge detection using anisotropic diffusion. *IEEE Trans. Pattern Anal. Mach. Intell.* **1990**, *12*, 629–639. [CrossRef]
14. Tomasi, C.; Manduchi, R. Bilateral filtering for gray and color images. In Proceedings of the International Conference on Computer Vision, Copenhagen, Denmark, 28–31 May 2002.
15. Aharon, M.; Elad, M.; Bruckstein, A. K-SVD: Design of dictionaries for sparse representation. *IEEE Trans. Signal Process.* **2006**, *54*, 4311–4322. [CrossRef]

16. Dabov, K.; Foi, A.; Katkovnik, V.; Egiazarian, K. Image denoising by sparse 3-D transform-domain collaborative filtering. *IEEE Trans. Image Process.* **2007**, *16*, 2080–2095. [CrossRef] [PubMed]
17. Wang, Y.; Song, X.; Gong, G.; Li, N. A Multi-Scale Feature Extraction-Based Normalized Attention Neural Network for Image Denoising. *Electronics* **2021**, *10*, 319. [CrossRef]
18. Ahmed, A.S.; El-Behaidy, W.H.; Youssif, A.A. Medical image denoising system based on stacked convolutional autoencoder for enhancing 2-dimensional gel electrophoresis noise reduction. *Biomed. Signal Process. Control* **2021**, *69*, 102842. [CrossRef]
19. Huang, C.; Hong, D.; Yang, C.; Cai, C.; Tao, S.; Clawson, K.; Peng, Y. A new unsupervised pseudo-siamese network with two filling strategies for image denoising and quality enhancement. *Neural Comput. Appl.* **2021**, *1*, 1–9. [CrossRef]
20. Wang, J.; Tang, Y.; Zhang, J.; Yue, M.; Feng, X. Convolutional neural network-based image denoising for synchronous measurement of temperature and deformation at elevated temperature. *Optik* **2021**, *241*, 166977. [CrossRef]
21. Usui, K.; Ogawa, K.; Goto, M.; Sakano, Y.; Kyougoku, S.; Daida, H. Quantitative evaluation of deep convolutional neural network-based image denoising for low-dose computed tomography. *Vis. Comput. Ind. Biomed. Art* **2021**, *4*, 21. [CrossRef]
22. Rajesh, C.; Kumar, S. An evolutionary block based network for medical image denoising using Differential Evolution. *Appl. Soft Comput.* **2022**, *121*, 108776. [CrossRef]
23. Gao, Q.W.; Li, B.; Xie, G.J.; Zhuang, Z.Q. An image de-noising method based on stationary wavelet transform. *J. Comput. Res. Dev.* **2002**, *39*, 1689–1694.
24. Donoho, D.L.; Johnstone, J.M. Ideal spatial adaptation by wavelet shrinkage. *Biometrika* **1994**, *81*, 425–455. [CrossRef]
25. Buades, A.; Coll, B.; Morel, J.M. A non-local algorithm for image denoising. In Proceedings of the 2005 IEEE Computer Society Conference on Computer Vision and Pattern Recognition (CVPR'05), San Diego, CA, USA, 20–26 June 2005.
26. Yi, Z.L.; Yin, D.; Hu, A.Z.; Zhang, R. SAR Image Despeckling Based on Non-local Means Filter. *J. Electron. Inf. Technol.* **2012**, *34*, 950–953.
27. Hu, M.-K. Visual pattern recognition by moment invariants. *IEEE Trans. Inf. Theory* **1962**, *8*, 179–187. [CrossRef]
28. Huynh-Thu, Q.; Ghanbari, M. Scope of validity of PSNR in image/video quality assessment. *Electron. Lett.* **2008**, *44*, 800–801. [CrossRef]
29. Wang, Z.; Bovik, A.C.; Sheikh, H.R.; Simoncelli, E.P. Image Quality Assessment: From Error Visibility to Structural Similarity. *IEEE Trans. Image Process.* **2004**, *13*, 600–612. [CrossRef]
30. Depeursinge, A.; Vargas, A.; Platon, A.; Geissbuhler, A.; Poletti, P.-A.; Müller, H. Building a reference multimedia database for interstitial lung diseases. *Comput. Med. Imaging Graph.* **2012**, *36*, 227–238. [CrossRef]

electronics

MDPI

Article

Image Denoising Based on GAN with Optimization Algorithm

Min-Ling Zhu [1], Liang-Liang Zhao [1] and Li Xiao [2,3,*]

[1] Computer School, Beijing Information Science and Technology University, Beijing 100101, China
[2] Key Laboratory of Intelligent Information Processing, Institute of Computing Technology Chinese Academy of Sciences, Beijing 100090, China
[3] Ningbo Huamei Hospital, University of Chinese Academy of Sciences, Ningbo 315010, China
* Correspondence: andrew.lxiao@gmail.com

Abstract: Image denoising has been a knotty issue in the computer vision field, although the developing deep learning technology has brought remarkable improvements in image denoising. Denoising networks based on deep learning technology still face some problems, such as in their accuracy and robustness. This paper constructs a robust denoising network based on a generative adversarial network (GAN). Since the neural network has the phenomena of gradient dispersion and feature disappearance, the global residual is added to the autoencoder in the generator network, to extract and learn the features of the input image, so as to ensure the stability of the network. On this basis, we proposed an optimization algorithm (OA), to train and optimize the mean and variance of noise on each node of the generator. Then the robustness of the denoising network was improved through back propagation. Experimental results showed that the model's denoising effect is remarkable. The accuracy of the proposed model was over 99% in the MNIST data set and over 90% in the CIFAR10 data set. The peak signal to noise ratio (PSNR) and structural similarity (SSIM) values of the proposed model were better than the state-of-the-art models in the BDS500 data set. Moreover, an anti-interference test of the model showed that the defense capacities of both the fast gradient sign method (FGSM) and project gradient descent (PGD) attacks were significantly improved, with PSNR and SSIM values decreased by less than 2%.

Keywords: image denoising; GAN; optimization algorithm; autoencoder; ResNet

Citation: Zhu, M.-L.; Zhao, L.-L.; Xiao, L. Image Denoising Based on GAN with Optimization Algorithm. *Electronics* **2022**, *11*, 2445. https://doi.org/10.3390/electronics11152445

Academic Editor: Byung Cheol Song

Received: 30 June 2022
Accepted: 3 August 2022
Published: 5 August 2022

1. Introduction

Image denoising is one of the hottest research topics in the field of image processing [1]. There are various traditional image denoising methods. Tang used an improved curvature filtering algorithm, where a projection operator was used to replace the minimum triangular tangent plane projection operator of the traditional curvature filtering [2]. Li proposed an adaptive matching and tracking algorithm. First, the sparse coefficients were calculated. Then the dictionary was trained to be an adaptive dictionary, which could reflect the image structure effectively by using the K singular value decomposition algorithm. Finally, the image was reconstructed by combining the sparse coefficients with the adaptive dictionary [3]. Dabov proposed block-matching and 3D filtering (BM3D), which made use of the self-similarity existing in natural images to match with adjacent image blocks, and then the similar blocks were integrated to form the denoised image through domain transformation [4]. Xu proposed a trilateral weighted sparse coding (TWSC) scheme for robust real image denoising [5]. Xie proposed a non-convex regular low rank sparse matrix decomposition method for image denoising [6]. Although the above traditional denoising methods achieved a good effect to a certain degree, there are highly time consuming and low robustness. Li proposed a new image denoising approach based on undecimated discrete wavelet transform (UDWT), which combines the technique of cone of influence (COI) analyzing and UDWT [7].

In recent years, with the rapid development of deep learning and remarkable achievements in the field of image processing, more and more people are applying deep learning to image denoising. For example, the convolutional neural network has two major characteristics, of local perception and parameter sharing, which have a good effect in image feature extraction and recognition. Wang proposed a gradient vector convolution (GVC) model for image denoising [8]. Wu proposed an interleaved cascade of shrinkage fields (CSF) to reduce noise and jointly restore the transmission diagram and scene radiance from a single noise image [9]. Zhang proposed a feedforward denoising convolutional neural network (DnCNN) model, which combined batch normalization and residual learning [10]. Yan proposed a self-consistent GAN network (SCGAN) to extract noise images directly from noisy images, to achieve unsupervised noise modeling [11]. Yu proposed a deep iterative down-up convolutional neural network (DIDN) for image denoising, which can process various noise levels using a single model, without input noise information as a solution [12]. Zhang proposed a fast and flexible denoising convolutional neural network (FFDNet), which used a noise estimation graph as input, balancing the suppression of uniform noise and the preservation of details [13]. Chen's proposed denoising method used GAN to model the noise distribution, to generate noise samples through the established model and form a training data set with clean image sets, and to train the denoising network model to perform blind denoising [14]. Dong proposed a convolutional neural network denoising method based on multi-scale redundancy of natural images [15]. Wang proposed a novel channel and spatial attention neural network for image denoising [16]. Cai proposed a new efficient image denoising scheme, where global structure and local similarity preservations combined method of optimal directions (MOD) with approximate K-SVD (AK-SVD) for dictionary learning [17]. Cai proposed a new development of non-local image denoising using fixed-point iteration for non-convex ℓp sparse optimization [18]. Although neural networks are widely applied in the field of image processing, they are vulnerable to adversarial attacks that lead to incorrect network outputs. In 2014, Szegedy Christian introduced the L-BFGS method, which induced the model to obtain a result completely deviating from the real value by adding slight disturbance to the input sample image of the model [19]. In 2015, Goodfellow Ian J proposed an adversarial sample generation algorithm based on the fast gradient sign method (FGSM), which sought the direction with the largest gradient change in the deep learning model and generated disturbances, to increase the loss of image classifiers in this direction [20]. Later, the FGSM derived project gradient descent (PGD) and other gradient-based attack algorithms. However, some current defense methods require a lot of manpower and material resources and have poor robustness [21].

In view of low robustness of traditional denoising methods and vulnerability of deep learning network under attacks, this paper introduces a simple and efficient method to improve the robustness of the denoising network. The whole backbone of the denoising-network is based on the GAN. Moreover, the denoised image is from the GAN. Random noise is added into the neural network and it is optimized through back propagation. The most important feature is that this method does not require additional resource consumption and can simultaneously improve the model's ability for denoising and defense against attack. Furthermore, an integrated image denoising network is designed. Finally, FGSM and PGD attack experiments were used to verify the anti-interference capability of the adversarial network.

2. Related Work

In this section, we briefly overview some of the basic network modules and loss functions that are involved in our design. First, we refer to the following three networks: The first is the autoencoder, which is a form of neural network and is composed of an encoder and decoder [22]. The encoder compresses the original data to obtain the features of the original data, and learns the features through other neural networks to reduce the burden of network generation. The decoder decompresses the learned features into original data. This is an unsupervised algorithm, and then the back propagation algorithm is used to train the network to make the output close to the standard image. The second is the residual module [23]. Although more features can be extracted, the training is also more difficult due to the increasing depth of the neural network. With the increase of depth, the original data information will be gradually lost in the process of convolution and pooling, and the error signal is prone to gradient dispersion during the back propagation. Therefore, the residual network is introduced to solve the training difficulties caused by increasing the network depth. The residual network uses jump connections to connect the features after convolution and pooling with the previous features, and the information representation is enhanced by the addition of both gradual and deep features. This method avoids the problem of image feature loss due to the increase of network depth, and solves the problem of gradient dispersion and ensures the stability of the network. The third aspect is the generative and adversarial network based on the two-person game idea, which is widely used in various aspects of the imaging field. A generative adversarial network is a method of unsupervised learning. It consists of a generator network and a discriminator network, and learns by playing two neural networks against each other. The generator network takes random samples from the latent space as input, and its output should imitate the real samples in the training set as much as possible. The input of the discriminator network is the real sample or the output of the generator network, and the purpose of the discriminator network is to distinguish the output of the generator network from the real sample as far as possible. The generator network tries to deceive the discriminator network as much as possible. The final purpose of the two networks is to make the discriminator network unable to judge whether the output result of the generator network is true or not [24].

Furthermore, we refer to three loss functions. The first is MSE loss [25]. The values of each pixel of the generated image and the original image are compared, and the mean square error of the generator network is represented by the loss of pixels. The second is GAN loss, which is mainly formed by the discrimination network to determine between the generated denoised image or the original real image [26]. The GAN loss ensures that the generator network generates an image as close to the real image as possible. Then the discriminator network is deceived, to achieve the optimal result of the generated image. The third is classification loss [27]. As the generated image may cause the loss of some features, it is necessary to analyze the generated image category. Then the generator network can generate the same image as the real image, as far as possible.

3. Network Structure Design and Optimization Algorithm

The whole network structure is based on GAN. The generator network uses an autoencoder for image generation. A discriminator network is used to discriminate between the generated images. When the discriminator network cannot discriminate the authenticity of the generated images, the generated images can be used as the input of a classification network, to further verify the denoising ability of the network for noisy images. On the other hand, Gaussian noise is added to the stochastic gradient estimates of the standard deviation path of each neural network neuron. In this way, the gradient estimates and the noise level are byproducts of back propagation.

3.1. Whole Network Structure Design

The network framework we proposed is shown in Figure 1. It consists of three subnetworks: a generator network (G), discriminator network (D), and classification network (C). The G inputs an image with noise and outputs an image with the same size as the original image, through feature extraction of the network; the D inputs the generated image and standard image, and outputs "0" or "1", which represent the similarity between the generated image and standard image; the C inputs generated images, to complete the classification of image content. In G and D, we apply the network optimization algorithm (OA) proposed in the following section, which improves the robustness of GAN networks. The MSE loss and GAN loss are used to update the iterative training parameters of the

GAN neural network; classification loss is used to update the iterative training parameters of the classification network. The training finally makes the network tend to be stable.

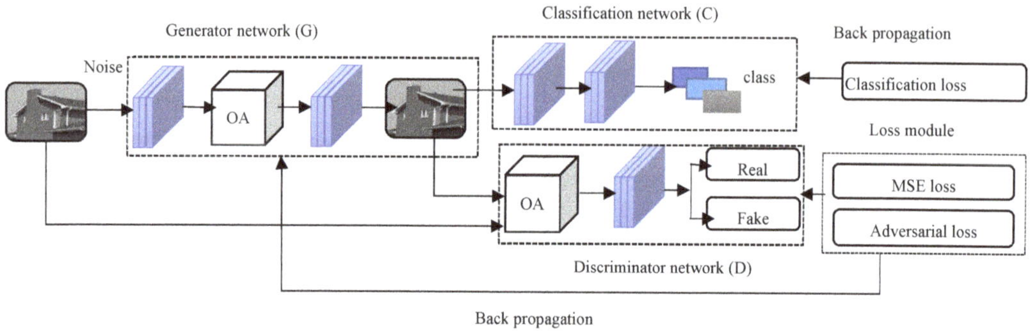

Figure 1. Whole network structure.

3.2. Optimization Algorithm

Here we deduce the OA in Figure 1. Let τ represent the layers of the neural network; m_t represents the number of neurons at layer t, $t \in 1, 2, \ldots, \tau$. The output of layer t is $x^{(t)} = [x_1^{(t)}, x_2^{(t)}, \ldots, x_{m_t}^{(t)}] \in R^{m_t}$, and $x^{(0)}$ is the input of the network.

Suppose the network has N inputs, denoted as $x^{(0)}(N)$, $N = 1, 2, \ldots, n$. For the n input, the i output of the t layer is Formulas (1) and (2).

$$x_i^{(t+1)}(n) = \varphi\left(v_i^{(t)}\right) \tag{1}$$

$$v_i^{(t)} = \sum_{j=0}^{m_t} \theta_{i,j}^{(t)} x_j^{(t)}(n) + z_i^{(t)}(n) \tag{2}$$

$x_j^{(t)}(n)$ is the j input of the n data in the t layer; $\theta_{i,j}^{(t)}$ is the weight of the i input in the t layer; $v_i^{(t)}$ is the i output of the t layer; φ is the activation function; $z_i^{(t)}(n)$ is the n data and independent random noise added to the i neuron in the t layer. Figure 2 shows a visualization of noise addition.

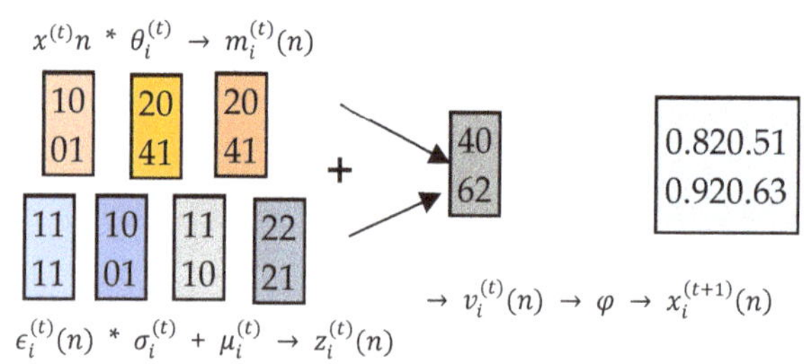

Figure 2. Optimization algorithm visualization.

L represents the loss function. For the n data $x^{(0)}(n)$ marked as $Y(n)$, $L(x^{(\tau)}(n), Y(n))$ represents the loss value. In our work, we tried to optimize the size of the noise level of the central normal random noise $\sigma_i^{(t)}$ of each neuron. $z_i^{(t)}(n) = \sigma_i^{(t)}\varepsilon_i^{(t)}(n)$, where $\varepsilon_i^{(t)}(n)$ is a standard normal random variable. The residual of the i neuron at the t layer of the n data propagates backward through the neural network and is defined as as Formula (3).

$$\delta_i^{(t)}(n) = \begin{cases} e_i^{(\tau)}(n)\varphi'\left(v_i^{(\tau-1)}(n)\right) & t = \tau \\ \varphi'\left(v_i^{(t-1)}(n)\right)\left(\sum_{j=0}^{m_k} \theta_{i,j}^{(t)}\delta_j^{(t+1)}(n)\right) & t < \tau \end{cases} \tag{3}$$

$e_i^{(\tau)}(n)$ is defined as formula (4):

$$e_i^{(\tau)}(n) = \frac{\partial L(x, Y(n))}{\partial x_i}\bigg|_{x=x^{(\tau)}(n)} \tag{4}$$

Back propagation essentially provides information about all parameters $\theta_{i,j}^{(t)}$ (t=1,2, … $\tau - 1$), path random derivative estimation of loss function L. As shown in Formula (5), $j \in \{0,\ 1, \ldots,\ m_t\}$, $i \in \{0,\ 1, \ldots, m_{t+1}\}$.

$$\frac{\partial L\left(x^{(\tau)}(n), Y(n)\right)}{\partial \theta_{i,j}^{(t)}} = \delta_j^{(t+1)}(n)x_j^{(t)}(n) \tag{5}$$

The algorithm flow is as follows:

(a) First input training data $P = \left\{\left(x^{(0)}(n),\ Y(n)\right)\right\}_{\mathrm{n}=1}^{\mathrm{N}}$, loss function L.
(b) Construct neural network.
(c) Use Formulas (1) and (2) to calculate the output $x^{(\tau)}(\mathrm{n})$.
(d) Calculate the loss function $\mathrm{L}\left(x^{(\tau)}(n),\ Y(n)\right)$.
(e) Use Formulas (3) and (5), respectively, to estimate the gradient of loss to weight and noise level.
(f) Update weights and noise levels.
(g) Repeat steps c to f until the parameters meet the requirements of the model.

3.3. Sub-Network Structure Design

The three sub-network structures proposed in this paper are shown in Figure 3.

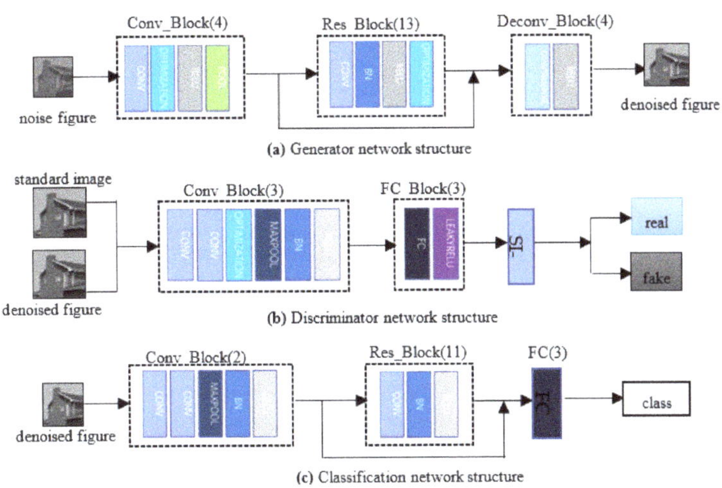

Figure 3. Sub-network structures.

Figure 3a shows the network structure of the generator network, which includes four convolution blocks, thirteen residual blocks, and four deconvolution blocks. Each one of four convolution blocks includes a convolution layer, optimization layer, relu layer, and pooling layer. In addition, each of thirteen residual blocks includes a convolutional layer, batch normalization layer, relu layer, and algorithm optimization layer. While, each one of the four deconvolution blocks includes a deconvolution layer and relu layer. The network outputs an image the same size as the standard image. The generator network is the core part of the whole network, and the image denoising effect largely depends on the ability of the generator network. Therefore, the neural network adopts encoding and decoding structures such as the autoencoder. A residual module jump connection is added in the middle, to enhance image feature representation, to avoid gradient dispersion, and to ensure the stability of the network.

Figure 3b shows the network structure of the discriminator network, which includes three convolution blocks, three linking blocks, and a sigmoid function layer. Each of three convolution blocks includes two convolution layers, an optimization layer, maximum pooling layer, batch normalization layer, and relu layer. Each of the three linking blocks includes a full link layer and leakyrelu layer. The sigmoid function layer outputs "0" or "1", which is used for the binary classification problem, to judge the difference between the positive and negative labels of the image. The discriminator network is designed based on the full convolution neural network, to discriminate the similarity between the standard image and the generated image.

Figure 3c shows the network structure of the classification network, which includes two convolution blocks, eleven residual blocks, and three full connection layers. Every two convolution blocks include a maximum pooling layer, batch normalization layer, and relu layer. Each of the eleven residual blocks includes a convolution layer, batch normalization layer, and relu layer. The final full connection layer outputs n categories to complete the classification of images. The classification network is used to classify the generated-images after the optimization of the generated network.

4. Experiments and Analyses

First, the proposed method was used to test the classification accuracy in the MNIST and CIFAR10 data sets. Then the method was compared with the DnCNN, BM3D, FFDNet, and IRCNN denoising methods, and the PSNR and SSIM values were calculated, which under the standard deviation of Gaussian noise were 25, 50, 75, and 100. Moreover, we performed a visual perception experiment. Finally, the network robustness was verified under FGSM and PGD attacks. The experiments illustrated that the method is effective.

4.1. Data Set and Parameter Setting

The MNIST data set is very well known. It consists of 60,000 training samples and 10,000 test samples, where each sample is a 28 $\times$ 28 pixel grayscale handwritten digital image. The Cifar-10 data set contains 50,000 training images and 10,000 test images, all of which are 3-channel color RGB images with a size of 32 $\times$ 32, including 10 categories in total. The two data sets were used to test the accuracy of model recognition under different noise conditions. Then we used the BDS500 data set to train and test the model. The peak signal to noise ratio (PSNR) and structural similarity (SSIM) were compared with other methods under different noise conditions.

The hardware platform of this experiment was a Tesla P100 with 16GB memory; software was Ubuntu18.04, CUDA10.02, python3.6; and the deep learning framework was Pytorch1.8; the batch processing was 128; the Adam algorithm was used to update the gradient; the initial learning rate was 0.001, and the learning rate decreased as the number of trainings increased; the momentum was 0.9.

4.2. Evaluation Index

The fidelity of image denoising is represented by the evaluation index, which is the error between the standard image and the denoised image, and the PSNR and SSIM are used for evaluation and analysis.

PSNR measures denoising performance, using the error between corresponding pixels of the denoising image and the standard image. PSNR is expressed as Formulas (6) and (7).

$$MSE = \frac{1}{mn}\sum_{i=0}^{m-1}\sum_{j=0}^{n-1}[I(i,j) - K(i,j)]^2 \tag{6}$$

$$PSNR = 10\lg\frac{MAX_I^2}{MSE} \tag{7}$$

where m and n represent the number of rows and columns of the image pixels, MAX_I is the maximum possible pixel value of the image. According to Formulas (6) and (7), the larger MSE is, the smaller $PSNR$ is, which indicates that the denoising effect is good and the denoised image is closer to the standard image.

SSIM is measured based on the luminance, contrast, and structure between the denoised image and standard image. The value ranges from "0" to "1", a larger value indicates a better denoising effect. SSIM is expressed as Formulas (8) and (9).

$$\begin{cases} l(x,y) = \dfrac{2\mu_x\mu_y + c_1}{\mu_x^2 + \mu_y^2 + c_1} \\ c(x,y) = \dfrac{2\sigma_x\sigma_y + c_2}{\sigma_x^2 + \sigma_y^2 + c_2} \\ s(x,y) = \dfrac{\sigma_{xy} + c_3}{\sigma_x\sigma_y + c_3} \end{cases} \tag{8}$$

$$SSIM(x,y) = \frac{(2\mu_x\mu_y + c_1)(2\sigma_{xy} + c_2)}{\left(\mu_x^2 + \mu_y^2 + c_1\right)\left(\sigma_x^2 + \sigma_y^2 + c_2\right)} \tag{9}$$

μ_x is the mean value of x; μ_y is the mean value of y; σ_x^2 is the variance of x; σ_y^2 is the variance of y; σ_{xy} is the covariance of x and y; $c_1 = (K_1L)^2, c_2 = (K_2L)^2$ which are constants that avoid zero; L is the range of pixel value; K_1=0.01 and $K_2 = 0.03$ are the default values.

4.3. Experimental Result and Analysis

4.3.1. Comparison of Classification Accuracy on Different Data Sets

In this paper, Gaussian noises with standard deviations of 25, 50, and 75 were added to the test set. The experimental results are shown in Figure 4.

From Figure 4a, we can see that under the influence of different noise environments the classification accuracy could reach more than 99%, and the experimental error remained within 0.005. This proves that the method is feasible for image denoising. It can resolve the classification problem of different noise levels and the images can be correctly classified under different noise levels.

Figure 4b shows the classification accuracy on CIFAR10, which could reach more than 90%. CIFAR10 is a rebuilt data set including RGB images with noise, so that the classification of CIFAR10 was harder. The experimental results showed the experimental error was stable within $\pm$0.1. This shows that the algorithm not only had a significant denoising effect for grayscale images, but also had a strong denoising ability for RGB color images, and it could realize the classification of color images and ensure the recognition accuracy of images. This paper mainly compared the accuracy gap between denoised images and standard images, without excessively pursuing the recognition accuracy of the data set. Therefore, the recognition of the data set did not achieved an optimal effect, which will be the next project.

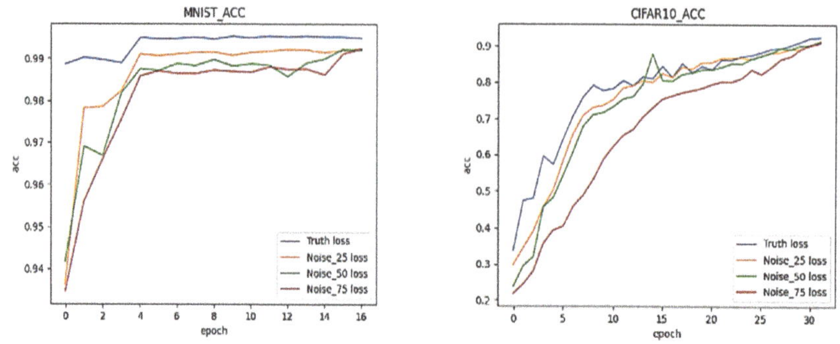

(a) (b)

Figure 4. MNIST and CIFAR10 classification accuracy. (**a**) classification accuracy on the MNIST data set, (**b**) classification accuracy on the CIFAR10 data set.

4.3.2. Comparison of PSNR and SSIM on the BDS500 Data Set among Different Methods

To compare the PSNR and SSIM values after denoising, Gaussian noises with standard deviations of 25, 50, 75, and 100 were added to the images from the BDS500 data set. Then the DnCNN, BM3D, FFDNet, IRCNN, LSLA-2, UDWT, and our method were tested. The results are shown in Tables 1 and 2.

Table 1. PSNR values of the different methods.

Noise (σ)	BM3D	UDWT	DnCNN	FFDNet	IRCNN	LSLA-2	This Paper
25	29.97	25.51	30.43	**30.44**	30.38	28.99	27.53
50	26.72	23.42	27.18	**27.32**	26.32	25.63	26.85
75	22.32	19.98	22.21	22.43	22.87	22.31	**24.49**
100	19.56	17.53	20.12	20.62	19.78	20.54	**24.71**

Table 2. SSIM values of the different methods.

Noise (σ)	BM3D	UDWT	DnCNN	FFDNet	IRCNN	LSLA-2	This Paper
25	0.8447	0.8053	**0.8597**	0.8582	0.8576	0.8286	0.8413
50	0.7659	0.7495	0.7865	0.7841	0.7853	0.7664	**0.8176**
75	0.7132	0.7054	0.7178	0.7232	0.7152	0.7143	**0.7868**
100	0.6856	0.6394	0.6871	0.6882	0.6725	0.6532	**0.7640**

It can be seen from Table 1 that the PSNR values of BM3D, DnCNN, FFDNet, IRCNN, UDWT, and LSLA-2 are slightly higher than this paper's method, when the standard deviation of Gaussian noise $\sigma = 25$, and the difference was almost the same when the standard deviation of Gaussian noise $\sigma = 50$, even being slightly higher than that of some methods. When the standard deviation of Gaussian noise was $\sigma > 50$, the proposed method was significantly higher than the other methods. When the standard deviation of Gaussian noise $\sigma > 50$, the PSNR of the proposed method was about 4 dB higher than the other methods.

Table 2 shows that the SSIM value of the proposed method was lower than that of other methods when $\sigma = 25$; and the SSIM value of the proposed method was significantly higher than that of the other methods when standard deviation of Gaussian noise was greater than 25.

4.3.3. Comparison of Visual Perception

In view of the evaluation index of visual perception difference, this paper selected a picture in the test set for visualization under different methods. The experimental results are shown in Figure 5. Where (a) is the standard image; (b) is the image with Gaussian noise; (d) is the image denoised by BM3D; © is the image denoised by DnCNN; (f) is the image denoised by FFDNet; and (g) is the image denoised by IRCNN. Although these methods also removed the noise of the image, the image looks partly fuzzy and some edge features have a fuzzy phenomenon. The image (c), denoised by the method proposed in this paper, has a more intuitive visual experience. The clarity of the denoised image is almost the same as that of the standard image, and the features of the image are relatively intact. The image in this paper is clearer.

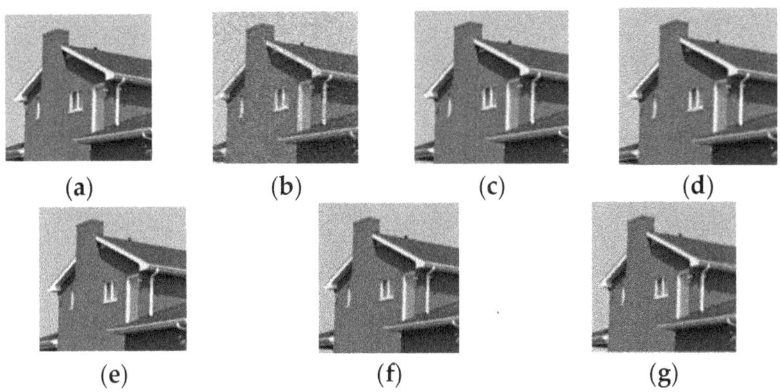

Figure 5. Image denoised using different methods. (**a**) original image, (**b**) noise image, (**c**) this paper, (**d**) MB3D, (**e**) DnCNN, (**f**) FFDNet, (**g**) IRCNN.

To sum up, when the noise level was low, the denoising effect of the method in this paper was equal to that of the other methods. However, when the noise standard deviation was greater than 25, the denoising ability and effect of the proposed method were better than the other methods, and both the values of PSNR and SSIM were higher than other methods. The test showed that when the noise environment was more complex, our method was more advantageous and had a stronger robustness and could effectively improve the image. This paper's method had little influence on the noise environment but its denoising ability was relatively stable in different environments.

4.3.4. FGSM Attack Result

FGSM is an algorithm based on gradient generation of adversarial samples and is a single-step, non-directional attack algorithm. Figures 6 and 7 show the comparison effect of SSIM and PSNR values between the generated images and the standard images under different attack degrees. The range of difference between the SSIM and PSNR values of the generated image and the standard image become smaller with a larger disturbance after FGSM attacks. Therefore, the method of adding random noise to the neurons of a neural network can improve the anti-interference ability of the network, which proved the superiority of our method in stability and robustness.

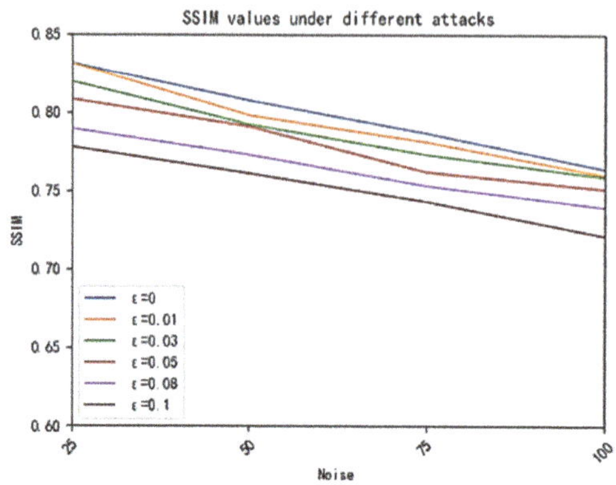

Figure 6. SSIM values under different levels of FGSM attacks.

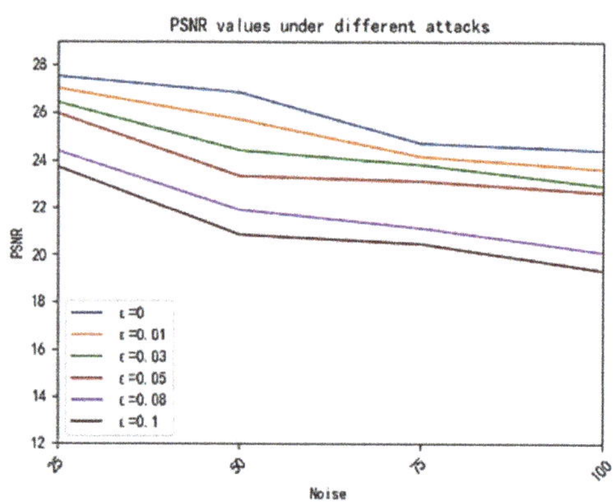

Figure 7. PSNR values under different levels of FGSM attacks.

4.3.5. Ablation Experiments and PGD Attack

In order to further verify the restoration ability of this paper's method with noisy images, an ablation experiment was carried out. First, the optimization algorithm (OA) was removed, to test the performance of the model. Gaussian noise with a standard deviation of 25, 50, 75, and 100 was added to the BDS500 dataset for the experiment. Comparing the PSNR and SSIM, the results are shown in Table 3. When OA was used in the generator network and discriminator network, it could optimize the network and achieve better results in the processing of noise images. This shows that our optimization method could improve the robustness of the network.

Table 3. Results of ablation experiments with no PGD (PSNR/SSIM).

	σ=25	σ=50	σ=75	σ=100
With OA (PSNR/SSIM)	27.53/0.8413	26.86/0.8176	24.49/0.7868	24.71/0.7640
Without OA (PSNR/SSIM)	21.13/0.6396	20.45/0.6034	19.12/0.5958	18.63/0.5756

Second, in order to further verify the robustness of this paper's method for the network, experiments with OA and without OA were performed, to test the defense performance of the model under different disturbance levels of PGD adversarial attack. The PGD attack is an iterative attack, which can be regarded as a copy of FGSM–K-FGSM (K represents the number of iterations). We performed a 10-step PGD adversarial training with a step size of

0.01, to verify the stability of the model under different disturbance levels. The results are shown in Table 4. The defense performance of the network against PGD attack decreased significantly without OA. With the increase of attack amplitude, the SSIM and PSNR values without OA decreased more than those of the network with OA. When $\epsilon = 0.05$, adding OA could even improve the SSIM and PSNR values by more than 100%. This proved that adding OA could improve the anti-interference ability and enhance the robustness of the network.

Table 4. Results of ablation experiments under PGD (PSNR/SSIM).

		σ=25	σ=50	σ=75	σ=100
With OA (PSNR/SSIM)	$\epsilon = \mathbf{0.01}$	26.93/0.8325	25.86/0.8123	23.91/0.7783	24.02/0.7601
	$\epsilon = \mathbf{0.02}$	26.52/0.8297	25.21/0.8043	23.42/0.7642	23.02/0.7554
	$\epsilon = \mathbf{0.05}$	26.36/0.8223	25.15/0.7931	22.97/0.7662	22.25/0.7510
Without OA (PSNR/SSIM)	$\epsilon = \mathbf{0.01}$	16.57/0.5217	15.50/0.5020	14.35/0.4715	13.36/0.4563
	$\epsilon = \mathbf{0.02}$	13.45/0.4570	12.62/0.4234	11.98/0.4044	10.52/0.3851
	$\epsilon = \mathbf{0.05}$	11.39/0.4178	10.84/0.3899	10.02/0.3620	9.15/0.3572

5. Conclusions

This paper proposed an image denoising method based on GAN network. In our method, a global residual is added into the autoencoder to extract and learn the features of the input image, preventing the loss of features in the process of denoising and preserving the details of the image features. Gaussian noise is added to the standard deviation path random estimation of each neuron in the neural network, to make it become a by-product of back propagation, which can effectively increase the robustness of the neural network and make it relatively stable in the case of noise environment fluctuations. MSE loss and adversarial loss are used to adjust the network, so that the network can achieve the best performance and have a better denoising effect. We compared our method with other methods. Although it was not as good as the other methods in the case of a low noise level, it was generally better than the other methods in the case of a high noise level. Both from the perspective of vision and quantitative objective evaluation, the denoising effect of the proposed method was remarkable in most scenes. The algorithm model provides help for target detection, recognition, and other applications, and it also has a good practicability. The future work after this paper is to further optimize the denoising effect in low noise environments, so as to achieve an optimal denoising effect in all noise environments

Author Contributions: Conceptualization: M.-L.Z. and L.X., methodology: M.-L.Z. and L.X., formal analysis: M.-L.Z. and L.-L.Z., investigation: M.-L.Z. and L.X., data curation: M.-L.Z. and L.-L.Z., writing– original draft preparation: M.-L.Z. and L.-L.Z., writing–review and editing: M.-L.Z. and L.-L.Z. All authors have read and agreed to the published version of the manuscript.

Funding: This research was funded by the Beijing Natural Science Foundation (No. 4202025), National Natural Science Foundation of China (No. 31900979) and Promoting the classified development of colleges and universities—the construction of the first level discipline of Computer Science and Technology (No. 5112211036).

Institutional Review Board Statement: Not applicable.

Informed Consent Statement: Not applicable.

Data Availability Statement: Not applicable.

Acknowledgments: We thank the company ZSE, a.s., for supporting the open-access publication of this paper.

Conflicts of Interest: The authors declare no conflict of interest.

References

1. Kumwilaisak, W.; Piriyatharawet, T.; Lasang, P.; Thatphithakkul, N. Image denoising with deep convolutional neural and Multi-Directional long Short-Term memory networks under poisson noise environments. *IEEE Access* **2020**, *8*, 86998–87010. [CrossRef]
2. Tang, C.; Xu, J.; Zhou, Z. Improved curvature filtering method for strong noise image denoising. *J. Image Graph.* **2019**, *24*, 26–36.
3. Li, G.; Li, J.; Fan, H. Adaptive matching pursuit image denoising algorithm. *Comput. Sci.* **2020**, *47*, 176–185.
4. Dabov, K.; Foi, A.; Katkovnik, V.; Egiazarian, K. Image denoising by Sparse 3-D transform-Domain collaborative Filtering. *IEEE Trans. Image Process.* **2007**, *16*, 2080–2095. [CrossRef] [PubMed]
5. Jun, X.; Lei, Z.; Zhang, D. A trilateral weighted sparse coding scheme for real-world image denoising. In Proceedings of the European Conference on Computer Vision (ECCV), Munich, Germany, 8–14 September 2018; Volume 9, pp. 20–36.
6. Xie, T.; Li, S.; Sun, B. Hyperspectral images denoising via nonconvex regularized Low-Rank and sparse matrix decomposition. *IEEE Trans. Image Process.* **2020**, *29*, 44–56. [CrossRef] [PubMed]
7. Li, Y.F. Image denoising based on undecimated discrete wavelet transform. In Proceedings of the 2007 International Conference on Wavelet Analysis and Pattern Recognition, Beijing, China, 2–4 November 2007; pp. 527–531.
8. Wang, Y.; Ren, W. Image denoising using anisotropic second and fourth order diffusions based on gradient vector convolution. *Comput. Sci. Inf. Syst.* **2012**, *9*, 1493–1511. [CrossRef]
9. Wu, Q.; Ren, W.; Cao, X. Learning interleaved cascade of shrinkage fields for joint image dehazing and denoising. *IEEE Trans. Image Process.* **2020**, *29*, 1788–1801. [CrossRef] [PubMed]
10. Zhang, K.; Zuo, W.; Chen, Y.; Meng, D.; Zhang, L. Beyond a gaussian denoiser: Residual learning of deep CNN for image denoising. *IEEE Trans. Image Process.* **2017**, *26*, 3142–3155. [CrossRef] [PubMed]
11. Yan, H.; Chen, X.; Tan, V.Y.F.; Yang, W.; Wu, J.; Feng, J. Unsupervised image noise modeling with Self-Consistent GAN. *arXiv* **2019**, arXiv:1906.05762v1.
12. Yu, S.; Park, B.; Jeong, J. Deep iterative Down-Up CNN for image denoising. In Proceedings of the 2019 IEEE/CVF Conference on Computer Vision and Pattern Recognition Workshops (CVPRW), Long Beach, CA, USA, 16–17 June 2017; Volume 6, pp. 2095–2103.
13. Zhang, K.; Zuo, W.; Zhang, L. FFDNet: Toward a fast and flexible solution for CNN based image denoising. *IEEE Trans. Image Process.* **2018**, *27*, 4608–4622. [CrossRef] [PubMed]
14. Chen, J.; Chen, J.; Chao, H.; Yang, M. Image blind denoising with generative adversarial network based noise modeling. In Proceedings of the IEEE Conference on Computer Vision and Pattern Recognition, Salt Lake City, UT, USA, 18–23 June 2018; Volume 6, pp. 3155–3164.
15. Dong, W.; Wang, P.; Yin, W.; Shi, G.; Wu, F.; Lu, X. Denoising prior driven deep neural network for image restoration. *IEEE Trans. Pattern Anal. Mach. Intell.* **2019**, *41*, 2305–2318. [CrossRef] [PubMed]
16. Wang, Y.; Song, X.; Chen, K. Channel and space attention neural network for image denoising. *IEEE Signal Process. Lett.* **2021**, *28*, 424–428. [CrossRef]
17. Cai, S.; Kang, Z.; Yang, M.; Xiong, X.; Peng, C.; Xiao, M. Image Denoising via Improved Dictionary Learning with Global Structure and Local Similarity Preservations. *Symmetry* **2018**, *10*, 167. [CrossRef]
18. Cai, S.; Liu, K.; Yang, M.; Tang, J.; Xiong, X.; Xiao, M. A new development of non-local image denoising using fixed-point iteration for non-convex ℓp sparse optimization. *PLoS ONE* **2018**, *13*, e0208503. [CrossRef] [PubMed]
19. Szegedy, C.; Zaremba, W.; Sutskever, I.; Bruna, J.; Erhan, D.; Goodfellow, I.; Fergus, R. Intriguing properties of neural networks. Computer Science. *arXiv* **2014**, arXiv:1312.6199v4.
20. Goodfellow, I.J.; Shlens, J.; Szegedy, C. Explaining and harnessing adversarial examples. Computer and Information Sciences. *arXiv* **2015**, arXiv:1412.6572v3.
21. Li, X.; Zhang, Z.; Peng, Y. Noise optimization for artificial neural networks. *arXiv* **2021**, arXiv:2102.04450v1.
22. Lin, W.; Gao, M.; Ruan, C.; Zhong, J. Denoising for intracranial hemorrhage images using autoencoder based on CNN. In Proceedings of the 2021 IEEE International Conference on Computer Science, Electronic Information Engineering and Intelligent Control Technology (CEI), Fuzhou, China, 24–26 September 2021; pp. 520–523.
23. Zhang, Y.; Tian, Y.; Kong, Y.; Zhong, B.; Fu, Y. Residual dense network for image restoration. *IEEE Trans. Pattern Anal. Mach. Intell.* **2021**, *43*, 2480–2495. [CrossRef] [PubMed]
24. Huang, Z.; Zhang, J.; Zhang, Y.; Shan, H. DU-GAN: Generative Adversarial Networks with Dual-domain U-Net based discriminators for Low-dose CT denoising. *IEEE Trans. Instrum. Meas.* **2022**, *71*, 4500512. [CrossRef]
25. Löhdefink, J.; Hüger, F.; Schlicht, P.; Fingscheidt, T. Scalar and vector quantization for learned image compression: A study on the effects of MSE and GAN loss in various spaces. In Proceedings of the 2020 IEEE 23rd International Conference on Intelligent Transportation Systems (ITSC), Rhodes, Greece, 20–23 September 2020; pp. 1–8.
26. Altakrouri, S.; Usman, S.B.; Ahmad, N.B.; Justinia, T.; Noor, N.M. Image to image translation networks using perceptual adversarial loss function. In Proceedings of the 2021 IEEE International Conference on Signal and Image Processing Applications (ICSIPA), Kuala Terengganu, Malaysia, 13–15 September 2021; pp. 89–94.
27. Cho, Y.S.; Kim, S.; Lee, J.H. Source model selection for transfer learning of image classification using supervised contrastive loss. In Proceedings of the 2021 IEEE International Conference on Big Data and Smart Computing (BigComp), Jeju, Korea, 17–20 January 2021; pp. 325–329.

electronics

MDPI

Article

An Improved U-Net for Watermark Removal

Lijun Fu [1], Bei Shi [2], Ling Sun [3], Jiawen Zeng [2], Deyun Chen [1,*], Hongwei Zhao [4] and Chunwei Tian [2,5]

[1] School of Computer Science and Technology, Harbin University of Computer Science and Technology, Harbin 150080, China
[2] School of Software, Northwestern Polytechnical University, Xi'an 710129, China
[3] School of Information and Electromechanical Engineering, Heilongjiang University of Industry and Business, Harbin 150025, China
[4] Shandong Baimeng Information Technology Co., Ltd., Weihai 264200, China
[5] Research & Development Institute, Northwestern Polytechnical University, Shenzhen 518000, China
* Correspondence: tg1950@hrbust.edu.cn

Abstract: Convolutional neural networks (CNNs) with different layers have performed with excellent results in watermark removal. However, how to extract robust and effective features via CNNs of black box in watermark removal is very important. In this paper, we propose an improved watermark removal U-net (IWRU-net). Taking the robustness of obtained information into account, a serial architecture is designed to facilitate useful information for guaranteeing performance in watermark removal. Taking the problem of long-term dependency into account, U-nets based simple components are integrated into the serial architecture to extract more salient hierarchical information for addressing watermark removal problems. To increase the adaptability of IWRU-net to the real world, we use randomly distributed blind watermarks to implement a blind watermark removal model. The experiment results illustrate that the proposed method is superior to other popular watermark removal methods in terms of quantitative and qualitative evaluations.

Keywords: serial architecture; U-net; blind watermark removal

Citation: Fu, L.; Shi, B.; Sun, L.; Zeng, J.; Chen, D.; Zhao, H.; Tian, C. An Improved U-Net for Watermark Removal. *Electronics* **2022**, *11*, 3760. https://doi.org/10.3390/electronics11223760

Academic Editor: Stefanos Kollias

Received: 16 October 2022
Accepted: 10 November 2022
Published: 16 November 2022

1. Introduction

To protect the copyright of files, added watermarks became a popular way to increase security of protected files [1]. To test the quality of added watermarks, watermark removal is an effective tool [2]. Integrating prior occurrences and the likelihood of cross-channel correlation can repair image information of the damaged channel in watermark removal [3]. Just-noticeable-distortion is used to estimate the energy to remove the watermark [4]. Taking into questions of scanned and back-lit pages in archaic documents account, a known lexicon of fragments is exploited to find watermarks and remove them for overcoming the effects of damaged files [5]. To improve the effect of removing watermarks, the discrete cosine transform domain and a key-based matrix are fused to remove visible watermarking [6]. Alternatively, authors use entropy and edge entropy as human visual system (HVS) characteristics to quickly extract watermark features and remove them [7]. To improve the generalization ability of the watermark removal algorithm, using wavelet transform can extract watermark features, the obtained features are used to revise singular values in order to test the robustness of the obtained watermark [8]. Ansari et al. adjusted block differences between HL and LH to automatically select the wavelet coefficient for improving watermark quality [9]. In addition, Fourier transform is effective for watermark removal. For instance, using the Fourier transform domain to remove watermarks of R, G and B in the watermark images is a good tool for blind color image watermark removal [10]. Taking robustness and imperceptibility into account, scholars ensure a signal-to-noise ratio that enables the host image to maximize the embedding strength for making a tradeoff between robustness and imperceptibility in the image watermarking removal [11]. Although these methods have performed well in watermark removal, they still suffer from the following drawbacks: (1) They use complex optimization methods to improve visual effect of watermark removal. (2) Manually choosing parameters is a good tool to improve the performance of watermark removal. To overcome these disadvantages, we use deep learning techniques, especially convolutional neural networks to deal with watermark removal.

Chen et al. proposed deep neural networks for watermark removal [12]. To improve the quality of watermark removal, Sai et al. used lower dimensional projections in the intermediate layers of a deep CNN to express the image content in watermark removal [13]. Haribabu et al. utilized an auto-encoder to deal with watermark images with two independent images for watermark removal [14]. To improve the robustness of watermarks, Chen et al. used an adaption of elastic weight consolidation and unlabeled data augmentation to better represent watermarks for watermark removal [15]. Using roughly localized and separate watermarks is a good tool for an image watermark [16]. Using a CNN, wavelet transform and residual regularization loss function, rather than down- and up-sampling operations, can improve the visual quality of watermark images [17].

Although these CNNs have obtained comparative results in watermark removal, the key question is how to extract effective features of CNNs with black box to better represent watermarks for more complex watermark removal. In this paper, we propose an improved watermark removal U-net (IWRU-net). To obtain more robust information, a serial architecture is presented to extract useful information to pursue better performance for watermark removal. To address a long-term dependency problem, U-net's base simple components are fused into the designed serial architecture to extract more salient hierarchical information for dealing with the watermark removal problem. Taking into the adaptability of IWRU-net in the real world account, we use randomly distributed blind watermarks to conduct a blind watermark removal model. The experiment's results show that the proposed method is effective in quantitative and qualitative evaluations.

This paper makes the following contributions.

1. A serial architecture is used to facilitate more useful information for improving the performance of watermark removal.
2. U-nets are gathered into a serial architecture to extract more salient hierarchical information to address the long-term dependency on deep CNNs for watermark removal.
3. To improve the adaptability of IWRU-net on mobile devices in the real world, randomly distributed watermarks with different types are used to train a blind watermark removal model.

The remaining parts of this paper have the following organizations. Section 2 represents related work on the proposed method. Section 3 lists the proposed method. Section 4 presents experiments. Section 5 offers conclusions.

2. Related Work

2.1. Deep CNNs for Watermark Removal

Due to their strong learning ability, CNNs are often exploited for image applications [18,19]. For example, CNNs are used to extract robust features for better representing watermarks for watermark removal [20]. To address different watermarks, a detector based on the CNN is used to detect the locations of watermarks and to remove watermarks [20]. To deal with blind watermark removal, Lee et al. used a pre-processing network, a watermark embedding network and a watermark extraction network to enhance the host image with super-resolution and the watermark invisibility for blind watermark removal [21]. To address artifacts and blurriness caused by opaque watermarks, dual convolutions are used for watermark removal [22]. That is, the first network is exploited to remove the watermark. The second network is utilized to optimize the second network for further filter watermarks. Due to problem of conventional perceptual hashing, perceptual hash is embedded into a CNN to obtain a weight way for verifying watermarking [23]. To address watermarks in remote operations, a novel zero-bit watermarking algorithm with adversarial model examples was used to extract the watermark and remove it [24]. In a tradeoff between robustness and transparency, a combination of water wave optimization, chaotic fruit fly optimization algorithm and CNN was developed for verifying watermarks [25]. To discriminate between attack watermarks and corrected watermarks, a generative adversarial network (GAN) was used for watermark removal [26]; that is, a generative network based on U-net architecture was used to extract high- and low-level features for generating images. Also, a discriminative network was used to distinguish between the truth of generated and original images, which can also be used to remove the watermark. To reduce the time of watermark removal, a lightweight CNN was deigned [27]. To improve watermark removal, a discrete cosine transformation and Harris hawks method were fused into a CNN to filter watermarks [28]. To improve the ability of watermark removal, a progressive pre-processing operation was gathered into a residual dense network to extract more low-frequency features and enhance the attack ability of the designed network for image watermark removal [29]. Motivated by that, we designed a novel CNN for image watermark removal.

2.2. *Cascaded Architectures for Image Applications*

To extract richer features, a cascaded architecture was designed to improve the performance in image applications [30,31]. For instance, Qin et al. trained a cascaded network composed simultaneously of a region proposal network and a fast R-CNN to improve the accuracy rate of facial detection [30]. To address undersampled data, a cascaded CNN was presented with MR images for overcoming undersampled data [32]. To overcome the effect of noise and artifacts, two identical networks were cascaded to enhance the classification results of medical images [33]. That is, the first network was used to remove noise; the second network was exploited to classify medical images. Taking into effect of different factors for image dehazing account, Li et al. used two sub-networks to address medium transmission and global atmospheric light to obtain more realistic effects, closer to the real world, to improve the practicality of the proposed method [34]. Alternatively, Yan et al. combined multi-frame geometry and jointed training to gather low- and high-frequency information to enhance the image quality of consumer depth cameras [35]. To improve the performance of image registration, each cascaded network can further deal with warped images to change the image quality [36]. To fully deal with the specific attributes of HSIs, a cascaded architecture based on two recurrent neural networks was used for hyper-spectral image classification [37]. Specifically, the first RNN was applied to remove redundant information from spectral bands. The second RNN can extract more extra information obtained from nonadjacent spectral bands. Two networks can improve the discriminative ability of the obtained classifier. Besides, a cascaded network can use a hierarchical architecture to extract more useful features to enhance the image quality [38]. Tian et al. designed a heterogeneous architecture and a stacked convolutional layer to mine richer low- and high-frequency features for addressing the unstable problem of a SR model [39]. To overcome the challenge of the low spatial resolution of hyper-spectral images, a network was implemented by cascading two sub-networks [40]. That is, the first network was used to obtain high resolution multispectral panchromatic images; the second network was exploited to predict abundance maps. Two networks can better deal with hyperspectral image resolution. The cascade network architecture was extended for facial expression recognition [41]. Combining group convolutional networks and stacked CNNs can be used to enhance the relationships of different channels for image super-resolution [42,43]. Following the above studies, we can see that cascading networks are useful for image applications. Inspired by that, we designed a cascading network architecture for image watermark removal.

3. The Proposed Method

3.1. *Network Architecture*

To extract robust and effective features for watermark removal, we propose an improved watermark removal U-net with 42 layers as well as an IWRU-net, as reported in Figure 1. To improve the robustness of obtained features, a serial architecture is implemented by cascading two sub-networks in order to obtain effective information for improving the performance of watermark removal. To address long-term dependency problem, U-nets base simple components are fused into the designed serial architecture to extract more salient hierarchical information for dealing watermark removal problem. To increase the adaptability of the IWRU-net to the real world, we use randomly distributed blind watermarks to implement a blind watermark removal model. To roughly express the above, we conduct the following equation.

$$\begin{aligned} I_c &= IWRUnet(I_w) \\ &= UnetBlock(UnetBlock(I_w)) \end{aligned} \tag{1}$$

where I_w denotes a watermark image and *IWRUnet* is a function of IWRU-net. I_c represents a clean image. *UnetBlock* expresses the function of the Unet Block. Also, *IWRUnet* is trained by the loss function illustrated in Section 3.2. Finally, each Unet Block is shown in Section 3.3.

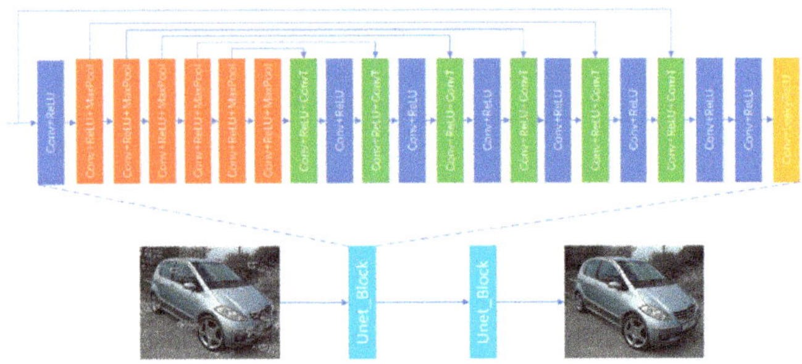

Figure 1. Network architecture of IWRU-net.

3.2. Loss Function

To improve the training efficiency, least absolute deviation (LAD) [44,45] is used to train the IWRU-net model for image watermark removal. That is, we use a watermark image and a clean image to act in an IWRU-net according to Equation (2) to train a watermark removal model.

$$D(\theta) = \frac{\sum_{j=1}^{t} \left| I_w^j - IWRUnet(I_w^j) \right|}{t} \tag{2}$$

where I_w^j is the jth watermark image, t expresses the total number of watermark images, D denotes a loss function for training a IWRU-net watermark removal model and θ is used to represent the parameters. Note that the parameters are optimized by Adam [46] when the IWRU-net is trained.

3.3. Each Unet Block

A serial architecture composed of two sub-networks is key to improving the robustness of the obtained features in the IWRU-net. Also, each 21-layer U-net filling in each sub-network is used to obtain more salient hierarchical information for watermark removal. That is, the 1st, 9th, 11th, 13th, 15th, 17th, 19th and 20th layers are composed of Conv+ReLU; the 2nd–7th layers include Conv+ReLU+MaxPool; the 8th, 10th, 12th, 14th, 16th and 18th layers contain Conv+ReLU+ConvT. Also, the last layer includes Conv+LeakyReLU. The mentioned Conv+ReLU is a combination of a convolutional layer and an activation function of ReLU [47]. The mentioned Conv+ReLU+MaxPool is a combination of a convolutional layer, an activation function of ReLU and a pooling function of max pooling [48]. Conv+ReLU+ConvT is a combination of a convolutional layer, ReLU and deconvolution

(ConvTranspose2d). In addition, a convolutional layer is used to obtain linear features. ReLU is exploited to map linear features into non-linear features. MaxPool is utilized to reduce the dimension of data to improve the efficiency of training a IWRU-net model. ConvTranspose2d is exploited to obtain our predicted results. LeakyReLU is exploited to map linear features onto non-linear features [49]. To enhance the memory ability of IWRU-net, we use concatenation operations to integrate shallow features to transmit deep layers. That is, features of the input and the 18th layer are fused by a concatenation operation as an input of the 19th layer. Features of the 2nd layer and 16th layer are merged by a concatenation operation as an input of the 16th layer. Features of the 3rd layer and 14th layer are gathered by a concatenation operation as an input of the 15th layer. Features of the 4th layer and 12th layer are fused by a concatenation operation as an input of the 13th layer. Features of the 5th layer and 10th layer are gathered by a concatenation operation as an input of the 11th layer. Features of the 6th layer and 8th layer are gathered by a concatenation operation as an input of the 9th layer. Also, each convolutional kernel size is 3×3. Input and output channel number of each layer are shown as follows. The 1st layer has 3 input channels and 48 output channels. The 2nd–8th layers have 48 input and output channels, respectively. The 9th, 10th, 12th, 14th, 16th and 18th layers have 96 input and output channels. The 13th, 15th and 17th have 144 input channels and 96 output channels, respectively.

The 19th layer has 99 input channel and 64 output channels. The 20th layer has 64 input channels and 32 output channels. The 21st layer has 32 input channels and 3 output channels.

To allow readers to understand illustrations the above visually, we conducted the following equations.

$$\begin{aligned} O^{i}{}_{F\_UnetBlock} &= UnetBlock(I_w) \\ &= CLR(CR(CR(C_o(CRC(CR(C_0(CRC(CR(C_0(CRC(CR(C_o(CRC(O_t),O_4))),O_3))),O_2))),I_w)))) \end{aligned} \tag{3}$$

$$O_t = CR(C_o(CRC(CR(C_o(CRC(6MCR(CR(I_w))),O_6))),O_5)) \tag{4}$$

$$O_6 = 5MCR(CR(I_w)) \tag{5}$$

$$O_5 = 4MCR(CR(I_w)) \tag{6}$$

$$O_4 = 3MCR(CR(I_w)) \tag{7}$$

$$O_3 = 2MCR(CR(I_w)) \tag{8}$$

$$O_2 = MCR(CR(I_w)) \tag{9}$$

where CR denotes a combination of a convolution and ReLU; $nMCR$ is n stacked MCR, where n varies from 1 to 6; CRC expresses a combination of a convolution, ReLU and Max pooling; CLR represents a combination of a convolution and LeakyReLU; O_j stands for output of the jth layer and $j = 2,3,4,5,6$; O_t is a temporary output; $O^{i}{}_{F\_UnetBlock}$ is the ith output of the Unet Block ($i = 1, 2$).

4. Experiments

4.1. Datasets

Training dataset. Following [50–52], we chose public large-scale visible watermarks (LVW) [20] for our training dataset to train our IWRU-net. The training dataset is composed of 60,000 watermarked images with 80 watermarks. Each watermark is embedded into 750 images. Also, we chose 3000 images without watermarks from the LVW. To enlarge the categories of training samples, we rotated seven conducted watermarks from -30° to 30°, scaled them from 70% to 100% and adjusted them to a transparency of 50% to 80% then randomly added them to the mentioned 3000 images to achieve a watermark coverage of 10% for constructing watermark images, where watermark coverage is the ratio of watermarking pixels to all the pixels in an entire image.

Test datasets. In order to fairly test the performance of our IWRU-net for watermark removal, we randomly selected 200 images from the LVW and colored large-scale watermark dataset (CLWD) [16] as test datasets. Specifically, 100 images were chosen from the LVW and the rest were chosen from the CLWD. Watermark images were rotated from -30° to 30°, underwent random scaling of 70% to 100% and random transparency adjustments of 50% to 80%. Additionally, a watermark was added into an image as a test image.

4.2. Experimental Settings

Our experiments were conducted on a PC. The PC has two CPUs of Intel(R) Xeon(R) Silver 4210 CPU@2.20GHz with RAM of 128 G and a Nvidia GeForce GTX 3090 GPU, where CUDN of 11.1 and CUDNN of 7.4.1 are used to accelerate the GPU. Our codes are run by PyTorch of 1.10.2 and Python of 3.8.12 on Ubuntu of 20.04.2. The initial learning rate is set to 1×10^{-4}. The number of epochs is 100. The learning rate has 0.5 times reduction each 200,000 iterations. All the training images and test images were scaled to 512×512 as inputs of the IWRU-net. Outputs of the IWRU-net need to be scaled the same as the original images. Other parameters are the same as in [53].

4.3. Experimental Analysis

Due to their hierarchical architecture, deep CNNs have obtained stronger learning abilities for image application, where hierarchical features have obtained features from different layers [54]. However, how to ensure obtained robust features is very important. Following Section 2.2, we can see that cascaded architectures are suitable to mine for more accurate features for image applications. Inspired by that, we designed an improved watermark removal U-net (IWRU-net) based on a cascaded architecture. That is, a serial architecture is used to facilitate useful information for guaranteeing performance in watermark removal. In this paper, we used two blocks (Unet_Blocks) in a serial way to form the serial architecture for image watermark removal in Figure 1. To address long-term dependency problems, we chose U-net as a simple component of each block (Unet_Block) to extract more salient hierarchical information to address watermark removal problems. To verify the effectiveness of the serial architecture, we used an IWRU-net and a single U-net on 100 images with seven watermarks from the LVW dataset to conduct experiments, where the settings of the chosen watermark images are the same as in Section 4.2. That is, the IWRU-net obtained higher peak signal-to-noise ratio (PSNR) [55] than that of a single U-net for image watermark removal as shown in Table 1, which shows the effectiveness of serial architecture in image watermark removal.

To increase the diversity of the obtained features, we used six up-sampling and down-sampling operations in each U-net to extract richer features. We used IWRU-net and IWRU-net without up- and down-sampling operations to test the superiority of up- and down-sampling operations in the IWRU-net for image watermark removal as reported in Table 1. To test the effectiveness of six up- and down-sampling operations in each U-net for image watermark, we chose the IWRU-net and the IWRU-net with four up- and down-sampling operations in each U-net to conduct comparative experiments in Table 1. That shows that the proposed IWRU-net exceeds IWRU-net with four up- and down-sampling operations in each U-net in terms of PSNR, which verifies the good performance of the six up- and down-sampling operations. Also, six up- and down-sampling operations can enhance the expressive ability of the designed IWRU-net. To discuss the effect of residual operations on serial architecture for image watermark removal, we used two residual operations to act the input and output of each Unet Block (block) for watermark removal. Due to the use of multiple concatenation operations in each U-net, two residual operations will result in the over-enhancement phenomenon of obtained features for image watermark removal. That is verified by both IWRU-net and IWRU-net with two extra residual operations in Table 1. To allow readers more easily to observe the effect of the IWRU-net, we chose one image from the LVW with one watermark randomly added onto the given clean image, rotated from -30° to 30°, scaled from 70% to 100% and adjusted to a transparency of 50% to 80% to conduct the visual watermark removal on the image. As presented in Figure 2, we can see that our IWRU-net has obtained more clearly detailed information than that obtained by the IWRU-net with two extra residual operations in the observation area. This shows that two extra residual operations have a native effect on the IWRU-net for image removal. In other words, the designed serial architecture is capable to deal with watermark removal from images.

Table 1. PSNR (dB) results of different methods on 100 watermark images from the LVW for image watermark removal.

Methods	PSNR
IWRU-net (ours)	44.85
IWRU-net with four up- and down-sampling operations in each U-net	34.75
IWRU-net without up- and down-sampling operations	43.18
A single U-net	43.71
IWRU-net with two extra residual operations	36.77

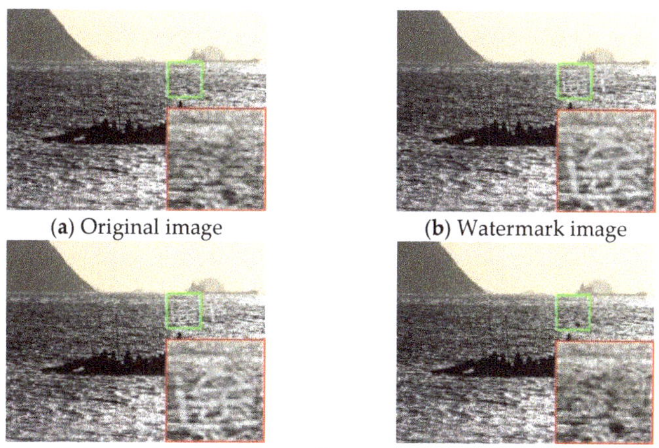

(**a**) Original image (**b**) Watermark image

(**c**) IWRU-net with two extra residual operations (**d**) IWRU-net (ours)

Figure 2. Visualization of different methods of watermark removal on an image from the LVW.

4.4. Experimental Results

Because image watermark removal is a low-level vision task, we chose a denoising convolutional neural network (DnCNN) [53], a fast and flexible denoising convolutional neural network (FFDNet) [56], a U-net [57], an attention-guided denoising convolutional neural network (ADNet) [58] and a robust deformed denoising CNN (RDDCNN) [31] as comparative methods to test the performance of image watermark removal in terms of qualitative and quantitative evaluations on the LVW and CLWD. For qualitative evaluation, we first used transparency rates of 1 and 0.5 to test the effects on the IWRU-net for image watermark removal. As shown in Table 2, our IWRU-net obtained a higher PSNR at a transparency of 1 than at 0.5 for image watermark removal, where one image is chosen from the LVW and its other setting is the same as in Section 4.2. This also shows that, when the transparency is lower, the IWRU-net has better results of watermark removal.

Table 2. PSNR (dB) results at different transparency rates of 100% and 50% with the IWRU-net for image watermark removal.

Transparency Rate	PSNR
100%	45.67
50%	41.32

Then, 100 randomly selected images from the LVW and CLWD datasets in Section 4.2 were used to test the effects of the transparency rate by varying it from 0.5, 0.6, 0.7 to 0.8 on the IWRU-net for image watermark removal. As illustrated in Tables 3 and 4, we can see that our IWRU-net is more effective than other methods, i.e., DnCNN, FFDNet and Unet for LVW and CLWD in terms of the PSNR and structural similarity (SSIM) [59] for blind watermark removal. This implies the robustness of our IWRU-net for image watermark removal.

Next, we also measured the complexity (parameters and flops [60]) of different methods, i.e., DnCNN, FFDNet, Unet, ADNet, RDDCNN and IWRU-net on an image with 512 × 512 from the LVW. Also, we used one image with 256 × 256, 512 × 512, and 1024 × 1024 from the LVW to test the running time of different methods, i.e., DnCNN, FFDNet, Unet, ADNet, RDDCNN and IWRU-net. As presented in Tables 5 and 6, we can see that our IWRU-net also his acceptably effective in terms of complexity and running time for image watermark removal. Compared with the cited methods, it is clear that our proposed IWRU-net is competitive for image watermark removal.

Table 3. Average PSNR (dB) and SSIM of different networks on LVW datasets for varying transparency rates of 0.5, 0.6, 0.7 and 0.8.

Methods	PSNR	SSIM
DnCNN [53]	42.95	0.9961
FFDNet [56]	38.48	0.9847
Unet [57]	43.71	0.9963
IWRU-net (ours)	44.85	0.9970

Table 4. Average PSNR (dB) and SSIM of different networks on CLWD datasets for varying transparency rates of 0.5, 0.6, 0.7 and 0.8.

Methods	PSNR	SSIM
DnCNN [53]	44.67	0.9753
FFDNet [56]	37.54	0.9912
Unet [57]	45.35	0.9972
RDDCNN [31]	46.25	0.9971
ADNet [58]	46.47	0.9972
IWRU-net (ours)	46.52	0.9975

Table 5. The complexity of different watermark removal methods.

Methods	Parameters	Flops
DnCNN [53]	0.5594 M	36.6582 G
FFDNet [56]	0.4945 M	8.1023 G
Unet [57]	1.0120 M	18.6813 G
RDDCNN [31]	0.5591 M	36.7060 G
ADNet [58]	0.5215 M	34.2393 G
IWRU-net (ours)	2.0240 M	37.3625 G

Table 6. Running time for different watermark removal methods on three image sizes.

Methods	256 × 256	512 × 512	1024 × 1024
DnCNN [53]	0.038228	0.154801	0.638453
FFDNet [56]	0.010732	0.037471	0.124227
Unet [57]	0.027889	0.097742	0.316260
RDDCNN [31]	0.057355	0.222245	1.559665
ADNet [58]	0.036286	0.147691	0.563838
IWRU-net (ours)	0.058375	0.199374	0.654419

To further test the performance of our IWRU-net, we used quantitative evaluation to conduct visual effects as follows. We chose four images from the LVW, adding different transparency rates of 0.5, 0.6, 0.7 and 0.8, respectively, to test the visual effects of the IWRU-net. Also, DnCNN, FFDNet and Unet were used as comparative methods. One chosen area of each predicted visual image was enlarged as an observation area. The observation area is clearer, so its corresponding method has better performance in image watermark removal. As shown in Figures 3–6, we can see that our IWRU-net has clearer areas for different transparency rates. It shows that our IWRU-net is more advantageous in terms of quantitative evaluation for image watermark removal. According to these findings, it is known that the IWRU-net is very suitable to image watermark removal for qualitative and quantitative evaluations.

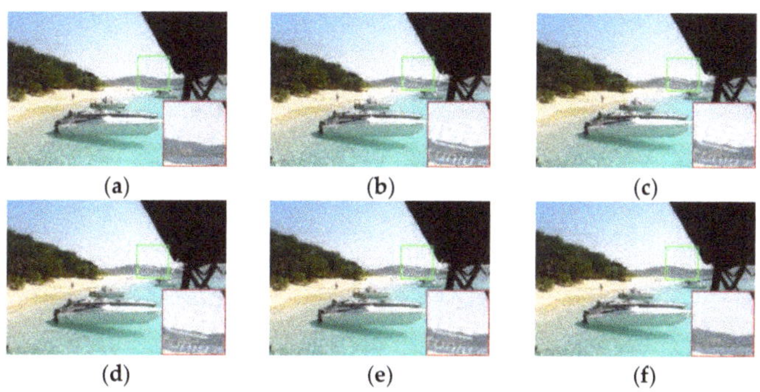

Figure 3. Visual effects of different methods with a transparency rate of 0.5 for image watermark removal. (**a**) Original image, (**b**) watermark image, (**c**) DnCNN, (**d**) FFDNet, (**e**) U-net and (**f**) IWRU-net (ours).

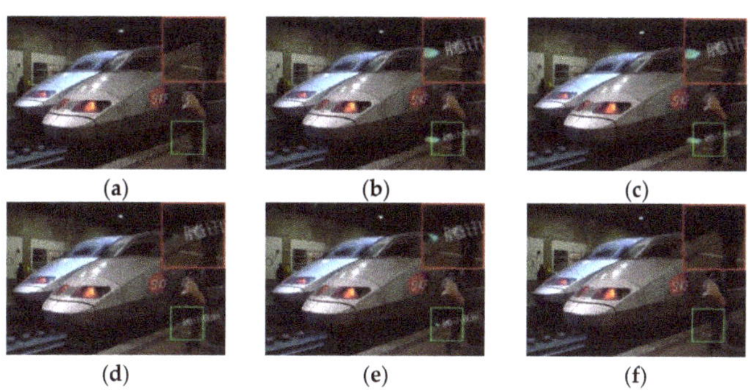

Figure 4. Visual effects of different methods with a transparency rate of 0.6 for image watermark removal. (**a**) Original image, (**b**) watermark image, (**c**) DnCNN, (**d**) FFDNet, (**e**) U-net and (**f**) IWRU-net (ours).

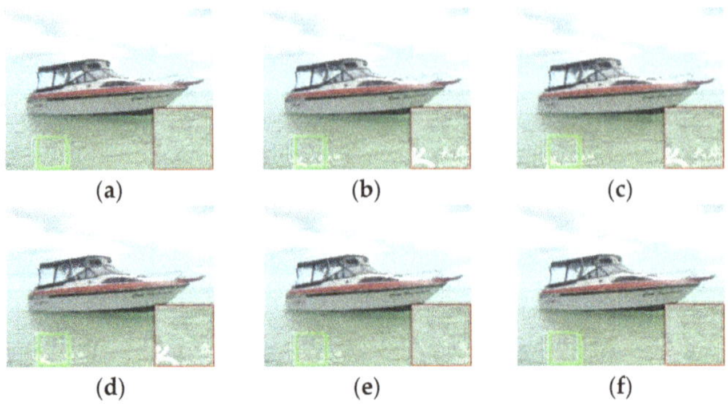

Figure 5. Visual effects of different methods with a transparency rate of 0.7 for image watermark removal. (**a**) Original image, (**b**) watermark image, (**c**) DnCNN, (**d**) FFDNet, (**e**) U-net and (**f**) IWRU-net (ours).

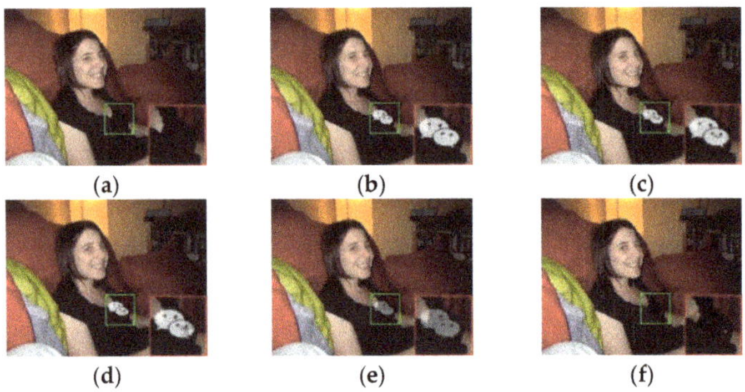

Figure 6. Visual effects of different methods with a transparency rate of 0.8 for image watermark removal. (**a**) Original image, (**b**) watermark image, (**c**) DnCNN, (**d**) FFDNet, (**e**) U-net and (**f**) IWRU-net (ours).

5. Conclusions

We propose an improved watermark removal U-net as IWRU-net. To improve the robustness of the obtained information, a serial architecture was used to facilitate more accurate information to guarantee the performance for watermark removal. To address long-term dependency problems, U-nets as simple components were integrated into the serial architecture to extract more salient hierarchical information for addressing watermark removal problems. To increase the adaptability of IWRU-net on mobile devices in the real world, randomly distributed watermarks of different types were used to train a blind watermark removal model. Our method is competitive with other popular watermark removal methods in terms of quantitative and qualitative evaluations. In the future, we will design lightweight CNNs for image watermark removal.

Author Contributions: Validation and part idea, L.F.; Data curation, writing and validation, B.S.; Investigation, L.S.; Part analysis, J.Z.; Part idea, visualization, writing (review), D.C.; Editing, H.Z.; Writing and Funding acquisition, C.T. All authors have read and agreed to the published version of the manuscript.

Funding: This work was supported in part by the Guangdong Basic and Applied Basic Research Foundation under Grant 2021A1515110079, in part by the Fundamental Research Funds for the Central Universities under Grant D5000210966.

Conflicts of Interest: The authors declare no conflict of interest.

References

1. Swanson, M.D.; Zhu, B.; Tewfik, A.H. Transparent robust image watermarking. In Proceedings of the 3rd IEEE International Conference on Image Processing, Lausanne, Switzerland, 19 September 1996; IEEE: Piscataway, NJ, USA, 1996; Volume 3, pp. 211–214.
2. Wong, P.W. A public key watermark for image verification and authentication. In Proceedings of the 1998 International Conference on Image Processing, ICIP98 (Cat. No. 98CB36269), Chicago, IL, USA, 7 October 1998; IEEE: Piscataway, NJ, USA, 1998; Volume 1, pp. 455–459.
3. Park, J.; Tai, Y.W.; Kweon, I.S. Identigram/watermark removal using cross-channel correlation. In Proceedings of the 2012 IEEE Conference on Computer Vision and Pattern Recognition, Providence, RI, USA, 16–21 June 2012; IEEE: Piscataway, NJ, USA, 2012; pp. 446–453.
4. Hsu, T.C.; Hsieh, W.S.; Chiang, J.Y.; Su, T. New watermark-removal method based on Eigen-image energy. *IET Inf. Secur.* **2011**, *5*, 43–50. [CrossRef]
5. Boyle, R.D.; Hiary, H. Watermark location via back-lighting and recto removal. *Int. J. Doc. Anal. Recognit. (IJDAR)* **2009**, *12*, 33–46. [CrossRef]
6. Yang, Y.; Sun, X.; Yang, H.; Li, C. Removable visible image watermarking algorithm in the discrete cosine transform domain. *J. Electron. Imaging* **2008**, *17*, 033008. [CrossRef]
7. Makbol, N.M.; Khoo, B.E.; Rassem, T.H. Block-based discrete wavelet transform-singular value decomposition image watermarking scheme using human visual system characteristics. *IET Image Process.* **2016**, *10*, 34–52. [CrossRef]
8. Ansari, I.A.; Pant, M. Multipurpose image watermarking in the domain of DWT based on SVD and ABC. *Pattern Recognit. Lett.* **2017**, *94*, 228–236. [CrossRef]
9. Huynh-The, T.; Banos, O.; Lee, S.; Yoon, Y.; Le-Tien, T. Improving digital image watermarking by means of optimal channel selection. *Expert Syst. Appl.* **2016**, *62*, 177–189. [CrossRef]
10. Fares, K.; Amine, K.; Salah, E. A robust blind color image watermarking based on Fourier transform domain. *Optik* **2020**, *208*, 164562. [CrossRef]
11. Huang, Y.; Niu, B.; Guan, H.; Zhang, S. Enhancing image watermarking with adaptive embedding parameter and PSNR guarantee. *IEEE Trans. Multimed.* **2019**, *21*, 2447–2460. [CrossRef]
12. Chen, X.; Wang, W.; Ding, Y.; Bender, C.; Jia, R.; Li, B.; Song, D.X. Leveraging unlabeled data for watermark removal of deep neural networks. In Proceedings of the 36th International Conference on Machine Learning, Long Beach, CA, USA, 10–15 June 2019; pp. 1–6.
13. Sharma, S.S.; Chandrasekaran, V. A robust hybrid digital watermarking technique against a powerful CNN-based adversarial attack. *Multimed. Tools Appl.* **2020**, *79*, 32769–32790. [CrossRef]
14. Haribabu, K.; Subrahmanyam, G.; Mishra, D. A robust digital image watermarking technique using auto encoder based convolutional neural networks. In Proceedings of the 2015 IEEE Workshop on Computational Intelligence: Theories, Applications and Future Directions (WCI), Kanpur, India, 14–17 December 2015; IEEE: Piscataway, NJ, USA, 2015; pp. 1–6.
15. Chen, X.; Wang, W.; Bender, C.; Ding, Y.; Jia, R.; Li, B.; Song, D.X. Refit: A unified watermark removal framework for deep learning systems with limited data. In Proceedings of the 2021 ACM Asia Conference on Computer and Communications Security, Hong Kong, China, 7–11 June 2021; pp. 321–335.
16. Liu, Y.; Zhu, Z.; Bai, X. Wdnet: Watermark-decomposition network for visible watermark removal. In Proceedings of the IEEE/CVF Winter Conference on Applications of Computer Vision, Waikoloa, HI, USA, 4–8 January 2022; pp. 3685–3693.
17. Lu, J.; Ni, J.; Su, W.; Xie, H. Wavelet-Based CNN for Robust and High-Capacity Image Watermarking. In Proceedings of the 2022 IEEE International Conference on Multimedia and Expo (ICME), Taipei, Taiwan, 18–22 July 2022; IEEE: Piscataway, NJ, USA, 2022; pp. 1–6.
18. Cao, L.; Liang, Y.; Lv, W.; Park, K.; Miura, Y.; Shinomiya, Y.; Yoshida, S. Relating brain structure images to personality characteristics using 3D convolution neural network. *CAAI Trans. Intell. Technol.* **2021**, *6*, 338–346. [CrossRef]
19. Jafarbigloo, S.K.; Danyali, H. Nuclear atypia grading in breast cancer histopathological images based on CNN feature extraction and LSTM classification. *CAAI Trans. Intell. Technol.* **2021**, *6*, 426–439. [CrossRef]
20. Cheng, D.; Li, X.; Li, W.; Lu, C.; Li, F.; Zhao, H.; Zheng, W. Large-scale visible watermark detection and removal with deep convolutional networks. In Proceedings of the Chinese Conference on Pattern Recognition and Computer Vision (PRCV), Guangzhou, China, 23–26 November 2018; Springer: Cham, Switzerland, 2018; pp. 27–40.
21. Lee, J.E.; Seo, Y.H.; Kim, D.W. Convolutional neural network-based digital image watermarking adaptive to the resolution of image and watermark. *Appl. Sci.* **2020**, *10*, 6854. [CrossRef]
22. Li, T.; Feng, B.; Li, G.; Li, X.; He, M.; Li, P. Visible Watermark Removal Based on Dual-input Network. In Proceedings of the 2021 ACM International Conference on Intelligent Computing and its Emerging Applications, Jinan, China, 28–29 December 2021; pp. 46–52.
23. Meng, Z.; Morizumi, T.; Miyata, S.; Kinoshita, H. An Improved Design Scheme for Perceptual Hashing based on CNN for Digital Watermarking. In Proceedings of the 2020 IEEE 44th Annual Computers, Software, and Applications Conference (COMPSAC), Madrid, Spain, 13–17 July 2020; IEEE: Piscataway, NJ, USA, 2020; pp. 1789–1794.
24. Le Merrer, E.; Perez, P.; Trédan, G. Adversarial frontier stitching for remote neural network watermarking. *Neural Comput. Appl.* **2020**, *32*, 9233–9244. [CrossRef]
25. Ingaleshwar, S.; Dharwadkar, N.V. Water chaotic fruit fly optimization-based deep convolutional neural network for image watermarking using wavelet transform. In *Multimedia Tools and Applications*; Springer: Berlin/Heidelberg, Germany, 2021; pp. 1–25.

26. Li, Q.; Wang, X.; Ma, B.; Wang, X.; Wang, C.; Gao, S.; Shi, Y. Concealed attack for robust watermarking based on generative model and perceptual loss. In *IEEE Transactions on Circuits and Systems for Video Technology*; IEEE: Piscataway, NJ, USA, 2021.
27. Dhaya, R. Light weight CNN based robust image watermarking scheme for security. *J. Inf. Technol. Digit. World* **2021**, *3*, 118–132.
28. Chacko, A.; Chacko, S. Deep learning-based robust medical image watermarking exploiting DCT and Harris hawks optimization. *Int. J. Intell. Syst.* **2022**, *37*, 4810–4844. [CrossRef]
29. Wang, C.; Hao, Q.; Xu, S.; Ma, B.; Xia, Z.; Li, Q.; Li, J.; Shi, Y.Q. RD-IWAN: Residual Dense based Imperceptible Watermark Attack Network. In *IEEE Transactions on Circuits and Systems for Video Technology*; IEEE: Piscataway, NJ, USA, 2022.
30. Qin, H.; Yan, J.; Li, X.; Hu, X. Joint training of cascaded CNN for face detection. In Proceedings of the IEEE Conference on Computer Vision and Pattern Recognition, Las Vegas, NV, USA, 1–26 July 2016; pp. 3456–3465.
31. Zhang, K.; Zuo, W.; Chen, Y.; Meng, D.; Zhang, L. Beyond a gaussian denoiser: Residual learning of deep cnn for image denoising. *IEEE Trans. Image Process.* **2017**, *26*, 3142–3155. [CrossRef] [PubMed]
32. Schlemper, J.; Caballero, J.; Hajnal, J.V.; Price, A.N.; Rueckert, D. A deep cascade of convolutional neural networks for MR image reconstruction. In Proceedings of the International Conference on Information Processing in Medical Imaging, Boone, CA, USA, 25–30 June 2017; Springer: Cham, Switzerland, 2017; pp. 647–658.
33. Wu, D.; Kim, K.; Fakhri, G.E.; Li, Q. A cascaded convolutional neural network for X-ray low-dose CT image denoising. *arXiv* **2017**, arXiv:1705.04267.
34. Li, C.; Guo, J.; Porikli, F.; Fu, H.; Pang, Y. A cascaded convolutional neural network for single image dehazing. *IEEE Access* **2018**, *6*, 24877–24887. [CrossRef]
35. Yan, S.; Wu, C.; Wang, L.; Xu, F.; An, L.; Guo, K.; Liu, Y. Ddrnet: Depth map denoising and refinement for consumer depth cameras using cascaded cnns. In Proceedings of the European Conference on Computer Vision (ECCV), Munich, Germany, 8–14 September 2018; pp. 151–167.
36. Zhao, S.; Dong, Y.; Chang, E.I.; Xu, Y. Recursive cascaded networks for unsupervised medical imageregistration. In Proceedings of the IEEE/CVF International Conference on Computer Vision, Seoul, Republic of Korea, 27 October–2 November 2019; pp. 10600–10610.
37. Hang, R.; Liu, Q.; Hong, D.; Ghamisi, P. Cascaded recurrent neural networks for hyperspectral image classification. *IEEE Trans. Geosci. Remote Sens.* **2019**, *57*, 5384–5394. [CrossRef]
38. Wu, J.; Ma, J.; Liang, F.; Dong, W.; Shi, G.; Lin, W. End-to-end blind image quality prediction with cascaded deep neural network. *IEEE Trans. Image Process.* **2020**, *29*, 7414–7426. [CrossRef]
39. Tian, C.; Xu, Y.; Zuo, W.; Zhang, B.; Fei, L.; Lin, C. Coarse-to-fine CNN for image super-resolution. *IEEE Trans. Multimed.* **2020**, *23*, 1489–1502. [CrossRef]
40. Lu, X.; Zhang, J.; Yang, D.; Xu, L.; Jia, F. Cascaded convolutional neural network-based hyperspectral image resolution enhancement via an auxiliary panchromatic image. *IEEE Trans. Image Process.* **2021**, *30*, 6815–6828. [CrossRef] [PubMed]
41. Xue, F.; Tan, Z.; Zhu, Y.; Ma, Z.; Guo, G. Coarse-to-fine cascaded networks with smooth predicting for video facial expression recognition. In Proceedings of the IEEE/CVF Conference on Computer Vision and Pattern Recognition, Vancouver, BC, Canada, 4 November 2022; pp. 2412–2418.
42. Tian, C.; Yuan, Y.; Zhang, S.; Lin, C.; Zuo, W.; Zhang, D. Image Super-resolution with An Enhanced Group Convolutional Neural Network. *arXiv* **2022**, arXiv:2205.14548. [CrossRef] [PubMed]
43. Tian, C.; Zhang, Y.; Zuo, W.; Lin, C.; Zhang, D.; Yuan, Y. A heterogeneous group CNN for image super-resolution. *arXiv* **2022**, arXiv:2209.12406. [CrossRef] [PubMed]
44. Bloomfield, P.; Steiger, W.L. *Least Absolute Deviations: Theory, Applications, and Algorithms*; Birkhäuser: Boston, MA, USA, 1983.
45. Pollard, D. Asymptotics for least absolute deviation regression estimators. *Econom. Theory* **1991**, *7*, 186–199. [CrossRef]
46. Kingma, D.P.; Ba, J. Adam: A method for stochastic optimization. *arXiv* **2014**, arXiv:1412.6980.
47. Agarap, A.F. Deep learning using rectified linear units (relu). *arXiv* **2018**, arXiv:1803.08375.
48. Murray, N.; Perronnin, F. Generalized max pooling. In Proceedings of the IEEE conference on computer vision and pattern Recognition, Columbus, OH, USA, 23–28 June 2014; pp. 2473–2480.
49. Xu, J.; Li, Z.; Du, B.; Zhang, M.; Liu, J. Reluplex made more practical: Leaky ReLU. In Proceedings of the 2020 IEEE Symposium on Computers and communications (ISCC), Rennes, France, 7–10 July 2020; IEEE: Piscataway, NJ, USA, 2020; pp. 1–7.
50. Li, X.; Lu, C.; Cheng, D.; Li, W.; Cao, M.; Liu, B.; Ma, J.; Zheng, W. Towards photo-realistic visible watermark removal with conditional generative adversarial networks. In Proceedings of the International Conference on Image and Graphics, Beijing, China, 23–25 August 2019; Springer: Cham, Switzerland, 2019; pp. 345–356.
51. Liang, J.; Niu, L.; Guo, F.; Long, T.; Zhang, L. Visible Watermark Removal via Self-calibrated Localization and Background Refinement. In Proceedings of the 29th ACM International Conference on Multimedia, Chengdu, China, 20 October 2021; pp. 4426–4434.
52. Cun, X.; Pun, C.M. Split then refine: Stacked attention-guided ResUNets for blind single image visible watermark removal. In Proceedings of the AAAI Conference on Artificial Intelligence, Virtually, 2–9 February 2021; Volume 35, pp. 1184–1192.
53. Zhang, Q.; Xiao, J.; Tian, C.; Chun Wei Lin, J.; Zhang, S. A robust deformed convolutional neural network (CNN) for image denoising. *CAAI Trans. Intell. Technol.* **2022**. [CrossRef]
54. Khan, A.; Sohail, A.; Zahoora, U.; Qureshi, A.S. A survey of the recent architectures of deep convolutional neural networks. *Artif. Intell. Rev.* **2020**, *53*, 5455–5516. [CrossRef]
55. Hore, A.; Ziou, D. Image quality metrics: PSNR vs. SSIM. In Proceedings of the 2010 20th International Conference on Pattern Recognition, Istanbul, Turkey, 23–26 August 2010; IEEE: Piscataway, NJ, USA, 2010; pp. 2366–2369.

56. Zhang, K.; Zuo, W.; Zhang, L. FFDNet: Toward a fast and flexible solution for CNN-based image denoising. *IEEE Trans. Image Process.* **2018**, *27*, 4608–4622. [CrossRef]
57. Ronneberger, O.; Fischer, P.; Brox, T. U-net: Convolutional networks for biomedical image segmentation. In Proceedings of the International Conference on Medical Image Computing and Computer-Assisted Intervention, Munich, Germany, 5–9 October 2015; Springer: Cham, Switzerland, 2015; pp. 234–241.
58. Tian, C.; Xu, Y.; Li, Z.; Zuo, W.; Fei, L.; Liu, H. Attention-guided CNN for image denoising. *Neural Netw.* **2020**, *124*, 117–129. [CrossRef] [PubMed]
59. Setiadi, D.R.I.M. PSNR vs SSIM: Imperceptibility quality assessment for image steganography. *Multimed. Tools Appl.* **2021**, *80*, 8423–8444. [CrossRef]
60. Dolbeau, R. Theoretical peak FLOPS per instruction set: A tutorial. *J. Supercomput.* **2018**, *74*, 1341–1377. [CrossRef]

electronics

MDPI

Article

Nighttime Image Dehazing Based on Multi-Scale Gated Fusion Network

Bo Zhao [1,2,*], Han Wu [1], Zhiyang Ma [2], Huini Fu [2], Wenqi Ren [3] and Guizhong Liu [1]

1 School of Information and Communication Engineering, Xi'an Jiaotong University, Xi'an 710049, China
2 China North Vehicle Research Institute, Beijing 100072, China
3 School of Cyber Science and Technology, Sun Yat-sen University, Shenzhen Campus, Shenzhen 528406, China
* Correspondence: whu_c2003@163.com

Abstract: In this paper, we propose an efficient algorithm to directly restore a clear image from a hazy input, which can be adapted for nighttime image dehazing. The proposed algorithm hinges on a trainable neural network realized in an encoder–decoder architecture. The encoder is exploited to capture the context of the derived input images, while the decoder is employed to estimate the contribution of each input to the final dehazed result using the learned representations attributed to the encoder. The constructed network adopts a novel fusion-based strategy which derives three inputs from an original input by applying white balance (WB), contrast enhancing (CE), and gamma correction (GC). We compute pixel-wise confidence maps based on the appearance differences between these different inputs to blend the information of the derived inputs and preserve the regions with pleasant visibility. The final clear image is generated by gating the important features of the derived inputs. To train the network, we introduce a multi-scale approach to avoid the halo artifacts. Extensive experimental results on both synthetic and real-world images demonstrate that the proposed algorithm performs favorably against the state-of-the-art dehazing for nighttime images.

Keywords: night image dehazing; encoder–decoder architecture; image fusion; multi-scale network

Citation: Zhao, B.; Wu, H.; Ma, Z.; Fu, H.; Ren, W.; Liu, G. Nighttime Image Dehazing Based on Multi-Scale Gated Fusion Network. *Electronics* **2022**, *11*, 3723. https://doi.org/10.3390/electronics11223723

Academic Editor: Gwanggil Jeon

Received: 22 October 2022
Accepted: 7 November 2022
Published: 14 November 2022

Publisher's Note: MDPI stays neutral with regard to jurisdictional claims in published maps and institutional affiliations.

1. Introduction

The single-image dehazing [1,2] aims to estimate the unknown clean scene given a hazy or foggy image. This is a classical image processing problem, which has received active research efforts in the computer vision communities [3]. Early dehazing methods focus on exploiting hand-crafted features based on the statistics of clean images, such as dark channel prior [1] and local max contrast [4]. To avoid hand-crafted priors, recent work [5–7] automatically learns haze-relevant features using convolutional neural networks (CNNs). In the dehazing literature, under the assumption of spatially invariant atmospheric light, the hazing process is usually modeled as [1],

$$\mathbf{I}(x) = \mathbf{J}(x)t(x) + \mathbf{A}(1 - t(x)), \tag{1}$$

where $\mathbf{J}(x)$ and $\mathbf{I}(x)$ denote the haze-free scene radiance and the observed hazy image, $\mathbf{A}$ is the global atmospheric light, and $t(x)$ is the scene transmission describing the portion of light that is not scattered and reaches the camera sensors. To recover the clear scene from a hazy input, most dehazing methods try to estimate the transmission $t(x)$ and the atmospheric light $\mathbf{A}$, given a hazy image.

Estimating transmission from hazy images is a severely ill-posed problem. Some approaches try to use visual cues to capture statistical properties of hazy images [8,9]. However, these transmission approximations are inaccurate, especially for the scenes where the colors of objects are inherently similar to those of atmospheric lights. Note that such an erroneous transmission estimation directly affects the quality of the dehazed image, resulting in undesired haze artifacts. Instead of using hand-crafted features, CNN-based approaches [5,7] are proposed to estimate the transmissions. However, these methods still follow the conventional dehazing methods in estimating atmospheric lights to recover clean images. Thus, if the transmission maps are not estimated well, they will interfere with the following airlight estimation and thereby lead to low-quality dehazed results.

In addition, even the state-of-the-art deep learning based methods need to compute the atmospheric light [5,7,10] or reformulated variables which are dependent on the atmospheric light [6,11]. These approaches suffer from important limitations on nighttime hazy scenes. This is mainly due to the multiple light sources that cause a strongly non-uniform illumination of the scene. However, we note that there are a few works to address nighttime dehazing.

To address the above issues, we propose a novel trainable neural network that does not explicitly estimate the transmission and atmospheric light. Thus, the artifacts arising from transmission and airlight estimation errors can be alleviated in the final restored results. The proposed neural network is built on a fusion strategy which aims to seamlessly blend several input images by preserving only the specific features of the composite output image.

We derive several inputs based on two major factors in nighttime hazy images that need to be dealt with. The first one is the color cast introduced by the environmental light. The second one is the lack of visibility due to attenuation. Therefore, we tackle these two problems by deriving three inputs from the original degraded image with the aim of recovering the visibility of the scene in at least one of them. The first input ensures a natural rendition (second column of Figure 1) of the output by eliminating chromatic casts caused by the atmospheric or environmental light. The second contrast-enhanced input generates a better holistic appearance but mainly in the thick hazy regions. However, the contrast-enhanced images are too dark in the light hazy regions. Hence, to recover the light hazy regions, we find that the gamma-corrected images restore information of the light hazy regions well. Consequently, the three derived inputs are gated by three confidence maps (fifth, sixth, and seventh columns of Figure 1), which aim to preserve the regions with good visibility. In addition, we propose to use the normalization (NM) of nighttime hazy images to provide detailed scene information by substituting gamma correction.

Figure 1. We exploit a multi-scale gated fusion network for nighttime haze removal. The first column gives degraded inputs. The second, third, and fourth columns show derived inputs for original images. The learned confidence maps for the derived inputs are shown in the fifth, sixth, and seventh columns, respectively. The last column shows our results by the proposed algorithm.

This paper is an extension of our preliminary version [12], which concentrates on daytime dehazing. In this paper, we first improve the network architecture (Section 3.2) and then adapt our network to work effectively on nighttime hazy scenes (Section 4). The contributions of this paper are summarized as follows:

- We propose a deep trainable neural network that restores clear images without assuming restrictions on scene transmission and atmospheric light.
- We demonstrate the effectiveness of a gated fusion network for single nighttime image dehazing by leveraging the derived inputs from an original input.
- We train the proposed model with a multi-scale approach to eliminate the halo artifacts that hurt image recovering.
- We show that the proposed algorithm can effectively process nighttime hazy images which are not well handled by most dehazing methods. We show that the proposed method performs favorably against the state-of-the-arts.

2. Related Work

2.1. Day-TIme Image Dehazing

Tang et al. [13] combined four types of haze-relevant features with Random Forest to estimate the transmission. Zhu et al. [14] introduced a linear model and learned the parameters of the model in a supervised manner under a color attenuation prior. However, these methods are still developed based on hand-crafted features.

Recently, CNNs have also been used for haze removal and related problems [15–18]. Cai et al. [5] proposed a DehazeNet and a BReLU layer to estimate the transmissions from hazy inputs. In [7], a coarse-scale network was first used to learn the mapping between hazy inputs and their transmissions, and then, a fine-scale network was exploited to refine the transmission. Zhang and Patel [10] proposed a densely connected encoder–decoder structure for joint estimating the transmission map and atmospheric light. Yang and Sun [11] build a deep architecture incorporating the prior learning for single image dehazing. In the recent level-aware progressive network (LAP-Net) model, an image is restored by fusing the results at various haze levels at different stages. However, one problem of these CNN-based methods [5,7] is that all these models require accurate transmission and atmospheric light estimation steps to restore clear images. Although the AOD-Net [6] method bypasses the estimation step, this approach still needs to compute an additional variable $\mathbf{K}(x)$ which integrates both transmission $t(x)$ and atmospheric light $\mathbf{A}$. Thus, the AOD-Net falls as one of the physics models as described in (1) that encounters issues with ill-posed problems. To alleviate these problems, several end-to-end networks [19–22] have recently been proposed to directly filter the input image.

Different from these CNN-based approaches, our proposed network is built on the principle of image fusion, and it is trained to produce the sharp image directly without estimating transmission and atmospheric light. The main idea of image fusion is to combine several images into a single one, retaining only the most significant features. This idea has been used in a number of applications such as image editing [23] and video super-resolution [24].

2.2. Nighttime Dehazing

Different from common image dehazing, nighttime hazy images often include visible man-made light sources with varying colors and non-uniform illumination [25]. These light sources may introduce noticeable amounts of glow that are not present in haze images taken in the daytime, which makes the estimation of atmospheric light inaccurate and causes some sharp images prior to becoming invalid. However, in recent years, the community has paid relatively less research attention to the nighttime haze removal problem.

Pei and Lee [26] estimate the ambient illumination and the haze thickness by transferring the hazy input into a grayish one; then, they recover the dehazed result using the refined DCP by a bilateral filter in local contrast correction. Zhang et al. [27] build a new imaging model for nighttime conditions; then, they remove the haze by using the DCP along with estimating the point-wise environmental light. Based on the proposed physics model, they estimate the ambient illumination and transmission by combining a maximum reflectance prior (MRP) [28]. However, MRP shares the common limitations of most statistical prior-based methods. When the scene objects are inherent with a solely distinct color, the maximum reflectance prior becomes invalid in nighttime scenes. In [29], Li et al. also introduce a nighttime haze model that is a linear combination of the direct transmission, airlight and glow. Using the physics model, the authors first reduce the effect of the glow and then recover the final dehazed result. Nevertheless, this approach tends to generate some halo artifacts in the dehazed results. Ancuti et al. [30] assume that the brightest pixels of local patches filtered by a minimal operator can capture the properties of atmospheric light, and they use the multi-scale fusion approach to obtain a visibility-enhanced image.

Similar to [25,30], we also propose a multi-scale fusion network for nighttime dehazing. Differently, without any tedious estimation of contrast, saturation, saliency, and airlight, we directly predict the weight maps for each derived input by the trainable network.

3. Multi-Scale Gated Fusion Network Architecture

This section presents the details of our multi-scale gated fusion network that employs an original degraded image and three derived images as inputs. We refer to this network as multi-scale GFN, or MSGFN, as shown in Figure 2. The central idea is to learn the confidence maps to combine several input images into a single one by keeping only the most significant features of them.

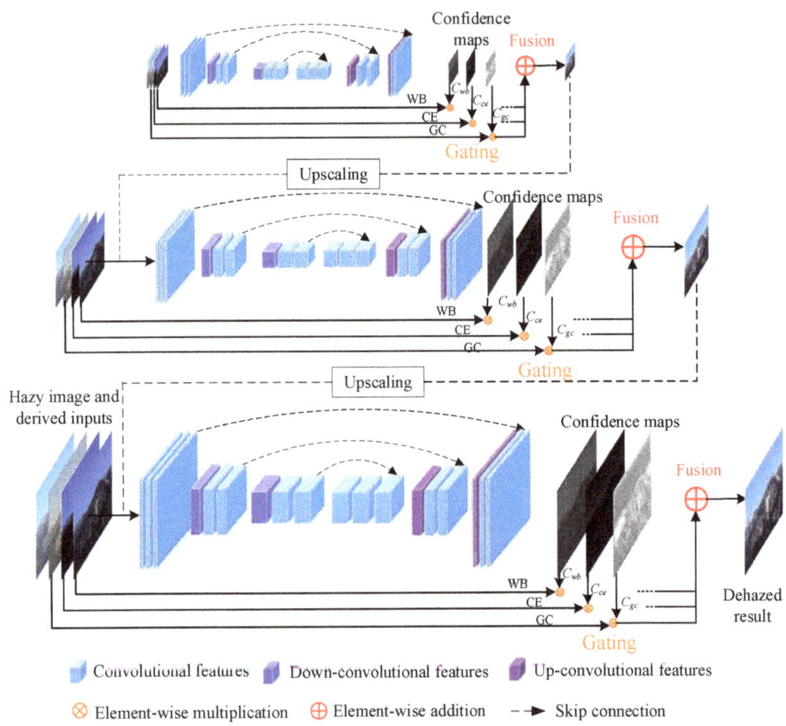

Figure 2. The architecture of the proposed multi-scale GFN, which takes a hazy image pyramid and the corresponding three enhanced versions as the input and outputs a latent image pyramid. These three derived inputs are weighted by the three confidence maps in each scale learned by our network, and the full-resolution output is the final dehazed result. The network contains layers of symmetric encoders and decoders. Skip shortcuts are connected from the convolutional feature maps to the deconvolutional feature maps.

3.1. Derived Inputs

We derive several inputs based on the following observations. The first one is that the colors in hazy images often change due to the influence of the atmospheric light. The second is the lack of visibility in distant regions due to scattering and attenuation phenomena. Based on these observations, we generate three inputs that recover the color and visibility of the entire image from the original hazy image. We first estimate the white balanced (WB) image $\mathbf{I}_{wb}$ of the hazy input $\mathbf{I}$ to recover the latent color of the scene. Then, we extract visible information including the contrast enhanced (CE) $\mathbf{I}_{ce}$ and the gamma corrected (GC) $\mathbf{I}_{gc}$ to generate better holistic quality.

White balanced input. Our first input is a white balanced image which aims to eliminate chromatic casts caused by the atmospheric color. In the past decades, a number of white balancing approaches [31,32] have been proposed. In this paper, we use the gray world assumption [33] based technique. Despite its simplicity, this low-level approach has shown to generate comparable results to those of more complex white balance methods [3]. The gray world assumption is that given an image with a sufficient quantity of color variations, the average value of the Red, Green and Blue components of the image should average out to a common gray value. This assumption is in general valid in any given real-world scene since the variations in colors are random and independent. It would be safe to say that given a large number of samples, the average should tend to converge to the mean value, which is gray. White balancing algorithms can make use of this gray world assumption by forcing images to have a uniform average gray value for the R, G, and B channels. For example, if an image is shot under a hazy weather condition, the captured image will have an atmospheric light **A** cast over the entire image. The effect of this atmospheric light cast disturbs the gray world assumption of the original image. By imposing the assumption on the captured image, we would be able to remove the atmospheric light cast and re-acquire the colors of our original scene. Figure 3b demonstrates such an effect.

Although white balancing could discard the color shifting caused by the atmospheric light, the results still present low contrast. To enhance the contrast, we introduce the following two derived inputs.

Contrast-enhanced input. Similar to prior dehazing methods [34,35], our second input is a contrast-enhanced image of the original hazy input. Ancuti [34] derived a contrast-enhanced image by subtracting the average luminance value $\tilde{I}$ of the entire image **I** from the hazy input and then using a factor μ to linearly increase the luminance in the recovered hazy regions as follows:

$$\mathbf{I}_{ce} = \mu(\mathbf{I} - \tilde{I}), \tag{2}$$

where $\mu = 2(0.5 + \tilde{I})$. Although $\tilde{I}$ is a good indicator of image brightness, there is a problem in this input, especially in denser haze regions. The main reason is that the negative values of $(\mathbf{I} - \tilde{I})$ may dominate the contrast-enhanced input as $\tilde{I}$ increases. As shown in Figure 3c, the dark image regions tend to be black after contrast enhancing.

(a) (b) (c) (d)

Figure 3. We derive three enhanced versions from nighttime hazy images. These derived inputs contain different important visual cues of the input hazy images. (**a**) Inputs; (**b**) WB; (**c**) CE; (**d**) NM.

3.2. Network Architecture

Only using one scale is subject to halo artifacts in the dehazed results, particularly for strong transitions within the confidence maps [34,35]. Hence, we perform estimation by varying the image resolution in a coarse-to-fine manner to prevent halo artifacts. The multi-scale approach is motivated by the fact that the human visual system is sensitive to local changes (e.g., edges) over a wide range of scales. As a merit, the multi-scale approach provides a convenient way to incorporate local image details over varying resolutions.

The proposed multi-scale GFN is shown in Figure 2. Finer level networks basically have the same structure as the coarsest network. However, the first convolutional layer takes the dehazed output from a previous stage as well as its own hazy image and derived inputs in a concatenated form. Each input size is twice the size of its coarser-scale network. As shown in Figure 2, there is an up-sampling layer to resize the coarser output before the next stage. At the finest scale, the original full-resolution image is recovered.

We use an encoder–decoder network in each scale, which has been shown to produce good results for a number of generative tasks. In particular, we choose a variation of the residual encoder–decoder block for image dehazing. We use skip connections between encoder and decoder halves of the network, where features from the encoder side are concatenated to be fed to the decoder. This significantly accelerates the convergence and helps generate a much clear dehazed image. In addition, we improve encoder–decoder modules by using residual blocks [36] after each convolution layer. We use shared weights in each scale, which operates in a way similar to using data multiple times [37] (i.e., data augmentation regarding scales) and reduces the number of parameters need to be learned.

We perform an early fusion by concatenating the original hazy image and three derived inputs in the input layer. Rectification layers are added after each convolutional or deconvolutional layer. The convolutional layers act as a feature extractor, which preserves the primary information of scene colors in the input layer, meanwhile eliminating the unimportant colors from the inputs. The deconvolutional layers are then combined to recover the weight maps of three derived inputs. In other words, the outputs of the deconvolutional layers are the *confidence maps* of the derived input images $\mathbf{I}_{wb}$, $\mathbf{I}_{ce}$ and $\mathbf{I}_{gc}$.

We use three down-convolutional blocks and three deconvolutional blocks in each scale. The stride for down-convolution layer is two, which down-samples feature maps to half size and doubles the channel of the previous layer. Each of the following ResBlocks contains two convolution layers. Each convolutional layer is of the same kernel size of 3×3 except the first layer. The first layer operates on the input image with kernel size of 5×5. In this work, we demonstrate that explicitly modeling confidence maps has several advantages. These are discussed later in Section 7.1. Once the confidence maps for the derived inputs are predicted, we fuse different inputs using the proposed gating method as illustrated in Figure 2,

$$J^k = \text{Gating}(\mathbf{I}^k_{wb}, \mathbf{I}^k_{ce}, \mathbf{I}^k_{gc}), \tag{3}$$

where J^k is the gated result at scale k. The gating function is defined by

$$\text{Gating}(x, y, z) = C_x \circ x + C_y \circ y + C_z \circ z, \tag{4}$$

where $\circ$ denotes element-wise multiplication, and $C_{(\cdot)}$ is the confidence map for the input.

The multi-scale approach desires that each scale output is a clear image of the corresponding scale. Thus, we train our network so that all intermediate dehazed images should form a pyramid of the sharp image. The MSE criterion is applied to every level of the pyramid. In particular, given a collection of N training pairs $\mathbf{I}_i$ and $\mathbf{J}_i$, where $\mathbf{I}_i$ is a hazy image and $\mathbf{J}_i$ is the clean version as the ground truth, the loss function at the k-th scale is defined as follows:

$$\mathcal{L}(\Theta, k) = \frac{1}{N} \sum_{i=1}^{N} \|\mathcal{F}(\mathbf{I}_{i,k}, \Theta, k) - \mathbf{J}_{i,k}\|^2, k \in \{1, 2, 3\}, \tag{5}$$

where Θ keeps the weights of the convolutional and deconvolutional kernels.

4. Nighttime Image Dehazing

Since nighttime scenes usually have artificial light sources that generate a glow effect in hazy images, most state-of-the-art dehazing methods based on (1) suffer from significant limitations on nighttime hazy scenes. Although several physics-based models [28,29] are developed to relax those strict constraints in (1) (e.g., homogeneous atmosphere illumination, unique extinction coefficient), a straightforward extension of common hazy image modeling to nighttime scenes cannot always hold in real cases. This is why our approach does not resort to an explicit inversion of the nighttime light propagation model in [28,29].

Fusion Process of Nighttime Dehazing

In this paper, we demonstrate that the proposed MSGFN can also effectively enhance nighttime hazy images. We employ the strategy described in Figure 2 to remove haze in nighttime images. For the derived inputs, we also use WB and CE to process a color correction step and visibility enhancement, respectively. However, there is another problem in nighttime hazy images that needs to be dealt with, i.e., non-uniform illumination caused by multiple light sources in the low-light environment. Therefore, we derive a third input, normalization (NM) of the nighttime hazy image, to obtain an illumination-balanced result and enhance the finest details in the nighttime scene.

The NM operation is obtained by linearly stretching all the pixel values in order to fit them into the interval [0, 1]. In this case, we achieve a better illumination result by contrast stretching the range of intensities of the hazy input. The main advantage of this operation is that we do not require any parameter to be tuned, and therefore, without information loss in the derived input. As shown in Figure 3d, the NM operation shifts and scales all the color pixel intensities of the input so that the pixel values cover the entire available dynamic range and obtain a balanced illumination.

Similar to the dehazing approach described in Section 3, we use the proposed MSGFN to predict three confidence maps for the derived inputs to ensure that regions of high contrast or high saliency will receive greater weights in the gated fusion process:

$$J^k = \text{Gating}(\mathbf{I}^k_{wb}, \mathbf{I}^k_{ce}, \mathbf{I}^k_{nm}), \tag{6}$$

where $\mathbf{I}^k_{nm}$ is the normalized version of the nighttime hazy input at scale k.

5. Nighttime Dehazing Results

We evaluate the proposed algorithm with nighttime configuration on real-world night hazy scenes, with comparisons to the state-of-the-art methods in terms of visual effect.

5.1. Training Data

Owing to the difficulty in obtaining realistic nighttime training data, we adopt the similar strategy as the daytime methods [38] to synthesize nighttime hazy scenes. Specifically, we select 4500 clear nighttime scenes in the KAIST dataset [39] and use the method proposed in [40] to estimate depth maps, which has been demonstrated to be effective for nighttime scene depth estimation. Then, we synthesize 4500 nighttime hazy images according to (1). Note that although some nighttime hazy imaging models are proposed [28,29] to account for artificial light sources, we found our synthesized nighttime hazy images based on (1) look natural as shown in Figure 4, since the proposed model in [28,29] is a generalization of (1) when the illumination is assumed to be a constant.

5.2. Quantitative Evaluation

For quantitative performance evaluation, we construct a new dataset of synthesized nighttime hazy images. We select 100 clear nighttime scenes (different from those that were used for training) from the KAIST dataset [39] to synthesize 500 hazy images (using different scattering coefficients to synthesize different haze concentrations). Figure 5 shows some dehazed images by the evaluated methods. The nighttime dehazing methods of MRP [28] and GMLC [29] generate the results with significant color distortions. The dehazed images by the deep learning approaches of MSCNN [7], GCAN [19], and GDN [41] still contain significant haze residuals. In contrast, our algorithm restores these images well. Overall, the dehazed results by the proposed algorithm are of higher visual quality and with fewer color distortions. The visual results in Figure 5 match the quantitative results shown in Table 1.

Figure 4. The proposed method for synthesizing nighttime hazy images. The first row shows original clear night scenes from [39],and the second row shows the synthesizing hazy images.

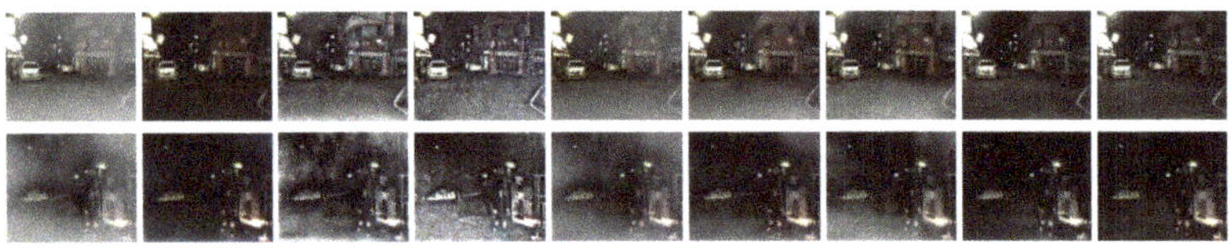

(a) (b) (c) (d) (e) (f) (g) (h) (i)

Figure 5. Dehazed results on synthetic nighttime images. The results by learning-based methods of MSCNN [7], GCAN [19], and GDN [41] have some remaining haze, while the nighttime dehazing methods of MRP [28] and GMLC [29] tend to generate some color distortions. In contrast, the dehazed results by our algorithm are close to the ground-truth images. (**a**) Hazy inputs; (**b**) DCP [42]; (**c**) MRP [28]; (**d**) GMLC [29]; (**e**) MSCNN [7]; (**f**) GCAN [19]; (**g**) GDN [41]; (**h**) Our results; (**i**) Ground truth.

Table 1. Average PSNR/SSIM of dehazed results by state-of-the-art dehazing methods on nighttime hazy images.

Input	DCP [42]	MSCNN [7]	MRP [28]	GMLC [29]	GCAN [19]	GDN [41]	MSGFN
13.70/0.6063	24.94/0.902	17.45/0.7113	16.49/0.6936	14.49/0.552	19.18/0.8133	21.03/0.8916	30.92/0.9492

5.3. Qualitative Evaluation

To demonstrate that the proposed method generalizes well in real-world nighttime hazy scenes, we use real-world hazy images for experiments against the state-of-the-art dehazing algorithms designed for nighttime scenes, i.e., Maximum Reflectance Prior (MRP) [28] as well as Glow and Multiple Light Colors (GMLC) [29], and daytime scenarios, i.e., DCP [42], MSCNN [7], GCAN [19], and GDN [41].

Figure 6b,c show the results by the recent nighttime dehazing methods, i.e., MRP [28] and GMLC [29]. The MRP method [28] tends to darken the hazy inputs in some regions. For example, the road regions of the first image are much darker than those obtained by other methods. In addition, the GMLC model [29] generates some artifacts in sky regions, e.g., the first and third images in Figure 6e. Figure 6d–g demonstrate the limitations of the daytime dehazing approaches, i.e., DCP [1], MSCNN [7], GCAN [19], and GDN [41] when applied to nighttime hazy inputs. Both the prior-based [1] and CNN-based [7,19,41] methods cannot recover colors well, and they only slightly remove the haze in these night scenes. In contrast, our algorithm generates dehazed results with clearer and sharper details and without artifacts in the sky regions as shown in Figure 6h.

(a) (b) (c) (d) (e) (f) (g) (h)

Figure 6. Qualitative comparison of different methods on real-world images. (**a**) Hazy inputs; (**b**) MRP [28]; (**c**) GMLC [29]; (**d**) DCP [42]; (**e**) MSCNN [7]; (**f**) GCAN [19]; (**g**) GDN [41]; (**h**) Our results.

6. Further Experiments

6.1. Comparison on O-Haze

In the main paper, we evaluate the proposed algorithm on all the 45 hazy images from the O-HAZE dataset [43] against the state-of-the-art methods. In this supplementary material, we retrain the proposed MSGFN using the same 40 training data as in the NTIRE 2018 challenge [2] and compare it with the winning methods in [2] on the five test images. As shown in Table 2, our proposed method performs favorably against the winning methods in the NTIRE 2018 challenge [2] and achieves the highest SSIM score.

Table 2. Average PSNR/SSIM of dehazed results on the 5 test images in the O-Haze [43] dataset. Although our algorithm ranks third in terms of PSNR, our method achieves the highest SSIM score.

Ranking in [2]	Methods	PSNR	SSIM
1	BJTU	24.598	0.777
2	KAIST-VICLAB [22]	24.232	0.687
–	Ours (MSGFN)	24.054	**0.787**
3	Scarlet Knights [21]	24.029	0.775
4	FKS	23.877	0.775
5	Dq-hisfriends	23.207	0.770
6	Ranjanisi [44]	23.180	0.705
7	Mt.Phoenix	23.124	0.755
8	Ranjanisi [44]	22.997	0.701
9	KAIST-VICLAB [45]	22.705	0.707
10	Mt.Phoenix	22.080	0.731
11	IVLab	21.750	0.717
12	CLEAR	20.291	0.683
13	CLFStudio	20.230	0.722
14	SiMiT-Lab [46]	19.628	0.674
15	AHappyFaceI	18.494	0.669
16	ASELSAN	18.123	0.675
17	Dehazing-by-retinex [47]	17.547	0.652
18	IMCL	16.527	0.616
	baseline (hazy images)	15.784	0.634

6.2. Mixed Training Strategy

To demonstrate the robustness of the proposed MSGFN on different training strategies, we train an additional network with all three datasets (daytime, nighttime, and underwater datasets) together. We refer to this network as "all-in-one" and refer to the original network in the main paper as "separate".

As shown in Table 3, the proposed model performs better on the daytime (SOTS and O-Haze) and nighttime datasets with the "separate" training strategy. Meanwhile, the performance on the underwater dataset becomes better with the "all-in-one" training strategy. Since the underwater inputs in the UIEB dataset are real-world images, the main reason may be that more types of training data benefit real-world image reconstruction.

Table 3. Comparison of MSGFN using different training strategies ("separate" vs. "all-in-one").

Dataset	Separate	All-in-One
Daytime (SOTS)	**25.37**/0.93	23.19/**0.94**
Daytime (O-Haze)	**21.21/0.76**	19.05/0.74
Nighttime	**30.92/0.95**	22.69/0.86
Underwater (UIEB)	17.61/0.86	**21.99/0.91**

7. Analysis and Discussions

7.1. Effectiveness of Fusion Strategy

Image fusion is a method to blend several images into a single one by retaining only the most useful features. To effectively blend the information of the derived inputs, we filter their important information by computing corresponding confidence maps. Consequently, in our gated fusion network, the derived inputs are gated by three pixel-wise confidence maps that aim to preserve the regions with good visibility. Our fusion network has two advantages: the first one is that it can reduce patch-based artifacts (e.g., dark channel prior [1]) by single pixel operations, and the other one is that it can eliminate the influence caused by transmission and atmospheric light estimation.

To show the effectiveness of fusion network, we also train an end-to-end network without a fusion process for the dehazing task. This network has the same architecture as MSGFN except the input is hazy image and output is dehazed result without confidence maps learning at each scale. In addition, we also conduct an experiment based on an equivalent fusion strategy, i.e., all the three derived inputs are weighted equally using 1/3. Figure 7 shows visual comparisons of on two real-world examples with different settings. In these examples, the approach without gating generates dark images in Figure 7b, and the method with an equivalent fusion strategy generates results with color distortion and dark regions as shown in Figure 7c. In contrast, our results contain most scene details and maintain the original colors which demonstrate the effectiveness of the learned confidence maps.

7.2. Effectiveness of Derived Inputs

We can design different inputs for different enhancement tasks. In practice, it is difficult to entirely remove the haze effects of hazy images by an enhancing approach. Therefore, the input generation process seeks to recover sharp regions in at least one of the derived inputs as analyzed in Section 3. They complement each other nicely to help dehazing by the gated fusion network as shown in Table 4.

Although we do not claim that these are the optimal inputs, our experiments show that the three derived inputs are the minimum inputs. Using two or fewer of them will not generate better results in the proposed network (Table 4) for nighttime image dehazing. In the future work, we will explore more effective derived inputs or directly learn the derived inputs in the fusion network. The network parameters comparison can be found in Table 5.

Figure 7. Effectiveness of the gated fusion network. (**a**) Hazy inputs; (**b**) w/o fusion; (**c**) Equivalent fusion; (**d**) MSGFN.

Table 4. Average PSNR/SSIM using different derived inputs. The method only using the original image means that we directly learn the mapping from degraded images to the clear ones.

Inputs				Dehazing
Original	**WB**	**CE**	**GC/NM**	**PSNR/SSIM**
√	×	×	×	22.38/0.90
√	√	√	×	24.83/0.92
√	√	×	√	23.54/0.92
√	×	√	√	23.96/0.89
√	√	√	√	**25.37/0.93**

Table 5. Comparison of MSGFN and state-of-the-art dehazing approaches with respect to parameters.

Model	Parameters
AOD-Net [6]	1.83×10^3
MSCNN [7]	8.01×10^3
DehazeNet [5]	8.24×10^3
Domain adaption [48]	2.27×10^5
PMS-Net [17]	2.44×10^5
GCAN [19]	7.03×10^5
EPDN [18]	1.74×10^7
DCPDN [10]	6.69×10^7
CGAN [16]	1.23×10^8
Ours	5.15×10^5

8. Conclusions

In this paper, we addressed the nighttime image dehazing via a multi-scale gated fusion network (MSGFN), a fusion based encoder–decoder architecture, by learning confidence maps for derived inputs. Compared with previous methods which impose restrictions on transmission and atmospheric light, our proposed MSGFN is easy to implement and reproduce since the proposed approach does not rely on the estimations of transmission and atmospheric/environmental light. In the approach, we first applied white balance to recover the scene color and then generated two contrast enhanced images for better visibility. Third, we carried out the MSGFN to estimate the confidence map for each derived input. Finally, we used the confidence maps and derived inputs to render the final result.

The experimental results on synthetic and real-world nighttime images demonstrate the effectiveness of the proposed approach.

Author Contributions: Supervision, W.R. and G.L.; Validation, Z.M.; Visualization, H.F.; Writing—original draft, B.Z.; Writing—review & editing, H.W. All authors have read and agreed to the published version of the manuscript.

Funding: This research received no external funding.

Data Availability Statement: Not applicable.

Conflicts of Interest: The authors declare no conflict of interest.

References

1. He, K.; Sun, J.; Tang, X. Single image haze removal using dark channel prior. In Proceedings of the IEEE Conference on Computer Vision and Pattern Recognition, Miami, FL, USA, 20–25 June 2009.
2. Ancuti, C.; Ancuti, C.O.; Timofte, R. Ntire 2018 challenge on image dehazing: Methods and results. In Proceedings of the IEEE Conference on Computer Vision and Pattern Recognition Workshops, Salt Lake City, UT, USA, 18–22 June 2018; pp. 891–901.
3. Li, Y.; You, S.; Brown, M.S.; Tan, R.T. Haze visibility enhancement: A survey and quantitative benchmarking. *Comput. Vis. Image Underst.* **2017**, *165*, 1–16. [CrossRef]
4. Tan, R.T. Visibility in bad weather from a single image. In Proceedings of the IEEE Conference on Computer Vision and Pattern Recognition, Anchorage, Alaska, USA, 24–26 June 2008.
5. Cai, B.; Xu, X.; Jia, K.; Qing, C.; Tao, D. Dehazenet: An end-to-end system for single image haze removal. *IEEE Trans. Image Process.* **2016**, *25*, 5187–5198. [CrossRef] [PubMed]
6. Li, B.; Peng, X.; Wang, Z.; Xu, J.; Feng, D. Aod-net: All-in-one dehazing network. In Proceedings of the IEEE International Conference on Computer Vision, Venice, Italy, 22–29 October 2017.
7. Ren, W.; Liu, S.; Zhang, H.; Pan, J.; Cao, X.; Yang, M.H. Single image dehazing via multi-scale convolutional neural networks. In Proceedings of the European Conference on Computer Vision, Amsterdam, The Netherlands, 11–14 October 2016.
8. Berman, D.; Avidan, S. Non-local image dehazing. In Proceedings of the IEEE Conference on Computer Vision and Pattern Recognition, Las Vegas, NV, USA, 27–30 June 2016.
9. Fattal, R. Single image dehazing. In Proceedings of the SIGGRAPH, Los Angeles, CA, USA, 11–15 August 2008.
10. Zhang, H.; Patel, V.M. Densely connected pyramid dehazing network. In Proceedings of the IEEE Conference on Computer Vision and Pattern Recognition, Salt Lake City, UT, USA, 18–23 June 2018; pp. 3194–3203.
11. Yang, D.; Sun, J. Proximal dehaze-net: A prior learning-based deep network for single image dehazing. In Proceedings of the European Conference on Computer Vision, Munich, Germany, 8–14 September 2018.
12. Ren, W.; Ma, L.; Zhang, J.; Pan, J.; Cao, X.; Liu, W.; Yang, M.H. Gated fusion network for single image dehazing. In Proceedings of the IEEE Conference on Computer Vision and Pattern Recognition, Salt Lake City, UT, USA, 18–23 June 2018; pp. 3253–3261.
13. Tang, K.; Yang, J.; Wang, J. Investigating Haze-Relevant Features in a Learning Framework for Image Dehazing. In Proceedings of the IEEE Conference on Computer Vision and Pattern Recognition, Columbus, OH, USA, 23–28 June 2014.
14. Zhu, Q.; Mai, J.; Shao, L. A Fast Single Image Haze Removal Algorithm Using Color Attenuation Prior. *IEEE Trans. Image Process.* **2015**, *24*, 3522–3533. [PubMed]
15. Zhang, H.; Patel, V.M. Density-aware Single Image De-raining using a Multi-stream Dense Network. In Proceedings of the IEEE Conference on Computer Vision and Pattern Recognition, Salt Lake City, UT, USA, 18–23 June 2018.
16. Li, R.; Pan, J.; Li, Z.; Tang, J. Single image dehazing via conditional generative adversarial network. In Proceedings of the IEEE Conference on Computer Vision and Pattern Recognition, Salt Lake City, UT, USA, 18–23 June 2018.
17. Chen, W.T.; Ding, J.J.; Kuo, S.Y. PMS-Net: Robust Haze Removal Based on Patch Map for Single Images. In Proceedings of the IEEE Conference on Computer Vision and Pattern Recognition, Long Beach, CA, USA, 15–20 June 2019; pp. 11681–11689.
18. Qu, Y.; Chen, Y.; Huang, J.; Xie, Y. Enhanced pix2pix dehazing network. In Proceedings of the IEEE Conference on Computer Vision and Pattern Recognition, Long Beach, CA, USA, 15–20 June 2019.
19. Chen, D.; He, M.; Fan, Q.; Liao, J.; Zhang, L.; Hou, D.; Yuan, L.; Hua, G. Gated context aggregation network for image dehazing and deraining. In Proceedings of the IEEE Winter Conference on Applications of Computer Vision, Waikoloa Village, HI, USA, 7–11 January 2019; pp. 1375–1383.
20. Hong, M.; Xie, Y.; Li, C.; Qu, Y. Distilling Image Dehazing With Heterogeneous Task Imitation. In Proceedings of the IEEE Conference on Computer Vision and Pattern Recognition, Seattle, WA, USA, 13–19 June 2020; pp. 3462–3471.
21. Zhang, H.; Sindagi, V.; Patel, V.M. Multi-scale single image dehazing using perceptual pyramid deep network. In Proceedings of the IEEE Conference on Computer Vision and Pattern Recognition Workshops, Salt Lake City, UT, USA, 18–22 June 2018; pp. 902–911.
22. Sim, H.; Ki, S.; Choi, J.S.; Seo, S.; Kim, S.; Kim, M. High-resolution image dehazing with respect to training losses and receptive field sizes. In Proceedings of the IEEE Conference on Computer Vision and Pattern Recognition Workshops, Salt Lake City, UT, USA, 18–22 June 2018; pp. 912–919.

23. Pérez, P.; Gangnet, M.; Blake, A. Poisson image editing. *ACM Trans. Graph.* **2003**, *22*, 313–318. [CrossRef]
24. Liu, D.; Wang, Z.; Fan, Y.; Liu, X.; Wang, Z.; Chang, S.; Huang, T. Robust video super-resolution with learned temporal dynamics. In Proceedings of the IEEE International Conference on Computer Vision, Venice, Italy, 22–29 October 2017.
25. Ancuti, C.; Ancuti, C.O.; De Vleeschouwer, C.; Bovik, A.C. Day and night-time dehazing by local airlight estimation. *IEEE Trans. Image Process.* **2020**, *29*, 6264–6275. [CrossRef] [PubMed]
26. Pei, S.C.; Lee, T.Y. Nighttime haze removal using color transfer pre-processing and dark channel prior. In Proceedings of the IEEE International Conference on Image Processing, Orlando, FL, USA, 30 September–3 October 2012.
27. Zhang, J.; Cao, Y.; Wang, Z. Nighttime haze removal based on a new imaging model. In Proceedings of the ICIP, Paris, France, 27–30 October 2014.
28. Jing, Z.; Yang, C.; Shuai, F.; Yu, K.; Chang, W.C. Fast haze removal for nighttime image using maximum reflectance prior. In Proceedings of the IEEE Conference on Computer Vision and Pattern Recognition, Honolulu, HI, USA, 21–26 June 2017.
29. Li, Y.; Tan, R.T.; Brown, M.S. Nighttime Haze Removal with Glow and Multiple Light Colors. In Proceedings of the IEEE International Conference on Computer Vision, Santiago, Chile , 7–13 December 2015.
30. Ancuti, C.; Ancuti, C.O.; De Vleeschouwer, C.; Bovik, A.C. Night-time dehazing by fusion. In Proceedings of the IEEE International Conference on Image Processing, Phoenix, AZ, USA, 25–28 September 2016.
31. Bekaert, P. Enhancing underwater images and videos by fusion. In Proceedings of the IEEE Conference on Computer Vision and Pattern Recognition, Providence, RI, USA, 16–21 June 2012.
32. Ancuti, C.O.; Ancuti, C.; De Vleeschouwer, C.; Sbert, M. Color channel compensation (3C): A fundamental pre-processing step for image enhancement. *IEEE Trans. Image Process.* **2019**, *29*, 2653–2665. [CrossRef] [PubMed]
33. Reinhard, E.; Adhikhmin, M.; Gooch, B.; Shirley, P. Color transfer between images. *Comput. Graph. Appl.* **2001**, *21*, 34–41. [CrossRef]
34. Ancuti, C.O.; Ancuti, C. Single image dehazing by multi-scale fusion. *IEEE Trans. Image Process.* **2013**, *22*, 3271–3282. [CrossRef] [PubMed]
35. Choi, L.K.; You, J.; Bovik, A.C. Referenceless prediction of perceptual fog density and perceptual image defogging. *IEEE Trans. Image Process.* **2015**, *24*, 3888–3901. [CrossRef] [PubMed]
36. He, K.; Zhang, X.; Ren, S.; Sun, J. Deep residual learning for image recognition. In Proceedings of the IEEE Conference on Computer Vision and Pattern Recognition, Las Vegas, NV, USA, 26–30 June 2016; pp. 770–778.
37. Tao, X.; Gao, H.; Shen, X.; Wang, J.; Jia, J. Scale-recurrent network for deep image deblurring. In Proceedings of the IEEE Conference on Computer Vision and Pattern Recognition, Salt Lake City, UT, USA, 18–23 June 2018; pp. 8174–8182.
38. Li, B.; Ren, W.; Fu, D.; Tao, D.; Feng, D.; Zeng, W.; Wang, Z. Benchmarking single-image dehazing and beyond. *IEEE Trans. Image Process.* **2018**, *28*, 492–505. [CrossRef]
39. Hwang, S.; Park, J.; Kim, N.; Choi, Y.; Kweon, I.S. Multispectral Pedestrian Detection: Benchmark Dataset and Baseline. In Proceedings of the IEEE Conference on Computer Vision and Pattern Recognition, Boston, MA, USA, 7–12 June 2015.
40. Li, Z.; Snavely, N. MegaDepth: Learning Single-View Depth Prediction from Internet Photos. In Proceedings of the IEEE Conference on Computer Vision and Pattern Recognition, Salt Lake City, UT, USA, 18–23 June 2018.
41. Liu, X.; Ma, Y.; Shi, Z.; Chen, J. GridDehazeNet: Attention-Based Multi-Scale Network for Image Dehazing. In Proceedings of the IEEE International Conference on Computer Vision, Seoul, Korea, 27 October–2 November 2019.
42. He, K.; Sun, J.; Tang, X. Single image haze removal using dark channel prior. *IEEE Trans. Pattern Anal. Mach. Intell.* **2011**, *33*, 2341–2353.
43. Ancuti, C.O.; Ancuti, C.; Timofte, R.; De Vleeschouwer, C. O-HAZE: A dehazing benchmark with real hazy and haze-free outdoor images. In Proceedings of the IEEE Conference on Computer Vision and Pattern Recognition Workshops, Salt Lake City, UT, USA, 18–22 June 2018; pp. 754–762.
44. Mondal, R.; Santra, S.; Chanda, B. Image dehazing by joint estimation of transmittance and airlight using bi-directional consistency loss minimized FCN. In Proceedings of the IEEE Conference on Computer Vision and Pattern Recognition Workshops, Salt Lake City, UT, USA, 18–22 June 2018; pp. 920–928.
45. Ki, S.; Sim, H.; Choi, J.S.; Kim, S.; Kim, M. Fully end-to-end learning based conditional boundary equilibrium gan with receptive field sizes enlarged for single ultra-high resolution image dehazing. In Proceedings of the IEEE Conference on Computer Vision and Pattern Recognition Workshops, Salt Lake City, UT, USA, 18–22 June 2018; pp. 817–824.
46. Engin, D.; Genç, A.; Kemal Ekenel, H. Cycle-dehaze: Enhanced cyclegan for single image dehazing. In Proceedings of the IEEE Conference on Computer Vision and Pattern Recognition Workshops, Salt Lake City, UT, USA, 18–22 June 2018; pp. 825–833.
47. Galdran, A.; Alvarez-Gila, A.; Bria, A.; Vazquez-Corral, J.; Bertalmío, M. On the duality between retinex and image dehazing. In Proceedings of the IEEE Conference on Computer Vision and Pattern Recognition, Salt Lake City, UT, USA, 18–22 June 2018; pp. 8212–8221.
48. Shao, Y.; Li, L.; Ren, W.; Gao, C.; Sang, N. Domain Adaptation for Image Dehazing. In Proceedings of the IEEE Conference on Computer Vision and Pattern Recognition, Seattle, WA, USA, 13–19 June 2020; pp. 2808–2817.

electronics

MDPI

Article

Semantic Segmentation of Side-Scan Sonar Images with Few Samples

Dianyu Yang, **Can Wang**, **Chensheng Cheng**, **Guang Pan** and **Feihu Zhang** *

School of Marine Science and Technology, Northwestern Polytechnical University, Xi'an 710072, China
* Correspondence: feihu.zhang@nwpu.edu.cn; Tel.: +86-15596611656

Abstract: Underwater sensing and detection still rely heavily on acoustic equipment, known as sonar. As an imaging sonar, side-scan sonar can present a specific underwater situation in images, so the application scenario is comprehensive. However, the definition of side scan sonar is low; many objects are in the picture, and the scale is enormous. Therefore, the traditional image segmentation method is not practical. In addition, data acquisition is challenging, and the sample size is insufficient. To solve these problems, we design a semantic segmentation model of side-scan sonar images based on a convolutional neural network, which is used to realize the semantic segmentation of side-scan sonar images with few training samples. The model uses a large convolution kernel to extract large-scale features, adds a parallel channel using a small convolution kernel to obtain multi-scale features, and uses SE-block to focus on the weight of different channels. Finally, we verify the effect of the model on the self-collected side-scan sonar dataset. Experimental results show that, compared with the traditional lightweight semantic segmentation network, the model's performance is improved, and the number of parameters is relatively small, which is easy to transplant to AUV.

Keywords: side-scan sonar; segmentation; CNN; SE-block; multi-channel

Citation: Yang, D.; Wang, C.; Cheng, C.; Pan, G.; Zhang, F. Semantic Segmentation of Side-Scan Sonar Images with Few Samples. *Electronics* **2022**, *11*, 3002. https://doi.org/10.3390/electronics11193002

Academic Editor: Byung Cheol Song

Received: 29 August 2022
Accepted: 19 September 2022
Published: 22 September 2022

1. Introduction

With the continuous improvement of the technical level, robot perception and recognition have begun to develop toward intelligence and automation in the underwater research field. Recognition and perception rely on front-end equipment capturing environmental features, which is sonar for the underwater environment. Therefore, correlation analysis and processing methods of sonar images have received extensive attention in recent years [1–3]. Side-scan sonar transmits sound waves and receives echoes from underwater objects to image underwater objects and calculate approximate distances [4]. The original sonar image has low resolution, serious noise interference, and a fuzzy target shape, which greatly complicates the recognition work of researchers [5].

However, achieving a lasting effect through a manually designed filtering algorithm in a complex and changeable underwater environment is not easy. If the judgment depends on experienced personnel, it will significantly increase the cost and reduce efficiency. Therefore, it is of great significance to design a feature extraction model for sonar images that can replace, or at least assist, human judgment.

Image processing models based on deep learning algorithms have made great progress recently. Among them, classical image classification models, such as VGG-net [6], GoogLeNet [7], and Resnet [8], have achieved good results on many camera image datasets. Image segmentation models represented by FCN [9], U-net [10], PSPNet [11] have also attracted the attention of many researchers. GAN networks are also widely used in machine learning data generation to solve the problem of insufficient data [12–14]. Given the good results of these algorithms, the researchers hope to apply them to underwater acoustic images, thereby advancing the field of underwater sensing and detection.

Song et al. [15] proposed a preliminary segmentation model of side-scan sonar image based on the FCN network model. Their model divides the image into the target area, shadow area, and seabed reverberation area. Finally, MRF is used to process the classification results to improve accuracy. Chen et al. [16] proposed a semi-supervised CNN network model, which uses many unlabeled or weakly labeled samples and a few densely labeled samples to segment the SAR images. Wu et al. [17] proposed a convolutional neural network model for side-scan sonar named ECNet. The network structure consists of an encoder and a decoder. The encoder obtains contextual features, and the decoder is used for image restoration. In addition, a single-stream deep neural network with multiple side outputs is added to optimize edge segmentation. Huo et al. [18] proposed a semi-synthetic sonar data generation method. For the input optical image, the CNN model combines image segmentation with intensity distribution simulation in different regions to generate synthetic sonar images of the plane and the drowning person to enrich the sonar image data set. Zhou et al. [19] added the Laplacian energy filter based on the CNN model, and the two-channel pulse-coupled neural network was used to fusion the side-scan sonar images and achieved good results. In the work of Połap et al. [20], a method based on a neural network model is proposed to search for target signals in ocean areas and restore areas with low image quality. Zhu et al. [21] used the convolutional neural network model to extract the target features of side-scan sonar images and input them into the trained SVM for classification.

Side-scan sonar is a kind of active imaging sonar. Its imaging principle is to send a short acoustic pulse with a slight horizontal opening angle (about 1 degree) and a large vertical opening angle to one or both sides of the vertical direction of the survey ship. After the pulse reaches the seabed, it is continuously reflected according to the distance from the seabed to the transducer. The sonar image with uneven gray level changes is drawn according to the strength of the reflected signal. Sonar images can be used to observe changes in the seafloor topography, whether there are obstacles to the navigation, and the type of seabed substrate. When the side-scanning sonar emission pulse propagates in water and meets the target, the target scatters the acoustic energy in all directions, and the transducer receives the backscattered echo. In contrast, the acoustic energy is difficult to reach the side and rear of the target (called the blind area).The sonar array moves forward with the carrier, and in the process of moving forward, sonar continues to transmit, receive and form sonar images [22]. As a result, the target (strong echo signal of the target) and its shadow (blind area behind the side of the target) appear at the corresponding position on the sonar image. It can be seen that the side-scan sonar reflects the echo intensity of the detected target so that the side-scan sonar image can be understood as a single-channel gray map, and the target with stronger reflection has greater brightness. However, the difference in brightness of most underwater targets is not apparent, so there must be a particular dimension of the color channel that contains most of the target information in the image.

On this basis, in this paper, a side-scan sonar image segmentation model is proposed based on the CNN network. Compared with camera images, side-scan sonar images are more challenging to acquire and have less data, so the network model needs to control the depth to avoid overfitting. In addition, due to the low color richness of side-scan sonar images, each channel contains a relatively large amount of information, so it is necessary to focus on the information in essential channels.

The main contributions of this paper are as follows:

(1) We introduced the SE module to increase channel attention in the feature extraction process and increase independent weight for each channel so that the more critical channels obtain a higher weight to improve the overall segmentation accuracy.
(2) We increased the convolution kernel size used from 3×3 to 7×7, which proved effective in sonar images with a larger size. Meanwhile, DW convolution was adopted to reduce the number of parameters given the increase in the number of parameters caused by the expansion of the convolution kernel size.

(3) Simply increasing the convolution kernel size cannot effectively improve the quality of feature extraction. Therefore, we constructed a parallel feature extraction channel using a small-size convolution kernel and concatenated its output with the leading network to achieve multi-scale feature extraction.
(4) We used a full convolution layer to restore the output of the decoder to the original image size and output the segmentation results. Then we conducted a contrast experiment with other lightweight CNN.

The rest of this paper is divided into five sections: Section 2 introduces the work of other researchers related to the model design; Section 3 presents the structure and details of the model; Section 4 uses the self-collected side-scan sonar data to verify the performance of the model; and Section 5 gives the conclusion.

2. Related Work

In this section, some essential concepts for model design are introduced, including the basic principle of the CNN network, the U-NET network's design idea, and the SE module's influence.

2.1. Principles of CNNs

Neural network models with CNN were completed by Lecun Y [23] and carried forward by AlexNet [24]. In the classical CNN model, data have two directions: forward propagation and backward propagation. Forward propagation realizes data feature extraction through the convolutional layer, pooling layer, activation function layer, and fully connected layer. The convolution layer is processed by multiple convolution checks to extract high-dimensional feature maps. The pooling layer compresses the parameters while preserving the main features. Finally, the activation function ensures the nonlinearity of the multi-layer network structure, and the last fully connected layer implements the mapping from image features to classification categories. According to the comparison between the output results of the forwarding propagation and label data, backpropagation performs gradient descent on network parameters layer by layer in reverse to improve the network performance. Finally, the network achieves due performance after multiple forward and backward propagation.

2.2. U-Net and FCN

There are many excellent models for semantic segmentation tasks, such as DeepLabV3 [25], hrnet [26], Transformer [27], etc. However, the original design concept of the segmentation model comes from FCN. The initial neural network model can only be applied to the classification task, and the emergence of FCN brought it into the field of image segmentation. Pioneering the model using the convolution layer instead of full connection as the last layer of the network's output solved the problem that the whole connection layer limits the input size. In addition, the model outputs from a one-dimensional probability vector into a two-dimensional probability matrix. That is, every pixel can be classified. FCN uses deconvolution and linear interpolation for image restoration and uses the feature fusion method of skip layer. It concatenates image features of high and low dimensions, which greatly impacts the design idea of the subsequent segmentation model. The structure of the FCN network model is shown in Figure 1.

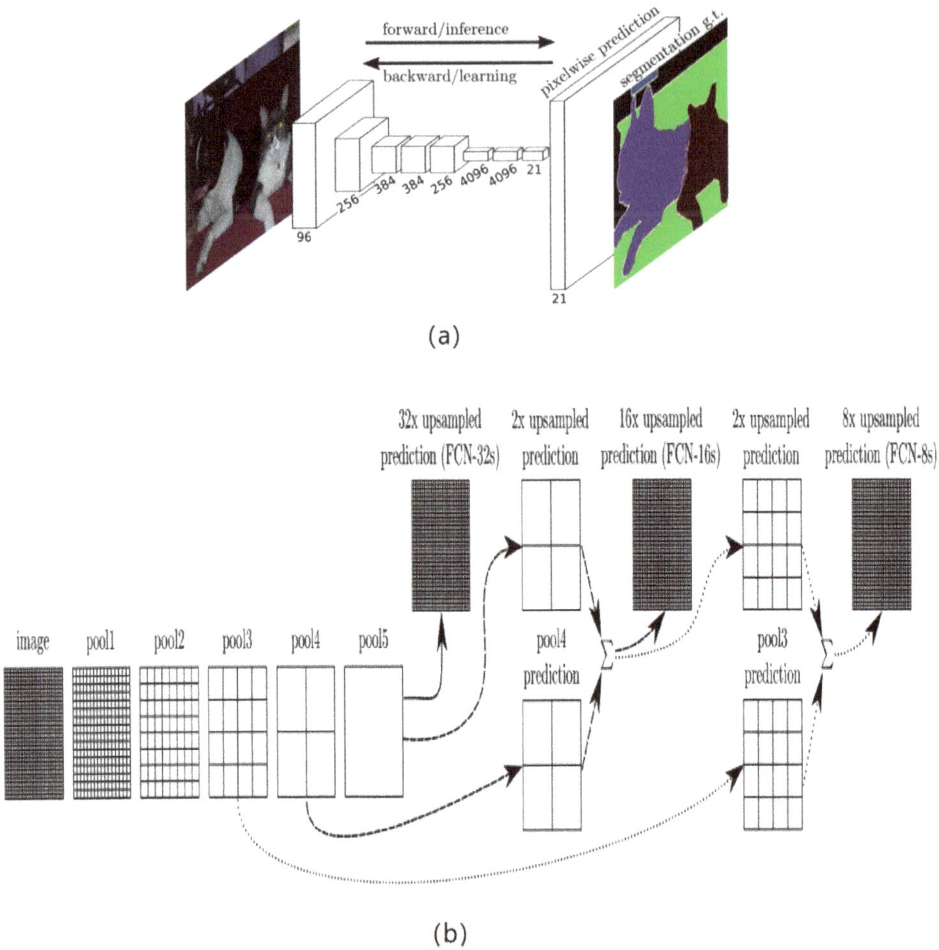

Figure 1. (**a**) FCN model with VGG as a backbone [9]. (**b**) Skip layer of FCN: There are three versions of the FCN network, namely FCN-8S, FCN-16S, and FCN-32S. The 32S version directly performs image restoration after a feature fusion, so the output quality is the lowest, but the number of parameters is the lowest. The 8S version can obtain the highest precision output after three times of feature fusion. 16S is relatively balanced.

U-net is an image segmentation network model that draws on the FCN model. The model still adopts the design idea of deconvolution restoration and full convolution instead of complete connection. However, it gives up using the VGG network as a backbone and designs a symmetric four-layer codec structure instead. At the same time, feature fusion is carried out between encoding and decoding structures at the same level, similar to skip layer. The u-net model is still the mainstream algorithm in all minor sample segmentation problems, such as medical image segmentation, due to its low depth, fewer parameters, and good segmentation effect. The U-NET model structure is shown in Figure 2.

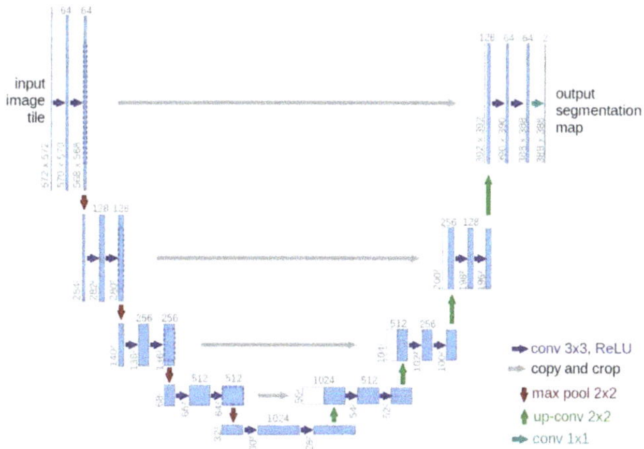

Figure 2. U-Net model [10]: classical symmetric codec structure with feature concatenate.

2.3. The Effect of SE-Block

SENet [28] is the ImageNet 2017 champion model. The SE-block structure is shown in Figure 3. Its full name is squeeze-and-excitation congestion networks. The main contribution is a channel attention extraction module called Se-block that can be added to any network structure.

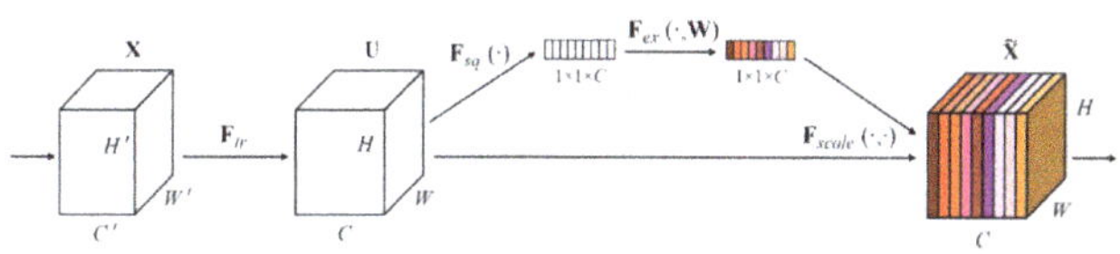

Figure 3. The structure of SE-block [28].

The module consists of two parts: the squeezing part, which compresses the original 3D data input into a one-dimensional vector, implemented mainly by global average pooling (this operation can extract the global features of each channel); and the crimping section, which uses a full connection layer to map the output of the compression module to a predicted weighting sequence, which is multiplied by all the channels for weighting. This module can effectively extract important channel features and ignore minor channels.

3. Method

The design ideas of our model are derived from U-NET, and we adopt a coding–decoding structure similar to U-NET and SENet, as well as the large convolution kernel and re-parameterizing mentioned in RepLKNet [29], but improve it for our downstream tasks. First, we added Se-block to the encoder, namely the feature extraction module, to obtain the weight of different feature channels. The network model will find the channel that significantly impacts the segmentation output result (the channel added after multiple convolutions, rather than the original RGB), increases the weight proportion of its corresponding parameters, and focuses on adjustment. Then, the large and small convolution kernels are used to capture features of different scales in parallel. Finally, after fusion and restoration, the image segmentation results are output.

3.1. Multi-Scale Feature Fusion

Due to the increasing complexity of images, multi-scale feature fusion has become a necessary capability for a qualified segmentation network. The skip layer of FCN, the codec information interaction of U-NET, and the ASPP module of the Deeplab model all belong to this kind of structure. The RepLKNet model proposes a structure-reparameterization method. The model uses a large convolution kernel (31 × 31) for feature extraction, and a parallel feature extraction channel using a conventional 3 × 3 small convolution kernel is added. After the parameter training of the convolution kernel is completed, the small convolution kernel is directly inserted into the large convolution kernel to realize the feature fusion of different levels of size and scale.

Due to the difficulty of obtaining side-scan sonar images, we cannot provide the massive amount of data required for training large convolutional kernels and deep networks, such as RepLKNet. Therefore, after slightly expanding the size of the convolution kernel, we did not insert the small convolution kernel directly into the large convolution kernel because this would destroy the feature extraction ability of the large convolution kernel itself. Instead, we use the concatenate method to incorporate features of different scales before restoring images using deconvolution.

3.2. Depthwise Separable Convolution

The concept of depthwise separable convolution was first proposed by MobileNet [30]. The standard convolution operation is decomposed into two steps: the first step is deep convolution, and the second step is point convolution. A specific example is used to compare the difference between this method and standard convolution: assuming that the size of the input image is 12 × 12 × 3 (3 represents three channels), and the desired output result is 8 × 8 × 128, so 128 5 × 5 × 3 convolution kernels are needed for convolution, and the number of operations in the whole process is 9600.

If deep convolution is used first, three 5 × 5 × 1 convolutions are used to convolve the three channels of the image, and the output result of 8 × 8 × 3 is obtained. Then point convolution is used, 128 1 × 1 × 3 convolution kernels (equivalent to one pixel containing three channels) are used to convolve the previous output results again, and finally, the output results of the same size are obtained. Still, the number of operations is reduced to 5 × 5 × 3 + 1 × 1 × 3 × 128 = 469.

The deep separable volume reduces the amount of network computation at the cost of increasing the depth of the network, which may affect the output results of the network while speeding up the calculation speed. Therefore, this practice may not play a positive role for networks mainly using small convolution kernels, but it is indispensable for our model.

3.3. Model Structure

The structure of our model is borrowed from the design of U-NET, and the central part is the four-layer codec, shown in Figure 4.

The encoder consists of four layers in total, and each layer contains an encode block. Each encode block uses a convolution kernel size of 7 × 7 (DW convolution is used to improve the operation rate while adding padding). The number of output channels in each layer is 32, 64, 128, and 256. Meanwhile, SE-blocks are added parallel to each layer to predict the channel weights.

Another parallel feature extraction channel uses a small-size convolution kernel; the main structure is similar to the central part.

The decoder input is the high-dimensional feature map extracted by the encoder, and the channel is 256. The decoder uses deconvolution to up-sample layer by layer. First, concatenate with the same dimensional features output by the feature channel using a small convolution kernel, then convolve twice and input to the next layer. After four repetitions, the image segmentation results are obtained through the full convolutional layer.

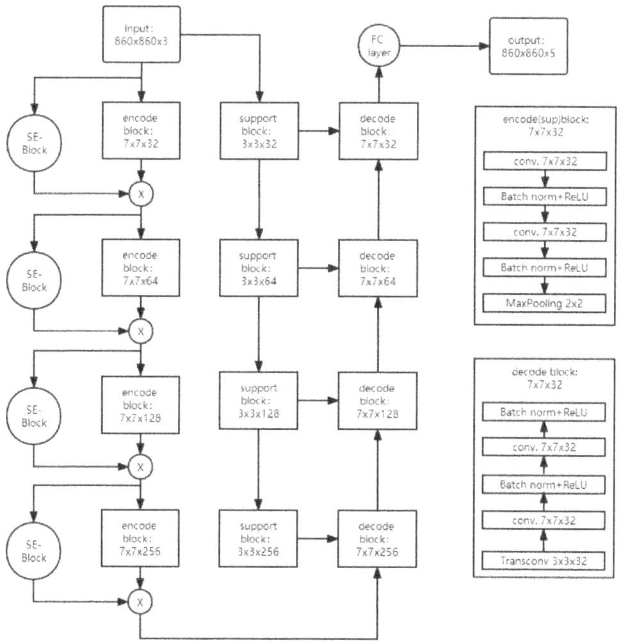

Figure 4. The structure of our model.

4. Experiment and Analysis

All experiments were conducted with Intel Core i9-10900F CPU@2.8 Ghz × 20, 64 GB RAM, Nvidia Geforce 3090 GPU, 24 GB of video memory, by CUDA Toolkit 11.3, CUDNN V8.2.1, Python 3.6, PyTorch-GPU 1.10.1, Ubuntu18.04.operating system.

4.1. Dataset Collection

We used Hydro 3060 dual-frequency side-scan sonar to collect sonar data needed for the experiment in the Lake District of Jiande, Hangzhou, China. The original image captured frame by frame was 960 × 960 pixels in size, and its effect is shown in Figure 5.

The side-scan sonar is mounted on an AUV and emits sound waves to both sides as the subject moves, collecting echoes from underwater objects to build an image. The bright parts of the image represent the targets with strong echoes, such as rocks and metals, while the parts without echoes will appear black, such as water bodies and blocked parts.

Our model is based on supervised learning, which requires manually annotated accurate data labels as training data. We annotated the data using LabelMe, open-source software on the Ubuntu platform. For the whole dataset, we divided the data into five categories (not every image contains labels from all five categories): (1) water; (2) the mountain part; (3) the land; (4) shaded part; and (5) unmarked area (background). The unlabeled area mainly refers to the debris area left after the first four types of image labeling. The labeled image is shown in Figure 6.

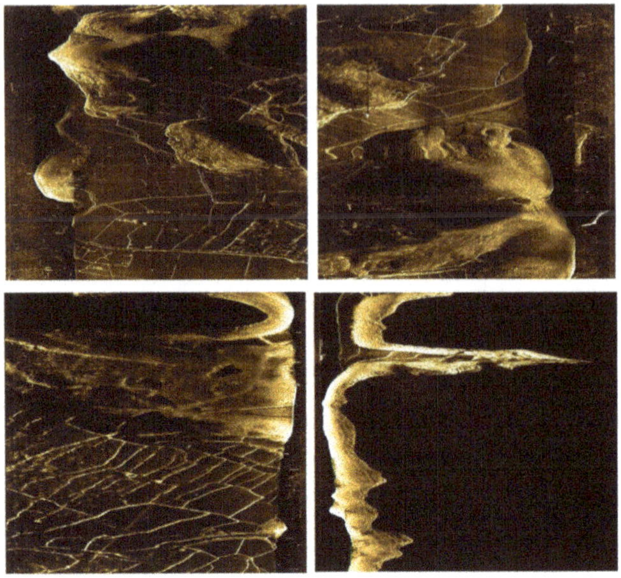

Figure 5. The original sonar image (each sonar image is cropped down the middle into two images).

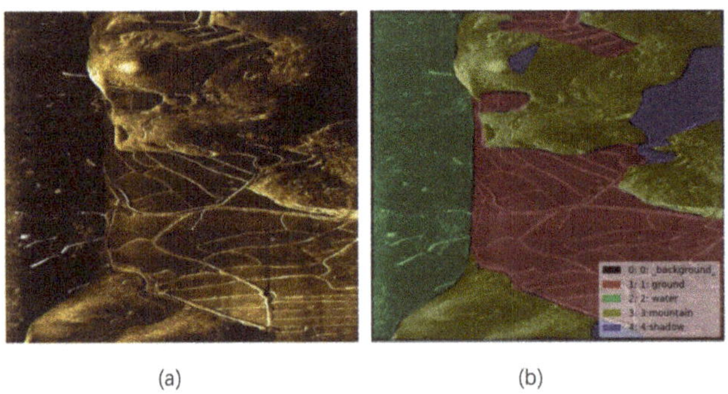

Figure 6. (**a**) Original image, (**b**) label.

4.2. Data Augmentation

As mentioned before, collecting side-scan sonar data is challenging, so the amount of data is not very rich. Therefore, we adopted the method of data amplification to increase the number of samples to ensure the training effect, and the method used is shown in Figure 7.

(1) The most common method is to flip the image at different angles, amplifying the data but also breaking the location correlation and making the network more generalized.
(2) Image translation is also a standard method, which controls the image translation in four directions by some random numbers, but not too much. Otherwise, it will destroy the feature structure of the image.
(3) By randomly clipping the original image, the size of the image can be reduced while the data are expanded, and the training can be accelerated.

The sonar image is less dependent on shape features but more on color features, so no color data amplification was carried out. The size of the original sonar data collected is 960 × 960, and the number is about 300. After data amplification, the data size is 860 × 860, and the number is increased by about four times. We randomly selected 60 percent as the training set, and the validation and test sets were 20 percent.

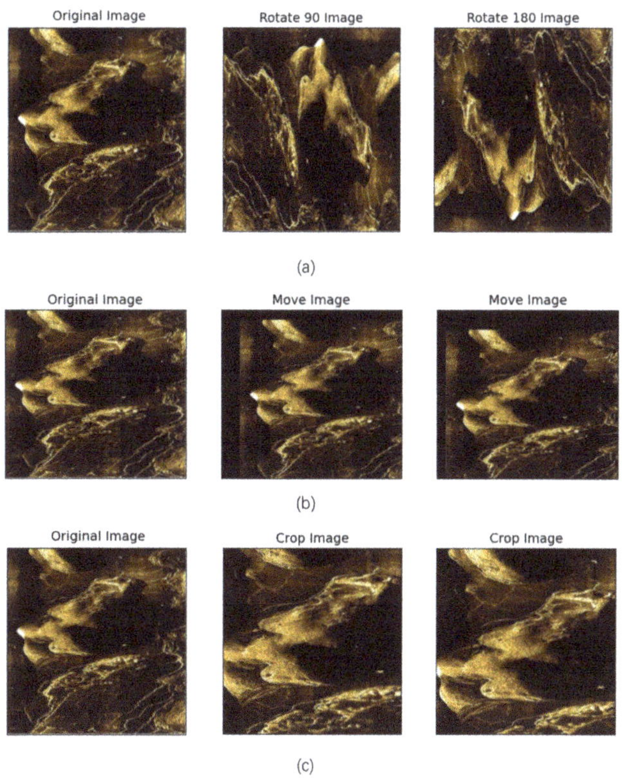

Figure 7. Data augmentation (**a**) image inversion, (**b**) image panning, (**c**) random crop.

4.3. Verification Indicators

We measure the model from two perspectives: the consumption of computing resources, and the model's accuracy. Computing resources are measured by the total number of network parameters and the FLOPs indicator, which refers to floating point operations. More FLOPs mean more computing resources consumed by the model. The calculation formula of the convolution layer FLOPs of the convolutional network is as follows:

$$FLOPs = (2c_{in}k^2 - 1)HWc_{out} \tag{1}$$

c_{in} and c_{out} represents the number of input and output channels in the convolution layer, and k represents the size of the convolution kernel. The size of the output feature graph is $H \times W$.

OA (overall accuracy) and MIoU (mean intersection over union) will measure the model accuracy. The calculation formula of OA is as follows:

$$OA = \frac{TP + TN}{TP + TN + FP + FN} \tag{2}$$

TP, TN, FP, FN mean true positive (positive sample is judged as a positive sample), true negative (negative sample is judged as a negative sample), false positive (negative sample is misjudged as a positive sample), and false negative (positive sample is misjudged as a negative sample).

The calculation formula of $MIoU$ is as follows:

$$MIoU = \frac{1}{k}\sum_{i=1}^{k}\frac{p \cap g}{p \cup g} \tag{3}$$

P means prediction, and G means ground truth.

4.4. Network Model Training

We use the processed sonar data for network training, and the hyperparameters used in the training process are listed in Table 1. The loss function used in the training process is the cross-entropy loss function, and the training process is shown in Figure 8.

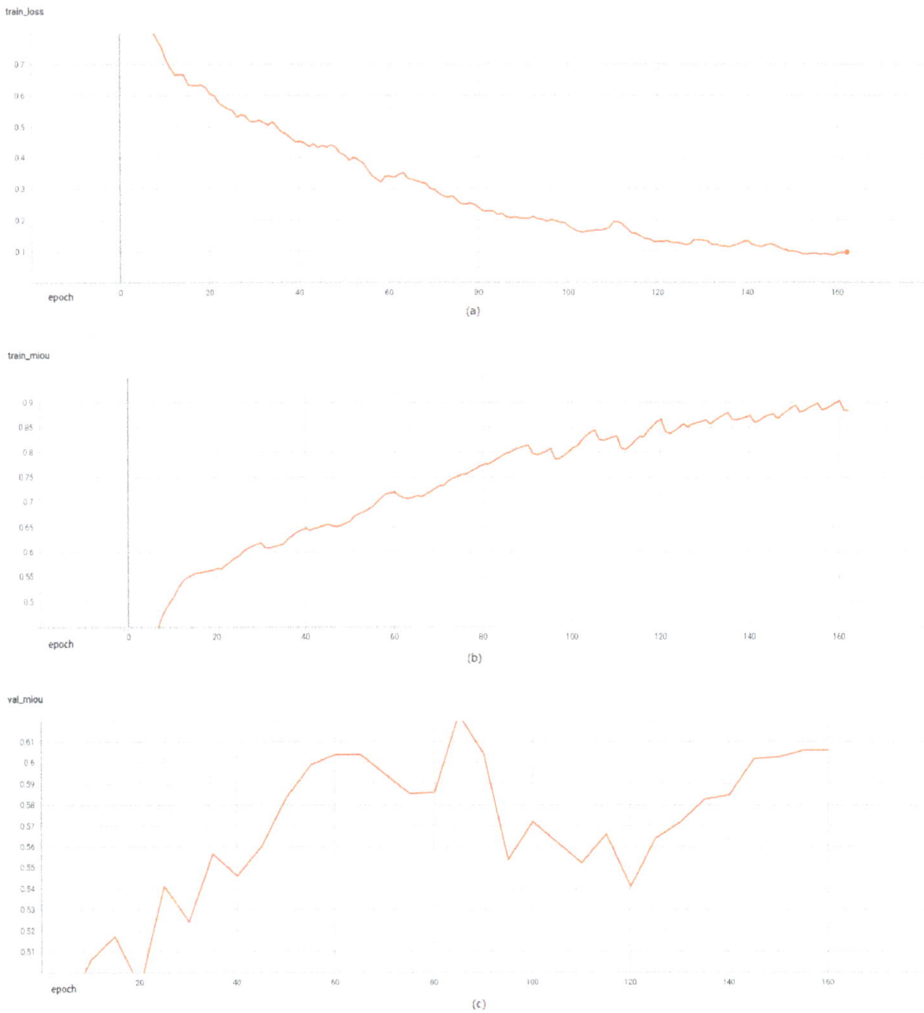

Figure 8. We use TensorBoard to draw the convergence curve of the training process, and the network has basically converged at 200 epochs. (**a**) train loss, (**b**) train MIoU, (**c**) val MIoU.

Table 1. Hyper-parameter.

Type	Value
num of workers	8
batch size	6
optimizer	SGD
learning rate	0.01
learning policy	poly
step size	10,000

$$H(p,q) = -\sum_{i=1}^{n} p(x_i)log(q(x_i)) \quad (4)$$

The cross-entropy function is used to measure the difference between two probability distributions. For example, machine learning tasks represent the difference between the network output and the label.

4.5. Performance and Comparison

In the experiment, the quantitative analysis of the segmentation results of U-Net, FCN, and PSPNet, which are typical lightweight networks, and our method is conducted. The comparison results are shown in the tables, and the recovered images are shown in Figure 9.

Figure 9. The segmentation results of the model output are shown in the figure. The original image, label, and output result are left to right. The colors in the picture are blue for water, gray for rocks, yellow for flat land, black for shadows, and white for fruitless areas (areas that are hard to distinguish).

Due to the small sample size, we used K-fold cross-validation on the dataset to calculate the model performance indicators. We set the value of K as 5, randomly divided all the data into five parts, and selected one of them as the validation set and the rest as the training set each time. Finally, the results obtained five times were averaged. The model indicators of five-fold cross-validation are shown in Table 2. The results shown in Table 3 show that the average OA and MIoU of our model in the dataset are 0.87159 and 0.67893, the highest of the four models. The total number of parameters is 21,340,813, which was above the average of the four models. The FLOPs are slightly higher because the currently used code and computing devices do not support DW convolution perfectly, and there is still room for further improvement.

Table 2. K-fold cross validation (K = 5).

K	OA	MIoU
1	0.869394	0.685063
2	0.856123	0.678424
3	0.854726	0.656946
4	0.884486	0.699488
5	0.856770	0.668848
avg	0.872299	0.677754

Table 3. Different model performance.

Model	OA	MIoU	Num of Para	FLOPs
FCN	0.864415	0.663187	18,643,845	212.4 G
U-Net	0.871427	0.674909	34,525,391	487.71 G
PSPNet	0.849124	0.651908	65,576,517	673.94 G
Ours	0.872299	0.677754	21,340,813	647.94 G

In order to test the effect of increasing the size of the convolution kernel, we carried out relevant comparative tests and adjusted the size of the convolution kernel from 3×3 to 11×11. The performance changes are shown in Table 4, and it can be found that the parameters currently used are the best ones.

Table 4. Model performance with different kernel size.

Size	OA	MIoU
3×3	0.864976	0.663328
5×5	0.862365	0.667896
7×7	0.872299	0.677754
9×9	0.866372	0.673241
11×11	0.862757	0.658241

5. Conclusions

This paper proposes a semantic segmentation model for side-scan sonar images based on the CNN network. The model uses a symmetric codec structure as the main body, adds a convolution kernel of different scales to extract multi-scale features, adds SE modules to focus on the weight of essential channels, and finally fuses at the output end. We verify the accuracy and reliability of the model on the self-collected sonar data and find that the model has a low computational cost and high portability. Our method achieves multiple classifications of side-scan sonar images at the semantic level. At the same time, most other researchers focus more on the recognition of objects with specific shapes or the simple binary classification of images. In addition, our model also has high portability. The large neural network model proposed by many researchers is inferior in real-time performance on AUV. After loading our model into the AUV control terminal, it can still complete the task and has low dependence on high-performance computers, which is also a significant advantage. In the future, we will consider further increasing the network depth and convolution kernel and find ways to make them effective in a small sample environment.

Author Contributions: D.Y. and F.Z. conceived the study and put forward the methodology. C.C. and C.W. performed the data collection and pre-processing. D.Y. carried out the software for the experiments and wrote the first draft of the manuscript. F.Z. and G.P. reviewed and edited the manuscript. All authors have read and agreed to the published version of the manuscript.

Funding: This study was supported by the National Natural Science Foundation of China (52171322), the National Key Research and Development Program (2020YFB1313200), and the Fundamental Research Funds for the Central Universities (D5000210944).

Acknowledgments: The authors would like to thank Songxiang Wang, Xijun Zhou, and Liyuan Chen et al. for their help during the experiment.

Conflicts of Interest: The authors declare no conflict of interest. The funders had no role in the design of the study.

References

1. Petrich, J.; Brown, M.F.; Pentzer, J.L.; Sustersic, J.P. Side scan sonar based self-localization for small autonomous underwater vehicles. *Ocean. Eng.* **2018**, *161*, 221–226. [CrossRef]
2. Reed, S.; Petillot, Y.; Bell, J. An automatic approach to the detection and extraction of mine features in sidescan sonar. *IEEE J. Ocean. Eng.* **2003**, *28*, 90–105. [CrossRef]
3. Acosta, G.G.; Villar, S.A. Accumulated ca–cfar process in 2-d for online object detection from sidescan sonar data. *IEEE J. Ocean. Eng.* **2015**, *40*, 558–569. [CrossRef]
4. Zhang, X.; Tan, C.; Ying, W. An imaging algorithm for multireceiver synthetic aperture sonar. *Remote Sens.* **2019**, *11*, 672. [CrossRef]
5. Wang, Z.; Guo, J.; Huang, W.; Zhang, S. Side-scan sonar image segmentation based on multi-channel fusion convolution neural networks. *IEEE Sens. J.* **2022**, *22*, 5911–5928. [CrossRef]
6. Simonyan, K.; Zisserman, A. Very deep convolutional networks for large-scale image recognition. *arXiv* **2014**, arXiv:1409.1556.
7. Szegedy, C.; Liu, W.; Jia, Y.; Sermanet, P.; Rabinovich, A. Going deeper with convolutions. In Proceedings of the 2015 IEEE Conference on Computer Vision and Pattern Recognition (CVPR), Boston, MA, USA, 7–12 June 2014.
8. He, K.; Zhang, X.; Ren, S.; Sun, J. Deep residual learning for image recognition. In Proceedings of the 2016 IEEE Conference on Computer Vision and Pattern Recognition (CVPR), Las Vegas, NV, USA, 27–30 June 2016.
9. Long, J.; Shelhamer, E.; Darrell, T. Fully convolutional networks for semantic segmentation. *IEEE Trans. Pattern Anal. Mach. Intell.* **2015**, *39*, 640–651.
10. Ronneberger, O.; Fischer, P.; Brox, T. U-net: Convolutional networks for biomedical image segmentation. In *International Conference on Medical Image Computing and Computer-Assisted Intervention*; Springer International Publishing: Cham, Switzerland, 2015.
11. Zhao, H.; Shi, J.; Qi, X.; Wang, X.; Jia, J. Pyramid scene parsing network. In Proceedings of the 2017 IEEE Conference on Computer Vision and Pattern Recognition (CVPR), Honolulu, HI, USA, 21–26 July 2016.
12. Tian, C.; Zhang, X.; Lin, J.C.W.; Zuo, W.; Zhang, Y. Generative Adversarial Networks for Image Super-Resolution: A Survey. *arXiv* **2022**, arXiv:2204.13620.
13. Tian, C.; Yuan, Y.; Zhang, S.; Lin, C.W.; Zuo, W.; Zhang, D. Image Super-resolution with an Enhanced Group Convolutional Neural Network. *arXiv* **2022**, arXiv:2205.14548.
14. Tian, C.; Xu, Y.; Zuo, W.; Lin, C.W.; Zhang, D. Asymmetric CNN for image superresolution. *IEEE Trans. Syst. Man Cybern. Syst.* **2021**, *52*, 3718–3730. [CrossRef]
15. Song, Y.; Zhu, Y.; Li, G.; Feng, C.; He, B.; Yan, T. Side scan sonar segmentation using deep convolutional neural network. In Proceedings of the OCEANS 2017, Anchorage, AK, USA, 18–21 September 2017.
16. Chen, J.; Summers, J.E. Deep convolutional neural networks for semi-supervised learning from synthetic aperture sonar (SAS) images. In Proceedings of the 173rd Meeting of Acoustical Society of America and 8th Forum Acusticum, Boston, MA, USA, 25–29 June 2017.
17. Wu, M.; Wang, Q.; Rigall, E.; Li, K.; Zhu, W.; He, B.; Yan, T. ECNet: Efficient convolutional networks for side scan sonar image segmentation. *Sensors* **2019**, *19*, 2009. [CrossRef] [PubMed]
18. Huo, G.; Wu, Z.; Li, J. Underwater object classification in sidescan sonar images using deep transfer learning and semisynthetic training data. *IEEE Access* **2020**, *8*, 47407–47418. [CrossRef]
19. Zhou, P.; Chen, G.; Wang, M.; Liu, X.; Chen, S.; Sun, R. Side-scan sonar image fusion based on sum-modified Laplacian energy filtering and improved dual-channel impulse neural network. *Appl. Sci.* **2020**, *10*, 1028. [CrossRef]
20. Połap, D.; Wawrzyniak, N.; Włodarczyk-Sielicka, M. Side-scan sonar analysis using roi analysis and deep neural networks. *IEEE Trans. Geosci. Remote Sens.* **2022**, *60*, 1–8. [CrossRef]
21. Zhu, P.; Isaacs, J.; Bo, F.; Ferrari, S. Deep learning feature extraction for target recognition and classification in underwater sonar images. In Proceedings of the 2017 IEEE 56th Annual Conference on Decision and Control (CDC), Melbourne, Australia, 12–15 December 2017.

22. Burguera, A.; Oliver, G. High-resolution underwater mapping using side-scan sonar. *PLoS ONE* **2016**, *11*, e0146396. [CrossRef] [PubMed]
23. Lecun, Y.; Bottou, L. Gradient-based learning applied to document recognition. *Proc. IEEE* **1998**, *86*, 2278–2324. [CrossRef]
24. Technicolor, T.; Related, S. Imagenet Classification with Deep Convolutional Neural Networks 2012. [50]. Available online: https://papers.nips.cc/paper/2012/file/c399862d3b9d6b76c8436e924a68c45b-Paper.pdf (accessed on 20 August 2022).
25. Chen, L.C.; Papandreou, G.; Schroff, F.; Adam, H. Rethinking atrous convolution for semantic image segmentation. *arXiv* **2017**, arXiv:1706.05587.
26. Wang, J.; Sun, K.; Cheng, T.; Jiang, B.; Xiao, B. Deep high-resolution representation learning for visual recognition. *IEEE Trans. Pattern Anal. Mach. Intell.* **2020**, *43*, 3349–3364. [CrossRef]
27. Strudel, R.; Garcia, R.; Laptev, I.; Schmid, C. Segmenter: Transformer for semantic segmentation. In Proceedings of the 2021 IEEE/CVF International Conference on Computer Vision (ICCV), Montreal, QC, Canada, 10–17 October 2021.
28. Hu, J.; Shen, L.; Albanie, S.; Sun, G.; Wu, E. Squeeze-and-excitation networks. *IEEE Trans. Pattern Anal. Mach. Intell.* **2019**, *42*, 2011–2023. [CrossRef]
29. Ding, X.; Zhang, X.; Zhou, Y.; Han, J.; Ding, G.; Sun, J. Scaling up your kernels to 31 × 31: Revisiting large kernel design in CNNs. *arXiv* **2022**, arXiv:2203.06717.
30. Howard, A.G.; Zhu, M.; Chen, B.; Kalenichenko, D.; Wang, W.; Weyand, T.; Andreetto, M.; Adam, H. Mobilenets: Efficient convolutional neural networks for mobile vision applications. *arXiv* **2017**, arXiv:1704.04861.

electronics

MDPI

Article

LSW-Net: A Learning Scattering Wavelet Network for Brain Tumor and Retinal Image Segmentation

Ruihua Liu [1,*], Haoyu Nan [1], Yangyang Zou [1], Ting Xie [2] and Zhiyong Ye [2]

1 School of Artificial Intelligence, Chongqing University of Technology, Chongqing 401135, China
2 College of Science, Chongqing University of Technology, Chongqing 400054, China
* Correspondence: lruih@cqut.edu.cn

Abstract: Convolutional network models have been widely used in image segmentation. However, there are many types of boundary contour features in medical images which seriously affect the stability and accuracy of image segmentation models, such as the ambiguity of tumors, the variability of lesions, and the weak boundaries of fine blood vessels. In this paper, in order to solve these problems we first introduce the dual-tree complex wavelet scattering transform module, and then innovatively propose a learning scattering wavelet network model. In addition, a new improved active contour loss function is further constructed to deal with complex segmentation. Finally, the equilibrium coefficient of our model is discussed. Experiments on the BraTS2020 dataset show that the LSW-Net model has improved the Dice coefficient, accuracy, and sensitivity of the classic FCN, SegNet, and At-Unet models by at least 3.51%, 2.11%, and 0.46%, respectively. In addition, the LSW-Net model still has an advantage in the average measure of Dice coefficients compared with some advanced segmentation models. Experiments on the DRIVE dataset prove that our model outperforms the other 14 algorithms in both Dice coefficient and specificity measures. In particular, the sensitivity of our model provides a 3.39% improvement when compared with the Unet model, and the model's effect is obvious.

Keywords: image segmentation; wavelet scattering; loss function; active contour; medical image

Citation: Liu, R.; Nan, H.; Zou, Y.; Xie, T.; Ye, Z. LSW-Net: A Learning Scattering Wavelet Network for Brain Tumor and Retinal Image Segmentation. *Electronics* **2022**, *11*, 2616. https://doi.org/10.3390/electronics11162616

Academic Editor: Giovanni Ramponi

Received: 15 July 2022
Accepted: 18 August 2022
Published: 20 August 2022

1. Introduction

Image segmentation is a class of image processing problems, and its task is to divide an image into two or more meaningful regions. The accuracy of image segmentation is particularly important in practical applications. In particular, biomedical image segmentation is prominent in clinical analysis, diagnosis, treatment planning, and the measurement of disease progression. Traditional image segmentation methods, such as the threshold method [1], region growing method [2], level set method [3–5], etc., have struggled to meet the need for accurate image segmentation in the context of big data.

In recent years, deep neural networks have made great progress in various artificial intelligence tasks including image recognition and image segmentation. A convolutional neural network (CNN) [6] introduces semantic information when segmenting objects; thereby, injecting new vitality into semantic segmentation research. Fully convolutional network models [7–9] based on CNN architecture have achieved excellent performance in automatic medical image segmentation, which further promotes the application of deep learning in image segmentation for applications such as brain tumor segmentation [10]. SegNet [11] adopts the encoder–decoder structure and transfers the pixel index value of the maximum pooling operation in the encoding process into the decoder, which not only retains the detailed information of the pixels but also improves the accuracy of semantic segmentation. The Attention Unet network (At-Unet) [12] adds an attention gating unit to the Unet model to provide pixel level attention for the feature map. The network tends to focus on feature points with more information and improves the feature extraction ability of the model. The Deeplab network [13,14] obtains multiscale context information by cascading atrous convolutions with different atrous rates, and then introduces a conditional random field to enhance the relevance of contextual semantic information, which in turn improves the segmentation accuracy. Although the above-mentioned network models have improved the image segmentation accuracy of some datasets, they are still unable to accurately extract the boundary features of brain tumors in images. This is due to the invasiveness of the imaging process and the ambiguity between biological forms, as is the case between tumors and adjacent organs or changes in lesions over different periods. This invasiveness and ambiguity can lead to the discontinuity of some segmentation boundaries, as shown in Figure 1a. When comparing Figure 1a,b, there are many discontinuous segmentations in Figure 1a.

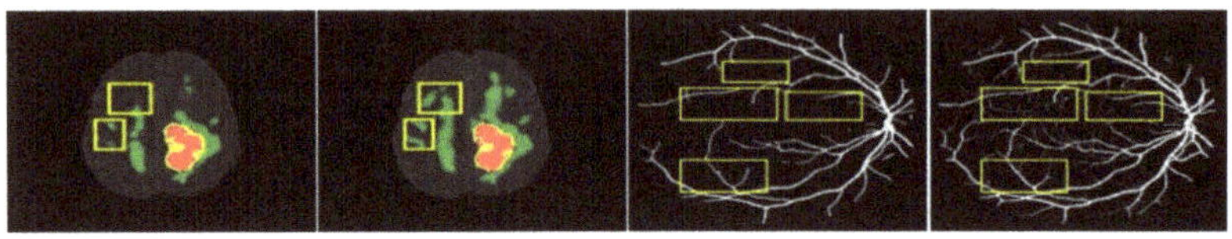

Figure 1. Medical image segmentation. (**a**) At-Unet; (**b**) Brain tumor ground truth; (**c**) AC-Loss; (**d**) Retinal ground truth, where the yellow boxes indicate the regions where the methods are mis-segmented compared to ground truth.

At the same time, many researchers are devoted to establishing the minimal loss energy function model. A level set function is proposed, that is, the region term of the CV energy function is used as the loss function, from which the CNN model can learn the spatial information of the image, which can improve the accuracy of image segmentation [14]. In order to solve the boundary error segmentation problem, an active contour loss function (AC-Loss) is constructed [15]. The AC-Loss function fully considers the internal and external areas of the segmented object, and the perimeter of the boundary. Unfortunately, some experiments have shown that when dealing with biomedical images with complex boundaries, such as retinal vessel images, because the AC-Loss function constrains the perimeter of the segmentation object boundary, it also limits the model's ability to segment small boundaries. The under-segmentation phenomenon of fine blood vessels is avoided in Figure 1c. By comparing Figure 1c,d, there are many small blood vessels that can be seen in Figure 1d that are not segmented in Figure 1c.

The problem of complex boundary contour features in medical images, also increases the difficulty of image boundary feature extraction and characterization in deep neural network learning. Inspired by the dual-tree wavelet scattering transform, we propose a boundary feature extraction module which can improve the network's ability to extract image boundary features. Specifically, the process can be described as follows: First, the dual-tree complex wavelet scattering transform is used to separate the high-frequency and low-frequency features of the feature map. Second, a convolution operator is adopted to extract low-frequency body features and high-frequency boundary features. Finally, the dual-tree complex wavelet scattering transform is then built into a fully convolutional network model, and a new learning scattering wavelet network (LSW-Net) semantic segmentation model is designed through end-to-end data-driven scattering learning transform features. In order to enhance its ability to extract image boundary contour information, the network utilizes the Unet network [8] as the backbone network and introduces the dual-tree complex wavelet scattering transform (DTCWT-Scat) during downsampling for boundary feature extraction. In order to further improve the network model's ability to extract complex boundary contours, an improved active contour loss function (IAC-Loss) is further constructed on the basis of the LSW-Net network. This loss function not only improves the network's sensitivity to small boundaries, it also better solves the problem of the under-segmentation of boundary contours.

Our main contributions are summarized as follows:

- In order to separate the high-frequency and low-frequency features of the feature map during downsampling, we introduce the DTCWT-Scat module into the Unet and innovatively propose the LSW-Net model.
- We design an improved active contour loss function, which can improve sensitivity to small boundaries and can better solve the problem of boundary under-segmentation.
- Through BraTS brain tumor segmentation experiments, our LSW-Net network has advantages when compared with traditional FCN, SegNet, At-Unet, and some advanced segmentation algorithms in terms of Dice coefficient, accuracy, sensitivity, and other indicators.
- Through the DRIVE retinal vessel segmentation experiments, the effectiveness and robustness of the LSW-Net + IAC-Loss model are illustrated.

2. Related Work

2.1. Dual-Tree Complex Scattering Wavelet Transform

Wavelet transform is a local waveform transform that can provide local representation of multiscale signals in both time and frequency domains. S. Mallat first proposed a wavelet scattering network with a non-feedback structure [16]. This network can not only present the image energy distribution in the frequency domain, but also maintain stability against small deformations. This partially makes up for the shortcomings of the CNN model, including small object segmentation and image boundary extraction capabilities. Some scholars have also actively tried to combine the wavelet algorithm with the CNN model. Oyallon [17] used a wavelet scattering network to replace the first layer of a residual network. The modified residual network produces roughly the same performance as the original residual network, but the training parameters are greatly reduced. Rodriguez [18] proposed a deep adaptive wavelet network to capture basic information from the input data for image classification. Through experiments on three image classification datasets, it was found that the model achieved high accuracy and also reduced training parameters. Recently, Cotter [19] proposed a dual-tree complex wavelet scattering network. After being combined with a CNN model, it achieves high accuracy in image classification tasks as well as fast inference ability. Figure 2 shows the output results of the first-order dual-tree complex wavelet scattering of brain tumor MRI images, including one low-frequency signal and high-frequency signals in six directions. The low-frequency signal is the main feature of the image, and the six high-frequency signals are the boundary feature of the image.

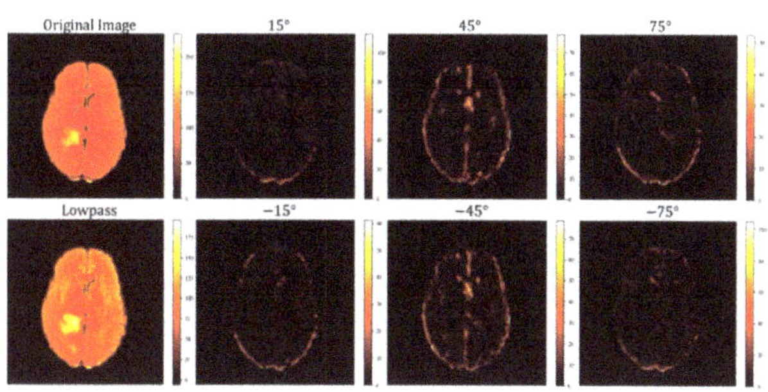

Figure 2. Visualization of the first-order dual-tree complex wavelet scattering transform of brain tumor MRI images.

2.2. Related Loss Functions

While the widely used cross-entropy loss function (CE-Loss) is not sensitive to the segmentation of small object boundaries, when the existing model is trained, the network model will optimize its parameters using a gradient descent method according to the loss function error. Figure 3 shows that the CE-Loss function does not perform very well for cases with small boundaries or a small number of misclassified boundaries. To solve this problem, Williams [15] et al. proposed the AC-Loss loss energy function, which can be described as; where the Region item is the area of the segmentation region, the AC item is the boundary length of the segmentation object, the item is the area of the segmentation region, and the item is the boundary length of the segmentation object. In order to reduce false boundary segmentation, this energy function is expected to minimize the area energy of the segmentation region and the energy of the segmentation target boundary length during model training.

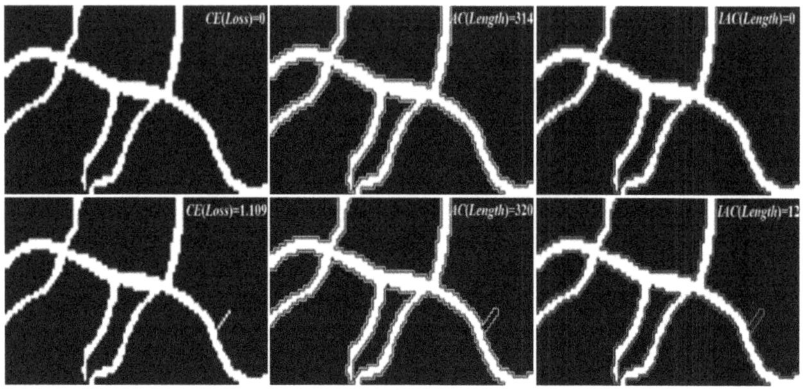

Figure 3. Comparison example of boundary contour segmentation. (**a**) CE-Loss; (**b**) AC-Loss; (**c**) IAC-Loss.

Unfortunately, it can be observed that the AC-Loss function is more sensitive than the CE-Loss term when there is a small boundary in complex medical images, such as the retinal vessel shown in Figure 3. However, when the target segmentation is completely correct, the AC-Loss term still maintains a high error value. As the model is further trained, this will further reduce the length of the segmentation target boundary resulting in under-segmentation for some boundaries.

3. Proposed Method

In this section, we first construct the DTCWT-Scat module and then propose a novel LSW-Net network model after introducing the DTCWT-Scat module into the Unet network. Furthermore, in order to solve the small target segmentation task, a new IAC-Loss function is designed. Finally, we document the LSW-Net algorithm and the IAC-Loss function calculation algorithm.

3.1. Learning Scattering Wavelet Network

Wavelet scattering can extract image texture features and boundary information but cannot make full use of contextual semantic information for image segmentation. FCN integrates multiscale contextual information through multilayer pooling and subsampling; however, it is still unable to distinguish boundary information from overall information. The natural solution of combining the two functional modules can not only enhance the complementarity between the boundary information and the global information but also improve the classification accuracy of image boundaries. Therefore, we designed a novel LSW-Net model that combines a wavelet scattering network and a fully convolutional network, which is based on the encoder–decoder structure of the fully convolutional network [20]. The LSW-Net framework can be described in detail as follows: First, the dual-tree complex wavelet scattering transform [19] is added during downsampling in order to effectively separate the high-frequency features and low-frequency features of the feature map. Second, the convolution operator is used to select the low-frequency main features and high-frequency boundary features of the feature map, respectively. Finally, we concentrate these features. The decoder is a process that uses a multilayer upsampling method to gradually restore its original resolution. The algorithm is shown in Algorithm 1.

Algorithm 1: Learning Scattering Wavelet Network

Input: Preprocess image, x;
  Num of encoder–decoder layers, $m = 4$;
  Kernel size, $k = 3$;
  Num of encoder kernels, $n^i = 64 \times 2^i$;
Num of decoder kernels, $n^j = 64 \times 2^{j-1}$;
Output: Predictive segmentation map, u;
initialization;
  $x^1 = F\left(x, k, n^i\right), (i = 0)$
Encoder:
for $i = 1$ to m do
    $z^{i+1} = \text{DTCWT\_}Scat(x^i)$
    $x^{i+1} = F(z^{i+1}, k, n \times 2^i)$
end
Decoder:
$d^{m+1} = x^{m+1}$
for $j = m + 1$ to 2 do
    $p^{j-1} = \text{concate}(x^{j-1}, \text{upsample}(d^j))$
    $d^{j-1} = F(p^{j-1}, k, n \times 2^j)$
end
$u = \text{softmax}(conv(d^1, 1))$
return u

The LSW-Net framework contains a convolutional feature extraction module that is followed by batch normalization [21] after each convolution. The purpose is to accelerate the convergence speed of the LSW-Net framework and reduce the correlation between layers, see Figure 4 for details. The details can be described as follows: First, we use a 3×3 size kernel for convolution and batch normalization. Then, we use the ReLU function to activate and to achieve the purpose of nonlinear transformation. Finally, the above process is repeated once. The mathematical expression is $F(\bullet, k, n) = [\text{ReLU}(norm(conv(\bullet, k, n)))]_2$, where $\bullet, k, n, [\bullet]_2$ indicates, respectively, the input map, the size of the convolution kernel, the number of convolution kernels, and the convolution feature extraction module which is executed twice.

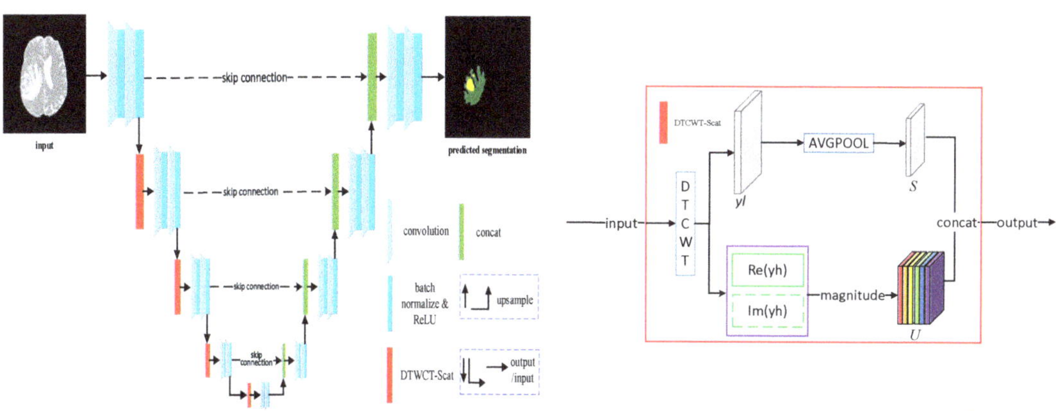

(a) (b)

Figure 4. LSW-Net model. (**a**) LSW-Net; (**b**) DTCWT-Scat module.

3.2. DTCWT-Scat Module

The DTCWT-Scat module can be described in detail as follows. First, the dual-tree complex wavelet transform is performed on the feature map. Then the low-frequency information is processed using average pooling low-pass filtering, that is, $S = Avgpool(yl, 2)$, and the magnitude (mag) of the high-frequency real and imaginary are also calculated, i.e., $U = mag(\mathrm{Re}(yh), \mathrm{Im}(yh)) = \sqrt{(\mathrm{Re}(yh))^2 + (\mathrm{Im}(yh))^2}$. Then, we merge S and U, as shown in Figure 4b, where $yl, yh, \mathrm{Re}(\bullet), \mathrm{Im}(\bullet)$ represents low-frequency information, high-frequency information, and real and imaginary operators, respectively.

The DTCWT-Scat module has two significant advantages. The first advantage is that it is able to perform a dual-tree complex wavelet transform on the input image. This transform supports the backpropagation of errors and can update the parameters so that the parameters of the previous convolution layer can be learned. Afterwards, the frequency domain features can be extracted. The second advantage is that the wavelet function has local waveform characteristics and is stable to local deformation. As a result, the LSW-Net model will be more stable and sensitive to small deformations in medical images such as tumors and will be more accurate for small feature extraction.

3.3. IAC-Loss Function

The flaws of CE-Loss and AC-Loss are acknowledged in Section 2.2. After absorbing the advantages of the AC-Loss function, we designed a contour segmentation minimum energy function, which can be written as follows,

$$\min_{c_1, c_2} Region \tag{1}$$

$$s.t. \int_C |\nabla u| ds = \int_C |\nabla v| ds. \tag{2}$$

where $Region = \int_\Omega ((c_1 - v)^2 - (c_2 - v)^2) \cdot u dx$, u, v, s, C, Ω represents the predicted image, segmentation image, curve arc length, segmentation contour curve, and image area, respectively. The variables c_1, c_2 are constant variables. Since $||\nabla u| - |\nabla v|| \leq |\nabla(u - v)|$, there is,

$$\int_C ||\nabla u| - |\nabla v|| ds \leq \int_C |\nabla(u - v)| ds. \tag{3}$$

Using the Lagrangian multiplier method, we construct a new contour segmentation energy function.

$$\min_{c_1, c_2, C} Loss_{IAC} = Region + \alpha \cdot IAC(Length), \tag{4}$$

$$IAC(Length) = \int_C |\nabla(u - v)| ds, \tag{5}$$

where α is the equilibrium coefficient. The first item is the area of the segmentation target area, and the second item is the difference between the target boundary length of the predicted image and the ground truth.

Figure 3 verifies the image segmentation advantages of the IAC-Loss energy function in complex backgrounds. It can be observed that when the segmentation target has no small target, $IAC(Length) = CE(Loss) = 0$, but $AC(Length) = 314$, which indicates that IAC-Loss and CE-Loss will stop during minimization, but AC-Loss will continue to decrease. When the segmentation has small targets, $IAC(Length) = 12$,$CE(Loss) = 1.109$, $AC(Length) = 320$, which shows that IAC-Loss not only improves the sensitivity to small boundaries but also better solves the problem of under-segmentation, see Figure 3.

In order to facilitate numerical calculation, the specific calculation discrete are also written,

$$Region = \sum_{\Omega}^{i=1,j=1} u_{i,j}(c_1 - v_{i,j})^2 + \sum_{\Omega}^{i=1,j=1} (1 - u_{i,j})(c_2 - v_{i,j})^2, \tag{6}$$

$$IAC(Length) = \sum_{\Omega}^{i=1,j=1} \left(\left| \nabla u_{x_{i,j}} - \nabla v_{x_{i,j}} \right| + \left| \nabla u_{y_{i,j}} - \nabla v_{y_{i,j}} \right| \right), \tag{7}$$

where α is the balance coefficient, $u_{ij} \in [0,1]$ is the predicted probability map, $v_{ij} \in \{0,1\}$ is the binary code of the ground truth, and c_1, c_2 can be defined as a constant of 1 or 0. $\nabla u_{x_{i,j}}, \nabla u_{y_{i,j}}, \nabla v_{x_{i,j}}, \nabla v_{y_{i,j}}$ are the differences of $u_{i,j}$ and $v_{i,j}$ in the horizontal and vertical directions, respectively. The algorithm flow is shown in Algorithm 2.

Algorithm 2: Improved AC-Loss function

Input: Predictive segmentation map, u;
Binary ground truth map, v;
Equilibrium coefficient, α;
Batch size, B; Channels, C;
Image width, W; Image height, H;
Output: IAC-Loss Error, $Loss_{IAC}$;
initialization;
$c_{in} = [1]_{B\times C\times W\times H}, c_{out} = [0]_{B\times C\times W\times H}$
$Region_{in} = u \times (v - c_{in})^2$
$Region_{out} = (1-u) \times (v - c_{out})^2$
$Region = Region_{in} + Region_{out}$
$\nabla h_x = h[:,:,1:,:] - h[:,:,:-1,:],\ h = u, v$
$\nabla h_y = h[:,:,:,1:] - h[:,:,:,:-1],\ h = u, v$
$IAC(Length) = |\nabla u_x - \nabla v_x| + |\nabla u_y - \nabla v_y|$
$Loss_{IAC} = Region + \alpha \cdot IAC(Length)$
return $Loss_{IAC}$

4. Experiments

In the following experiments, we use the DRIVE [22] and MICCAI-BraTS2020 [23] datasets. In the experimental results, the BraTS brain tumor segmentation evaluation metrics were recorded when $epoch = 200$, and the DRIVE retinal blood vessel segmentation evaluation metrics were recorded when $epoch = 10$.

All models are trained on an i7-10750H, NVIDIA RTX 2070 GPU with 8G RAM. The Python language is used for programming and the deep learning framework used is Pytorch.

4.1. Data Preprocessing and Evaluation Metrics

The BraTS2020 brain tumor dataset has 369 patient samples, and each patient contains 4 modalities of MRI image data. After splicing and slicing the four-modal data, slices are obtained. In this experiment, 297 samples are randomly selected as the training set and validation set, and the remaining 72 samples are reserved as the test set. After removing the slices without lesions there are still 19,874 slices, of which 80% of the slices are randomly selected as the training set and 20% of the slices are selected as the validation set. In the DRIVE retinal dataset, the first 20 images are selected as the training set and validation set and the last 20 images are used as the test set.

In the DRIVE retinal dataset, the first 20 images are selected as the training set and validation set, and the last 20 images are used as the test set. In this experiment, since there are only a small number of sample sets in the DRIVE retinal dataset, we preprocess the images of the training set according to image rotation, horizontal flip, vertical flip, translated, and random cropping, in order to expand the sample size of the training set.

In this paper, our model quality is evaluated in terms of the standard evaluation metrics such as precision, Dice coefficient, sensitivity, specificity, and accuracy, which are shown in Table 1. TP, FP, FN, TN represent true positives, false positives, false negatives, and true negatives, respectively.

Table 1. Evaluation metrics.

Metric	Description
Pre (Precision)	$\frac{TP}{FP+TP}$
Dice (Dice coefficient)	$\frac{2 \cdot TP}{2 \cdot TP+FP+FN}$
Sen (Sensitivity)	$\frac{TP}{TP+FN}$
Spe (Specificity)	$\frac{TN}{FP+TN}$
Acc (Accuracy)	$\frac{TP+TN}{FP+FN+TP+TN}$

4.2. Experiment 1: BraTS Brain Tumor Segmentation

In this subsection, we will compare the evaluation metrics of the LSW-Net model with FCN, SegNet, and At-Unet models, as well as the 3D visualization. We use a combination of binary cross entropy (BCE) and Dice loss to train the LSW-Net. The loss is formulated as:

$$loss_{BraTs} = loss_{Dice} + 0.5 \cdot loss_{BCE}, \tag{8}$$

where $loss_{Dice} = 1 - \frac{2\sum y_i \cdot \hat{y}_i}{\sum y_i + \sum \hat{y}_i}$, $loss_{BCE} = -\frac{1}{N}\sum[y_i \cdot \log(\hat{y}_i) + (1 - y_i) \cdot \log(\hat{y}_i)]$, $y_i \in \{0, 1\}$ is the binary-coded value of the ground truth, and $\hat{y}_i \in [0, 1]$ is the predicted value.

In Table 2, the evaluation metrics of the LSW-Net model are recorded, where ET, TC, WT, and AVG represent the enhanced tumor area, tumor core, the entire tumor area, and the average metric, respectively. After comparison to the classic FCN, SegNet, and At-Unet models, it can be observed that the Dice coefficient, accuracy, and sensitivity of the LSW-Net model are all excellent. The LSW-Net model improved the Dice coefficient, accuracy, and sensitivity by at least 3.51%, 2.11%, and 0.46%, respectively.

Table 2. Comparison of LSW-Net model with classical segmentation algorithms on BraTS2020.

Method	Pre				Dice				Sen			
	ET	TC	WT	AVG	ET	TC	WT	AVG	ET	TC	WT	AVG
FCN [7]	0.7650	0.6554	0.7831	0.7345	0.7656	0.6802	0.8125	0.7528	0.8197	0.7904	0.8722	0.8274
SegNet [11]	0.7748	0.7076	0.8669	0.7831	0.7316	0.6984	0.8448	0.7583	0.7615	0.7754	0.8464	0.7944
At-Unet [12]	0.7764	0.7235	0.8791	0.7930	0.7646	0.7312	0.8600	0.7853	0.8080	0.8240	0.8665	0.8328
LSW-Net (Ours)	**0.8319**	**0.7447**	**0.9077**	**0.8281**	**0.7947**	**0.7448**	**0.8797**	**0.8064**	**0.8125**	**0.8308**	**0.8690**	**0.8374**

Figure 5 shows the 2D visualization comparison of the segmentation results of four brain tumor samples between the LSW-Net model and the classic FCN, SegNet, At-Unet models. After comparison with the ground truth, it can be observed that the LSW-Net model performs better than the three classical models in terms of segmentation and is more suitable for BraTS brain tumor dataset image segmentation. In the segmentation results in line one of Figure 5, it can be observed that the other three classic models have misclassified in the enhanced tumor area and the edema area. Conversely, the LSW-Net model has a clear and complete outline, which also shows the validity of the LSW-Net model.

The LSW-Net model segmentation results have fewer outliers and mis-segmented blocks, so they are closer to ground truth when compared to the 3D visualization of the classical FCN, SegNet, and At-Unet segmentation results. The 3D visualization in Figure 6 shows that the LSW-Net model has achieved a good overall segmentation effect. This contributes to a clearer understanding and judgment of tumor size, boundary, and location.

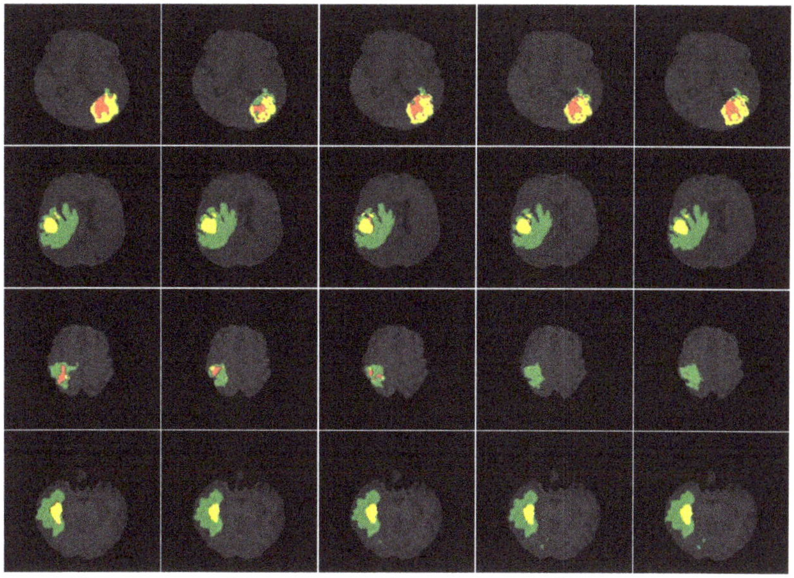

(**a**) (**b**) (**c**) (**d**) (**e**)

Figure 5. Comparison of segmentation results of four brain tumor samples, in which red, yellow, and green indicate necrotic area, enhanced tumor area, and edema area, respectively. (**a**) FCN; (**b**) SegNet; (**c**) At-Unet; (**d**) LSW-Net (Ours); (**e**) Ground truth.

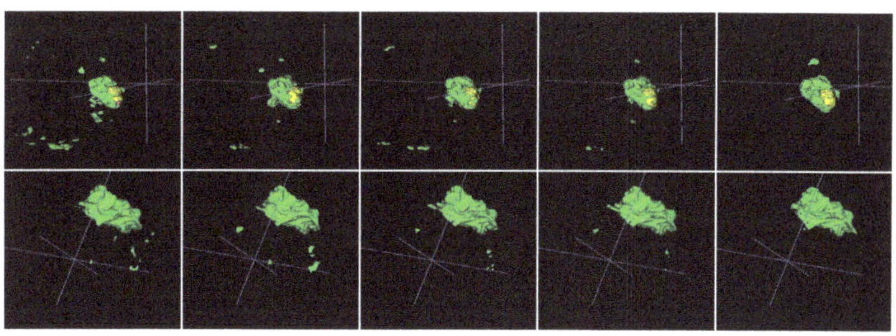

(**a**) (**b**) (**c**) (**d**) (**e**)

Figure 6. 3D visualization comparison of the segmentation results of two brain tumors, in which red, yellow, and green indicate the necrotic area, the enhanced tumor area, and the edema area, respectively. (**a**) FCN; (**b**) SegNet; (**c**) At-Unet; (**d**) LSW-Net (Ours); (**e**) Ground truth.

In addition to the experimental comparison of the LSW-Net with the classical segmentation algorithms, the performance of the LSW-Net model, using the BraTS brain tumor dataset, is assessed here against several advanced segmentation algorithms developed in recent years by researchers such as Zhang et al. [24], Li et al. [25], Feng et al. [26], Latif et al. [27] and Hao et al. [28], see Table 3. The comparison assesses performance in terms of the Dice coefficients for ET, TC, WT, and average (AVG). The evaluation metrics in Table 3 show that the LSW-Net model and these advanced segmentation algorithms have advantages and disadvantages in ET, TC, and WT indicators. However, the most important evaluation indicator is the average indicator of the Dice coefficient, that is, the AVG indicator. The AVG index of the LSW-Net model is the highest, which also shows that our model has better segmentation performance on the BraTS2020 dataset.

Table 3. Comparison of LSW-Net model with some advanced algorithms on BraTS2020.

Method	Year	Dice			
		ET	TC	WT	AVG
Zhang et al. [24]	2019	0.7070	0.7380	0.8850	0.7767
Li et al. [25]	2019	0.7450	**0.8080**	0.8650	0.8060
Feng et al. [26]	2020	0.7100	0.7300	**0.9000**	0.7800
Latif et al. [27]	2021	0.7180	0.7460	0.8960	0.7860
Hao et al. [28]	2021	0.7926	0.7465	0.8764	0.8051
LSW-Net (Ours)		**0.7947**	0.7448	0.8797	**0.8064**

4.3. Experiment 2: DRIVE Retinal Segmentation

In this subsection, we will verify the effectiveness of the IAC-Loss function for segmentation on the DRIVE retina dataset. In the experiment, the LSW-Net model will be used as the backbone network and the loss function will be the IAC-Loss function, denoted as the LSW-Net + IAC-Loss model. Finally, the LSW-Net + IAC-Loss model segmentation results are compared with other 14 models, including Cheng et al. [29], Azzopardi et al. [30], Roychowdhury et al. [31], DRIU [32], HED [33], Unet [34], Recurrent Unet [34], R2Unet [34], Guo et al. [35], Du et al. [36], Arias et al. [37], Zou et al. [38], and MD-Net [39] models. In addition, two examples of segmentation effects are shown in terms of overall and local details. Compared with the other 14 models in Table 4, it can be seen that the LSW-Net + IAC-Loss model is higher than the other 14 algorithms in terms of Dice coefficient and specificity; it is second only to the MD-Net [39] model in the accuracy index. Compared with the segmentation results of the classic Unet model, the sensitivity of the LSW-Net model offers an improvement of 3.39%, which is an obvious improvement and has a significant effect. These advantages indicate that our model performs well.

Table 4. Comparison of LSW-Net + IAC-Loss model with some advanced models on DRIVE.

Method	Year	Dice	Sen	Spe	Acc
Cheng et al. [29]	2014	-	0.7252	0.9798	0.9474
Azzopardi et al. [30]	2015	-	0.7655	0.9704	0.9442
Roychowdhury et al. [31]	2016	-	0.7250	0.9830	0.9520
DRIU [32]	2016	0.6701	**0.9696**	0.9115	0.9165
HED [33]	2017	0.6400	0.9563	0.9007	0.9054
Unet [34]	2019	0.8142	0.7537	0.9820	0.9553
Recurrent Unet [34]	2019	0.8155	0.7751	0.9816	0.9556
R2Unet [34]	2019	0.8171	0.7792	0.9813	0.9556
Guo et al. [35]	2020	0.8215	0.8283	0.9726	0.9542
Du et al. [36]	2021	-	0.7814	0.9810	0.9556
Arias et al. [37]	2021	-	0.8597	0.9690	0.9563
Zou et al. [38]	2021	0.8129	0.7761	0.9792	0.9519
MD-Net [39]	2021	0.8099	0.8065	0.9826	**0.9676**
MFE-Net [40]	2022	0.8204	0.7853	0.9812	0.9563
LSW-Net + IAC-Loss (Ours)		**0.8216**	0.7876	**0.9837**	0.9565

The comparison experiment, using the segmentation results from the DRIVE retinal blood vessel dataset, is shown in Figure 7. In the first and third rows of Figure 7 it can be observed that the segmentation results of DRIU [32] and HED [33] have obvious over-segmentation. In lines two and four of Figure 7, it can be seen that the segmentation results of the LSW-Net + IAC-Loss model have less noise and clearer contours. Through this experimental comparison, it can be shown that the LSW-Net + IAC-Loss model has better segmentation effectiveness.

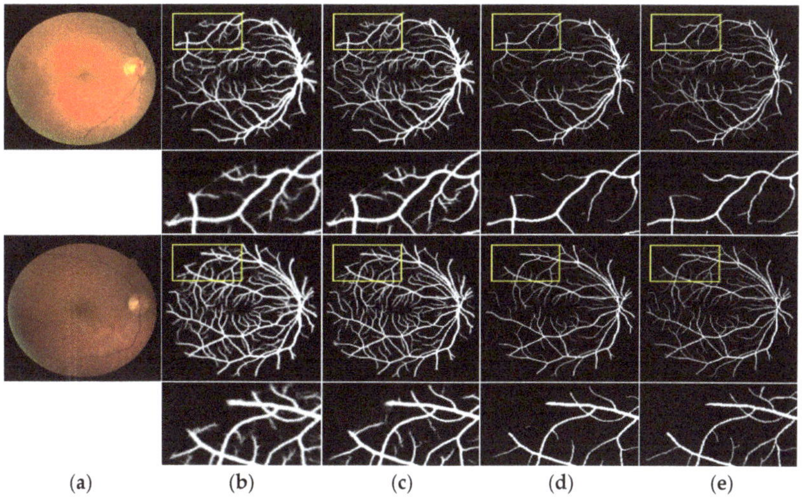

Figure 7. Comparison of LSW-Net + IAC-Loss with advanced segmentation models results on DRIVE. Lines 2 and 4 are the local details. (**a**) Test Image; (**b**) HED; (**c**) DRIU; (**d**) LSW-Net + IAC-Loss (Ours); (**e**) Ground truth.

4.4. Experiment 3: Discussion on Equilibrium Coefficient α

To evaluate the effect of the balance coefficient α in the LSW-Net + IAC-Loss model, ablation experiments are performed on the DRIVE dataset in this subsection.

In the experiment, the evaluation metrics of the LSW-Net + IAC-Loss model are recorded in Table 5, where $\alpha \in [0.1, 0.5]$. Table 5 shows that the differences in each metric are not obvious; however, they all reach a high level, which also shows that the LSW-Net + IAC-Loss model has better robustness to the balance coefficient α. When $\alpha = 0.3$, the specificity is the highest while the Dice coefficient and sensitivity are relatively low. Alternatively, when $\alpha = 0.1$ or $\alpha = 0.5$, the Dice coefficient and sensitivity index values increase. Therefore, we suggest that the metrics can be fine-tuned by controlling the balance coefficient α according to actual needs.

Table 5. Influence of α balance coefficient on LSW-Net + IAC-Loss model segmentation result indicators.

α	Pre	Dice	Sen	Spe	Acc
0.1	0.8525	**0.8231**	0.9565	**0.7957**	0.9799
0.2	0.8542	0.8222	0.9564	0.7925	0.9802
0.3	0.8588	0.8216	0.9565	0.7876	**0.9837**
0.4	**0.8602**	0.8221	**0.9566**	0.7873	0.9813
0.5	0.8571	0.8227	**0.9566**	0.7909	0.9807

4.5. Experiment 4: IAC-Loss Effectiveness Evaluation

In this subsection, we further evaluate the advantages of the IAC-Loss function. For the segmentation of the DRIVE retina dataset, the LSW-Net is used as the backbone network and the loss functions are the CE-Loss, AC-Loss, and IAC-Loss functions, denoted as +CE-Loss, +AC-Loss, and +IAC-Loss models, respectively. The segmentation results are shown in Table 6.

Table 6. Comparison of IAC-Loss with related loss function metrics.

	Dice	Sen	Spe	Acc
+AC-Loss	0.7875	0.7147	**0.9853**	0.9509
+CE-Loss	0.8182	0.7920	0.9789	0.9551
+IAC-Loss (α = 0.1)	**0.8231**	**0.7957**	0.9799	**0.9565**

Compared with the +AC-Loss model, the +IAC-Loss model improves on the Dice coefficient and sensitivity by 3.56% and 8.1%, respectively. The accuracy is also increased by 0.56%. This illustrates the effectiveness of the IAC-Loss function for image segmentation, see Table 6.

After comparing the enlarged details of lines two and four in Figure 8, it can be seen that the boundary contour segmentation of the +IAC-Loss model is the best. Through the comparative experiments above, it can be determined that the IAC-Loss function has greater advantages for complex image boundary contours.

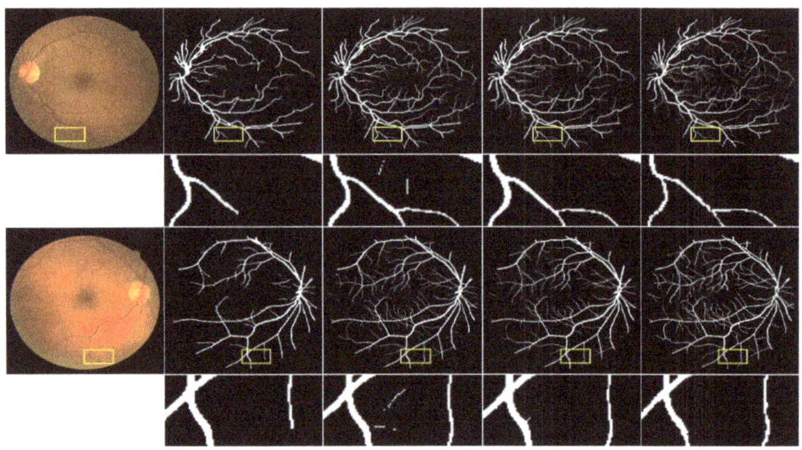

(a) (b) (c) (d) (e)

Figure 8. Comparison of +IAC-Loss with related loss function results. Lines 2 and 4 are the local details respectively. (**a**) Test Image; (**b**) +AC-Loss; (**c**) +CE-Loss; (**d**) +IAC-Loss; (**e**) Ground truth.

5. Conclusions

In this research, we have proposed the LSW-Net model for the BraTS2020 dataset, which achieved good experimental simulation results on the segmentation discontinuity problem. We have constructed an LSW-Net + IAC-Loss model in order to solve the weak boundary problem of small blood vessels in the DRIVE retinal vessel dataset. After introducing the dual-tree complex wavelet transform, the experimental results show that the LSW-Net has the ability to extract features and achieve better segmentation results. In the future we will further integrate the attention mechanism and the transformer method to design a better image segmentation network model.

Author Contributions: Methodology, R.L.; software, H.N.; validation, H.N. and Y.Z.; formal analysis, R.L.; resources, R.L., T.X. and Z.Y.; project administration, R.L.; funding acquisition, R.L. and Z.Y.; data curation, T.X. and Z.Y.; writing—review and editing, R.L.; visualization, H.N. All authors have read and agreed to the published version of the manuscript.

Funding: This work was supported in part by: Chongqing Natural Science Foundation under Grants No. cstc2019jcyj-msxmX0500 and No. cstc2019jcyj-msxmX0240; Technology Research Program of Chongqing Education Commission under Grand No. KJQN202001129.

Data Availability Statement: The DRIVE and BraTS2020 datasets are publicly available [22,23], https://drive.grand-challenge.org/. https://www.med.upenn.edu/cbica/brats2020/.

Conflicts of Interest: The authors declare no conflict of interest.

References

1. Kaur, T.; Saini, B.S.; Gupta, S. A novel fully automatic multilevel thresholding technique based on optimized intuitionistic fuzzy sets and tsallis entropy for MR brain tumor image segmentation. *Australas. Phys. Eng. Sci. Med.* **2018**, *41*, 41–58. [CrossRef]
2. Sukanya, A.; Rajeswari, R.; Murugan, K.S. Region based coronary artery segmentation using modified frangi's vesselness measure. *Int. J. Imaging Syst. Technol.* **2020**, *30*, 716–730. [CrossRef]
3. Chen, Y.; Chen, G.; Wang, Y.; Dey, N.; Sherratt, R.S.; Shi, F. A distance regularized level-set evolution model based MRI dataset segmentation of brain's caudate nucleus. *IEEE Access* **2019**, *7*, 124128–124140. [CrossRef]
4. Chan, T.F.; Vese, L.A. Active contours without edges. *IEEE Trans. Image Process.* **2001**, *10*, 266–277. [CrossRef] [PubMed]
5. Lie, J.; Lysaker, M.; Tai, X.-C. A binary level set model and some applications to mumford-shah image segmentation. *IEEE Trans. Image Process.* **2006**, *15*, 1171–1181. [CrossRef] [PubMed]
6. Krizhevsky, A.; Sutskever, I.; Hinton, G.E. Imagenet classification with deep convolutional neural networks. *Adv. Neural Inf. Process. Syst.* **2012**, *25*, 84–90. [CrossRef]
7. Long, J.; Shelhamer, E.; Darrell, T. Fully convolutional networks for semantic segmentation. In Proceedings of the IEEE Conference on Computer Vision and Pattern Recognition, Boston, MA, USA, 7–12 June 2015; pp. 3431–3440.
8. Ronneberger, O.; Fischer, P.; Brox, T. Unet: Convolutional networks for biomedical image segmentation. In Proceedings of the International Conference on Medical Image Computing and Computer-Assisted Intervention, Munich, Germany, 5–9 October 2015; Springer: Berlin, Germany, 2015; pp. 234–241.
9. Cicek, O.; Abdulkadir, A.; Lienkamp, S.S.; Brox, T.; Ronneberger, O. 3d Unet: Learning dense volumetric segmentation from sparse annotation. In Proceedings of the International Conference on Medical Image Computing and Computer-Assisted Intervention, Athens, Greece, 17–21 October 2016; Springer: Berlin, Germany, 2016; pp. 424–432.
10. Shrestha, H.; Dhasarathan, C.; Kumar, M.; Nidhya, R.; Shankar, A.; Kumar, M. A deep learning based convolution neural network-DCNN approach to detect brain tumor. In Proceedings of the Academia-Industry Consortium for Data Science, Wenzhou, China, 19–20 December 2020; Springer: Singapore, 2022; pp. 115–127.
11. Badrinarayanan, V.; Kendall, A.; Cipolla, R. Segnet: A deep convolutional encoder-decoder architecture for image segmentation. *IEEE Trans. Pattern Anal. Mach. Intell.* **2017**, *39*, 2481–2495. [CrossRef] [PubMed]
12. Schlemper, J.; Oktay, O.; Schaap, M.; Heinrich, M.; Kainz, B.; Glocker, B.; Rueckert, D. Attention gated networks: Learning to leverage salient regions in medical images. *Med. Image Anal.* **2019**, *53*, 197–207. [CrossRef]
13. Chen, L.C.; Papandreou, G.; Kokkinos, I.; Murphy, K.; Yuille, A.L. Deeplab: Semantic image segmentation with deep convolutional nets, atrous convolution, and fully connected CRFs. *IEEE Trans. Pattern Anal. Mach. Intell.* **2017**, *40*, 834–848. [CrossRef]
14. Kim, Y.; Kim, S.; Kim, T.; Kim, C. CNN-based semantic segmentation using level set loss. In Proceedings of the 2019 IEEE Winter Conference on Applications of Computer Vision (WACV), Waikoloa Village, HI, USA, 7–11 January 2019; IEEE: New York, NY, USA, 2019; pp. 1752–1760.
15. Chen, X.; Williams, B.M.; Vallabhaneni, S.R.; Czanner, G.; Williams, R.; Zheng, Y. Learning active contour models for medical image segmentation. In Proceedings of the IEEE/CVF Conference on Computer Vision and Pattern Recognition, Long Beach, CA, USA, 15–20 June 2019; pp. 11632–11640.
16. Bruna, J.; Mallat, S. Invariant scattering convolution networks. *IEEE Trans. Pattern Anal. Mach. Intell.* **2013**, *35*, 1872–1886. [CrossRef]
17. Oyallon, E.; Belilovsky, E.; Zagoruyko, S. Scaling the scattering transform: Deep hybrid networks. In Proceedings of the IEEE International Conference on Computer Vision, Venice, Italy, 22–29 October 2017; pp. 5618–5627.
18. Rodriguez, M.X.B.; Gruson, A.; Polania, L.; Fujieda, S.; Prieto, F.; Takayama, K.; Hachisuka, T. Deep adaptive wavelet network. In Proceedings of the IEEE/CVF Winter Conference on Applications of Computer Vision, Snowmass Village, CO, USA, 1–5 March 2020; pp. 3111–3119.
19. Cotter, F. Uses of Complex Wavelets in Deep Convolutional Neural Networks. Ph.D. Thesis, University of Cambridge, Cambridge, UK, 2020.
20. Pathak, D.; Krahenbuhl, P.; Donahue, J.; Darrell, T.; Efros, A.A. Context encoders: Feature learning by inpainting. In Proceedings of the IEEE Conference on Computer Vision and Pattern Recognition, Las Vegas, NV, USA, 27–30 June 2016; pp. 2536–2544.
21. Ioffe, S.; Szegedy, C. Batch normalization: Accelerating deep network training by reducing internal covariate shift. In Proceedings of the International Conference on Machine Learning, Lille, France, 6–11 July 2015; pp. 448–456.
22. Staal, J.; Abramoff, M.D.; Niemeijer, M.; Viergever, M.A.; Van Ginneken, B. Ridge-based vessel segmentation in color images of the retina. *IEEE Trans. Med. Imaging* **2004**, *23*, 501–509. [CrossRef] [PubMed]
23. Menze, B.H.; Jakab, A.; Bauer, S.; Cramer, J.K.; Farahani, K.; Kirby, J.; Burren, Y.; Porz, N.; Slotboom, J.; Wiest, R.; et al. The multimodal brain tumor image segmentation benchmark (brats). *IEEE Trans. Med. Imaging* **2014**, *34*, 1993–2024. [CrossRef] [PubMed]

24. Zhang, C.; Shen, X.; Cheng, H.; Qian, Q. Brain tumor segmentation based on hybrid clustering and morphological operations. *Int. J. Biomed. Imaging* **2019**, 1–11. [CrossRef]
25. Li, H.; Li, A.; Wang, M. A novel end-to-end brain tumor segmentation method using improved fully convolutional networks. *Comput. Biol. Med.* **2019**, *108*, 150–160. [CrossRef]
26. Bowen, F.; Qi, L.X.; Yu, G.; Qing, L.; Yang, L. Three-dimensional parallel convolution neural network brain tumor segmentation based on dilated convolution. *Laser Optoelectron. Prog.* **2020**, *57*, 141009. [CrossRef]
27. Latif, U.; Shahid, A.R.; Raza, B.; Ziauddin, S.; Khan, M.A. An end-to-end brain tumor segmentation system using multi-inception-Unet. *Int. J. Imaging Syst. Technol.* **2021**, *31*, 1803–1816. [CrossRef]
28. Hao, K.; Lin, S.; Qiao, J.; Tu, Y. A generalized pooling for brain tumor segmentation. *IEEE Access* **2021**, *9*, 159283–159290. [CrossRef]
29. Cheng, E.; Du, L.; Wu, Y.; Zhu, Y.J.; Megalooikonomou, V.; Ling, H. Discriminative vessel segmentation in retinal images by fusing contextaware hybrid features. *Mach. Vis. Appl.* **2014**, *25*, 1779–1792. [CrossRef]
30. Azzopardi, G.; Strisciuglio, N.; Vento, M.; Petkov, N. Trainable cosfire filters for vessel delineation with application to retinal images. *Med. Image Anal.* **2015**, *19*, 46–57. [CrossRef]
31. Roychowdhury, S.; Koozekanani, D.D.; Parhi, K.K. Blood vessel segmentation of fundus images by major vessel extraction and sub-image classification. *IEEE J. Biomed. Health Inform.* **2014**, *19*, 1118–1128. [CrossRef]
32. KManinis, K.; Tuset, J.P.; Arbelaez, P.; Gool, L.V. Deep retinal image understanding. In Proceedings of the International Conference on Medical Image Computing and Computer-Assisted Intervention, Athens, Greece, 17–21 October 2016; Springer: Berlin, Germany, 2016; pp. 140–148.
33. Xie, S.; Tu, Z. Holistically-nested edge detection. In Proceedings of the IEEE International Conference on Computer Vision, Santiago, Chile, 7–13 December 2015; pp. 1395–1403.
34. Alom, Z.; Yakopcic, C.; Hasan, M.; Taha, T.M.; Asari, V.K. Recurrent residual Unet for medical image segmentation. *J. Med. Imaging* **2019**, *6*, 014006. [CrossRef] [PubMed]
35. Guo, X.; Chen, C.; Lu, Y.; Meng, K.; Chen, H.; Zhou, K.; Wang, Z.; Xiao, R. Retinal vessel segmentation combined with generative adversarial networks and dense Unet. *IEEE Access* **2020**, *8*, 194551–194560. [CrossRef]
36. XDu, X.-F.; Wang, J.-S.; Sun, W.-Z. Unet retinal blood vessel segmentation algorithm based on improved pyramid pooling method and attention mechanism. *Phys. Med. Biol.* **2021**, *66*, 175013.
37. Arias, M.E.G.; Santos, D.M.; Borrero, I.P.; Vazquez, M.J.V. A new deep learning method for blood vessel segmentation in retinal images based on convolutional kernels and modified Unet model. *Comput. Methods Programs Biomed.* **2021**, *205*, 106081. [CrossRef]
38. Zou, B.; Dai, Y.; He, Q.; Zhu, C.; Liu, G.; Su, Y.; Tang, R. Multi-label classification scheme based on local regression for retinal vessel segmentation. *IEEE/ACM Trans. Comput. Biol. Bioinform.* **2020**, *18*, 2586–2597. [CrossRef] [PubMed]
39. Shi, Z.; Wang, T.; Huang, Z.; Xie, F.; Liu, Z.; Wang, B.; Xu, J. MD-Net: A multi-scale dense network for retinal vessel segmentation. *Biomed. Signal Process. Control* **2021**, *70*, 102977. [CrossRef]
40. Yan, H.; Xie, J.; Yue, X.; Wang, J.; Guo, S. MFE-Net: Multi-type feature enhancement net for retinal blood vessel segmentation. In Proceedings of the 2022 5th International Conference on Artificial Intelligence and Big Data (ICAIBD), Chengdu, China, 27–30 May 2022; pp. 51–56.

electronics

MDPI

Article

An Image Style Diversified Synthesis Method Based on Generative Adversarial Networks

Zujian Yang and Zhao Qiu *

School of Computer Science and Technology, Hainan University, Haikou 570228, China; 20081200210006@hainanu.edu.cn
* Correspondence: qiuzhao@hainanu.edu.cn

Abstract: Existing research shows that there are many mature methods for image conversion in different fields. However, when the existing methods deal with images in multiple image domains, the robustness and scalability of images are often limited. We propose a novel and scalable approach, using a generative adversarial networks (GANs) model that can transform images across multiple domains, to address the above limitations. Our model can be trained on image datasets with different domains in a single network, with the ability to translate images and the ability to flexibly translate input images to any desired target domain. Our model is mainly composed of a generator, discriminator, style encoder, and a mapping network. The datasets use the celebrity face dataset CelebA-HQ and the animal face dataset AFHQ, and the evaluation criteria use FID and LPIPS to evaluate the images generated by the model. Experiments show that our model can generate a rich variety of high-quality images, and there is still some room for improvement.

Keywords: generative adversarial networks; multiple domains; translate images

Citation: Yang, Z.; Qiu, Z. An Image Style Diversified Synthesis Method Based on Generative Adversarial Networks. *Electronics* **2022**, *11*, 2235. https://doi.org/10.3390/electronics11142235

Academic Editor: Gemma Piella

Received: 22 June 2022
Accepted: 14 July 2022
Published: 17 July 2022

Publisher's Note: MDPI stays neutral with regard to jurisdictional claims in published maps and institutional affiliations.

1. Introduction

Diversifying the style of an image essentially edits the attributes of the image, so that the image has the attributes that people require to meet their different needs. Diversifying image attributes can also increase the diversity of data, resulting in more datasets. Compiling the properties of images is a challenging problem for vision applications. Generative adversarial networks (GANs) [1] can be an important tool for generating images of people's desired attributes. After the use of GANs, the task of compiling and generating images has been greatly improved. There are applications from text to images [2–5]; unsupervised image-to-image compilation of two different domains [6,7]; and multi-domain image compilation [8], etc.

We design a model where the generator takes the input image as conditional input, and the style encoding of the target domain as the label input, and transforms the input image to the target domain indicated by the input label, where the style encoding is provided by the mapping network or the style encoder. The mapping network outputs the style code corresponding to the target image domain by randomly sampling the latent vector z and domain y. The style encoder can output the input image x and the corresponding domain into the style code s corresponding to the target domain. The mapping network consists of multi-layer perceptions (MLPs) and has multiple branches outputting, each of which can generate a style code for a specific image domain. In addition, the style encoding network consists of residual modules, MLPs and convolutional layers, and it also has multiple outputting branches, consisting of multi-layer perceptions. Style encoders and mapping networks benefit from a multi-task learning setting that can generate diverse style codes.

Our generator encodes the input picture and style, and then goes through a downsampling module, an intermediate layer module, and an upsampling module, all of which consist of residual units and convolutional attention units. The residual unit of the upsampling module contains adaptive instance normalization (AdaIN) [9]. The style encoding is combined with the input image through AdaIN, and the scale and shift vectors are provided by learning affine transformation. The residual module is followed by a convolutional attention module to enhance the effective features in the output feature map, while some irrelevant noises are suppressed. Repeatedly superimposing the attention module can gradually improve the expressive ability of the network.

2. Related Work

2.1. Generative Adversarial Networks

The generative adversarial network (GAN) model was originally proposed to generate images from random noise. Its structure generally consists of a generator and a discriminator, and is trained in an adversarial manner. GAN has many advantages, it can train any kind of generator network, and its design also does not need to follow any kind of factorization model, nor does it need to use Markov chains to repeatedly sample, and it does not need to infer during the learning process, but the GAN has the problem that the network is difficult to converge. Therefore, in [10,11], it is suggested that Wasserstein-1 distance and gradient penalty is used to improve the stability of the optimization process. Conditional GANs (cGANs) [2,12] take conditional variables as inputs to the generator and discriminator, to generate images with desired properties.

2.2. Image-to-Image Translation

Good results have been achieved in image-to-image style transfer research [12–15]. Pix2pix [12] uses cGAN [2] to train the model in a supervised manner, combining adversarial loss and L1 loss, so paired samples are required. In order to solve the problem that the data need to be paired, the unpaired image transformation framework has been proposed [13–15]. UNIT [14] proposes to add VAE [16] to CoGAN [17] for unsupervised image-to-image translation, which builds two encoders sharing the same latent space, and sharing weights, to learn the joint distribution of images across domains. CycleGAN [14], NICE-GAN [15], and DiscoGAN [13] preserve key properties between input and translated images by exploiting cycle consistency loss, but they can only learn the relationship between two different domains at a time. To address this problem, StarGAN [8] proposes the use of a generator that can generate images of all domains. Instead of just taking images as conditional input, StarGAN also takes the label of the target domain as input, and the generator is used to transform the input image to the target domain indicated by the input label. Moreover, DualStyleGAN [18] adds an external style control module on the basis of StyleGAN [19], and learns external styles on small-scale data through a progressive transfer learning method, which can effectively imitate the style of artistic portraits and achieve sample-based high-definition stylized faces. RAMT-GAN [20] realizes cross-domain image conversion based on a dual input/output network constructed by the BeautyGAN [21] architecture, and introduces identity preservation loss and background invariance loss to ensure that the generated facial makeup images are accurate and realistic. Different from the above methods, our framework not only uses a single model to learn the relationship between multiple domains, but also introduces an attention module to make the features of the generated images more obvious and improve the quality of the generated images.

3. The Proposed Method

In this section, we describe our proposed framework and its training objective function.

3.1. The Proposed Framework

X and Y represent the image set and the domain of the image, respectively. Given an image $x \in X$ and an arbitrary domain $y \in Y$, where y is the encoding of the field Y, our goal is to train a generator G, that can generate different images of different domains y corresponding to images x. The generator G adopts an encoder–decoder structure. Spatial pooling is necessary to extract high-level abstract representations of image features, but spatial pooling reduces the spatial resolution and fine details of the image feature map, and the features map will easily lose details when it is restored later. In order to improve the quality of the coded image, some skip connections are applied between the encoder and the decoder to prevent the important features of the image from being lost. Skip connections are added between every two corresponding encoder layers and decoder layers. When the network is very deep, the decoder layer cannot complete the restoration of image details. The skip connections pass the information of the feature map through the convolutional layer to the corresponding part in the decoder layer. In addition, spatial attention and channel attention modules are applied in the encoder and decoder, so that the important features of the image are enhanced and the unimportant features are suppressed. Our framework mainly consists of three modules:

- Mapping network (F) and style encoder (E). The schematic diagram of their structure is shown in Figures 1 and 2 respectively. For a given latent code z and a domain y, or a given image x and the corresponding image domain y, the style code s = $F_y(z)$ generated by the mapping network, where $F_y(z)$ represents the F output of the corresponding domain y. The output of the style encoder is s = $E_y(x)$, where $E_y(x)$ represents the E output for the corresponding domain y. E consists of MLPs, convolutional neural net- works (CNNs) and residual blocks with multiple output branches, which can provide a variety of style encodings, and the number of output branches is determined by the number of image domains. F consists of multiple MLPs with output branches, and the number of output branches is also determined by the number of domains.
- Generator (G). As shown in Figure 3, this is a schematic diagram of our residual downsampling module, and Figure 4 is the schematic diagram of our AdaIN-residual upsampling module. Figure 5 is the generator structure diagram. The generator transforms an input image x into an output image G(x, s), where s is a domain-specific style code provided by a mapping network F or a style encoder E. Our generator consists of four downsampling blocks, two intermediate blocks and four upsampling blocks, all of which have pre-activated residual units. We apply adaptive normalization (AdaIN) [9,19] to the upsampling module in the generator, which can inject s into G. AdaIN receives two sources of information: the content input x and the style input s, and matches the channel-wise mean and standard deviation of x to the channel-wise mean and standard deviation of s. As shown in the following Equation (1). Simply speaking, AdaIN realizes style transfer by changing the data distribution of features at the feature map level, with small computational and storage costs, and is easy to implement, additionally, there are skip connections between the encoder and decoder, which can effectively avoid some important features of loss. In addition, the generator also adds the attention mechanism module (CBAM) [22] of the convolution module, to make the important features of the feature map more obvious and suppress the unimportant features.

$$AdaIN(x,s) = \sigma(s)(\frac{x-\mu(x)}{\sigma(x)}) + \mu(s) \quad (1)$$

- Discriminator (D). Our discriminator D is a multi-task discriminator [23,24]. As shown in Figure 6. It contains multiple output branches, as well as multiple preactivated residual blocks. Each branch D_y learns a binary classification to determine whether an image x is a real image of its domain y or a fake image G(x, s) generated by Generator.

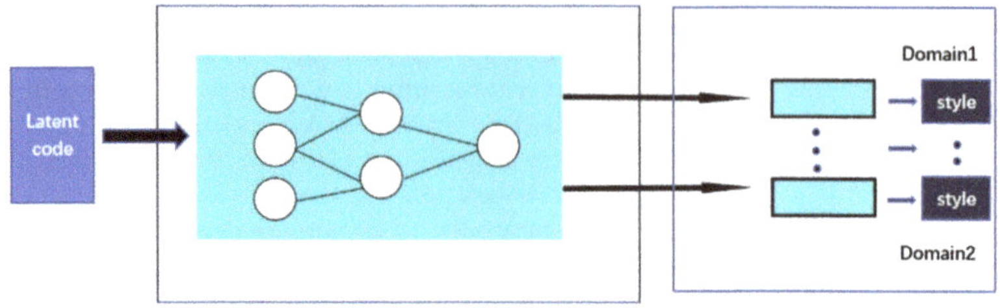

Figure 1. Mapping network.

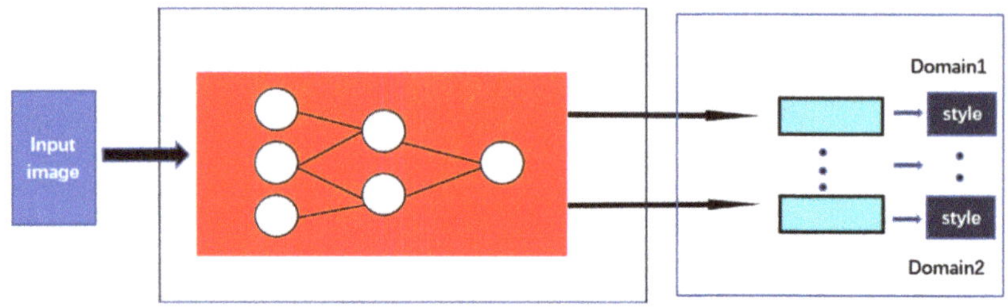

Figure 2. Style encoder.

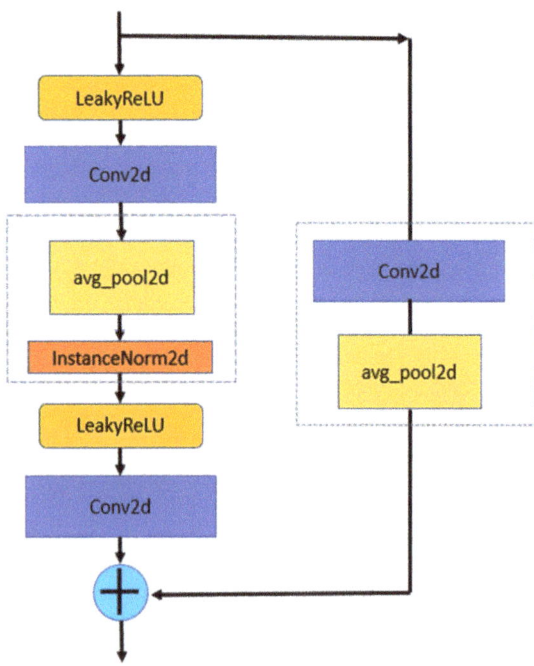

Figure 3. Residual block (the part within the dashed line indicates whether the judgment is executed or not).

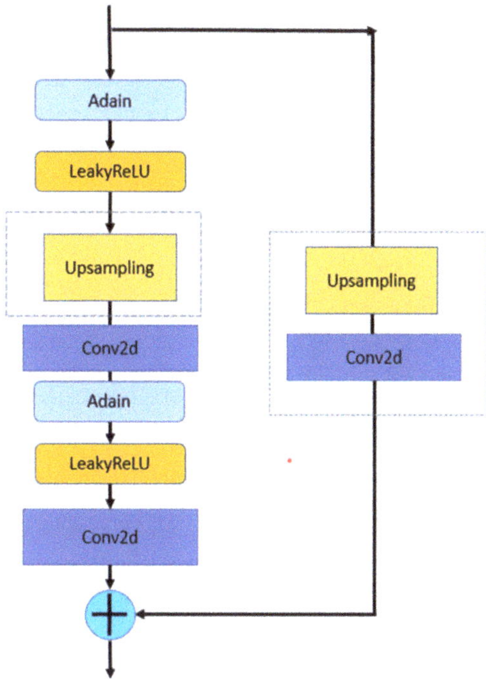

Figure 4. AdainResBlk block (the part within the dashed line indicates whether the judgment is executed or not).

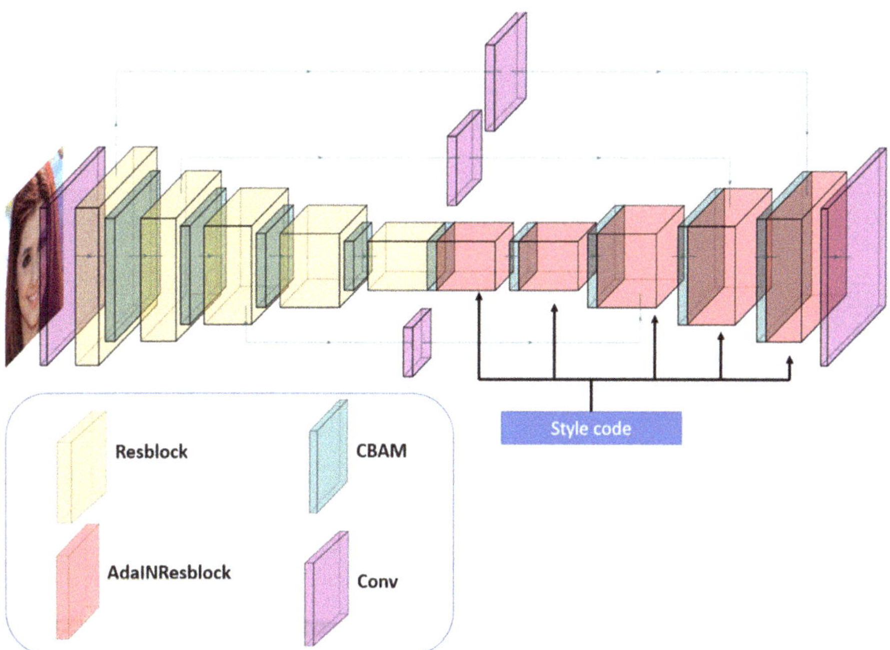

Figure 5. Generator.

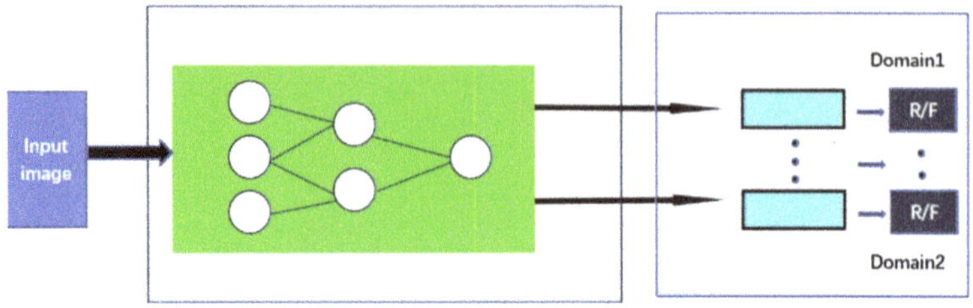

Figure 6. Discriminator.

3.2. The Function of Training

Our goal is to train a generator G that can learn mappings between multiple domains. For a given image x and y, we train as follows:

Adversarial loss: To make the generated image indistinguishable from the real image, an adversarial loss is used:

$$\mathcal{L}_{adv} = \mathbb{E}_{x,y}[\log D_y(x)] + \mathbb{E}_{x,\tilde{y},z}\left[\log\left(1 - D_{\tilde{y}}(G(x,s))\right)\right] \tag{2}$$

x is the input image, y is the image source domain, s is generated by the randomly sampled code $z \in Z$ and the image target domain $\tilde{y} \in Y$ through the mapping network $s = F_{\tilde{y}}(z)$ or style encoder $s = E_{\tilde{y}}(x)$. The generator G takes the image x and the style code s as input to generate a picture G(x, s). Dy (·) represents the D output corresponding to the domain y. Generator scholars use s to generate images G(x, s) that are indistinguishable from real images of the domain $\tilde{y}$.

Style reconstruction: In order for the generator G to use the style code when generating images and train a style encoder E to learn the output of different domains, the learned style encoder E allows G to transform the input image to reflect the style code of the reference image. Our style reconstruction loss is:

$$\mathcal{L}_{sty} = \mathbb{E}_{x,\tilde{y},z}\left[\left|\left|s - E_{\tilde{y}}(G(x,s))\right|\right|_1\right] \tag{3}$$

z is the latent code generated by random noise, $\tilde{y}$ is the given image domain, x is the real picture, s is generated by z and $\tilde{y}$ through the mapping network F or x and $\tilde{y}$ through the style encoder E. Similar approaches have also been used by previous methods [25–27] with this loss. Most of them use multiple encoders to train different pictures to their latent codes; we only train one encoder to learn to map pictures of different domains to their latent codes.

Diverse styles: This loss is derived from MSGAN [28], regularizing the generator with a diversity-sensitive loss [28,29]:

$$\mathcal{L}_{ds} = \mathbb{E}_{x,\tilde{y},z_1,z_2}[||G(x,s_1) - G(x,s_2)||_1] \tag{4}$$

where the style code s_1, s_2 is generated by two random latent codes z_1, z_2 through the mapping network $F\left(s_i = F_{\tilde{y}}(z_i), i = 1,2\right)$. Compared with MSGAN, the denominator image is removed, making the training more stable. Maximizing regularization also enables the generator to discover more style features and generate diverse images.

Cycle consistency loss: This loss is derived from CycleGAN [6]. The purpose is to make the generated images properly maintain the characteristics of the original image:

$$\mathcal{L}_{cyc} = \mathbb{E}_{x,y,\tilde{y},z}[||x - G(G(x,s),\hat{s})||_1] \tag{5}$$

Among them, where $\tilde{s}$ is generated by the style encoder E from the source domain y corresponding to the input images x and x ($\tilde{s} = E_y(x)$). By letting the generator G use the style code s to reconstruct the input image x, the generator G can retain the original features of x when changing the style of x.

Total loss function: Our overall objective function is as follows:

$$\min_{G,F,E}\max_{D} \mathcal{L}_{adv} + \lambda_{sty}\mathcal{L}_{sty} - \lambda_{ds}\mathcal{L}_{ds} + \lambda_{cyc}\mathcal{L}_{cyc} \tag{6}$$

where λ_{sty}, λ_{ds}, λ_{cyc} are the hyperparameters of each loss function. We train our model with the above objective function. We use in all experiments λ_{cyc} = 1, λ_{sty} = 1 and λ_{ds} = 1.

4. Experiments

In this section, we first compare recent image attribute transfer methods with our framework, through research. Next, we conduct classification experiments on image attribute transfer and synthesis. Finally, we show empirical results on image-to-image attribute transfer learned by our framework from several datasets. For different datasets, these models need to be trained separately for each dataset.

4.1. Baseline Model

We use MUNIT [25], DRIT [30] and MSGAN [28], as our baselines, all of which learn multimodal implicatures between two or more domains. For multi-domain comparisons, we train these models multiple times for the image domain.

MUNIT [25] reduces the image dimension of the dataset into two types of low-dimensional codes: content code and style code. It combines the content code with the style code of another image domain to generate style transfer and uses the decoder to increase the dimension of the newly combined code to generate the resulting image, before the generated image is decomposed into two codes again. For the original code to calculate the error back propagation, the c encoder and the s encoder should be well integrated in the decoder. Adaptive instance normalization (AdaIN) is used, along with an MLP network, which is used to generate parameters to assist residential blocks to generate high-quality images.

DRIT [30] proposes a decoupled representation-based method that can produce various outputs without paired training images. Embedding images into two spaces: an invariant domain content space that captures information shared across domains, and a domain-specific attribute space, using decoupled features as input, can greatly reduce mode collapse and achieve diversity. Introduce cross-cycle consistency loss to handle unpaired training data.

MSGAN [28] proposes an effective and simple pattern search regularization method that solves the pattern collapse problem of cGAN. The method is easy to add to existing models and has been proven to generalize well and effectively.

4.2. Dataset

We use CelebA-HQ [31] and the AFHQ dataset for experiments. We divide CelebA-HQ into two domains, male and female, and the AFHQ dataset is a dataset divided into three domains, cat, dog, and wildlife, each providing 5000 high-quality images at 512 × 512 resolution. We do not use additional information (other than domain labels), and learn information (such as styles) without supervision. For a fair comparison, we resize all images to a resolution of 256 × 256 for experiments.

4.3. Training

All models are trained using Adamw [32], β = 0, β = 0.99. For data addition, we simply crop the image and do some normalization. For all models, we set the learning rate to 1e-4 and the weight decay to le-4. Trained on 4 pieces of 2080Ti for about 3 days.

4.4. Evaluation Metrics

We use Frechet inception distance (FID) [33] to evaluate the quality of generated images, with lower scores indicating high correlation with higher-quality images; using learned perceptual image patch similarity (LPIPS) [34] to evaluate the diversity of generated images, with higher scores indicating better diversity of generated images.

4.5. Comparison of Image Synthesis

In this section, we evaluate the performance of the silent framework on image synthesis from two aspects: latent-guided synthesis and reference-guided synthesis.

Latent-guided synthesis. Figure 7 shows a qualitative comparison with related methods on the CelebA-HQ dataset. Each method uses random noise in the latent space to produce a diverse picture output. Figures 8 and 9 are qualitative comparisons on the AFHQ dataset.

Qualitative comparison of latent guided image synthesis results on CelebA-HQ and AFHQ datasets is undertaken. Each method uses a randomly sampled latent code to transform the source image (top row) into the target domain. To learn meaningful styles, we transform latent codes, z, into domain-specific style codes, s, through a mapping network, M. After injecting style codes into a generator, E, we use a style reconstruction loss that allows the generator to generate different images in field style.

Figure 7. Using random noise to guide generation of images in CelebA-HQ.

Figure 8. Using random noise to guide generation of images in AFHQ.

Figure 9. Using random noise to guide generation of images in AFHQ.

Table 1 shows that our method outperforms some methods. We outperform comparative methods in FID vs. LPIPS in CelebA-HQ dataset, but perform worse than the best comparison method MSGAN in dataset. We speculate that it may be that when the decoder adds attention mechanism to suppress the unimportant features of the image, it may also limit the development of image diversity, so that image diversity cannot generate pictures well, according to the provided style codes s.

Table 1. Quantitative comparison of latent-guided synthesis. (The bold number indicates the best result.)

Method	CelebA-HQ		AFHQ	
	FID	LPIPS	FID	LPIPS
MUNIT	31.6	0.365	43.6	0.501
DRIT	52.3	0.176	95.4	0.328
MSGAN	33.5	0.375	61.6	**0.517**
Ours	**18.6**	**0.423**	**28.6**	0.412

Reference-guided synthesis. Figure 10 is the reference guide image synthesis result on CelebA-HQ. The source and reference images in the first row and column are real images, and the rest are images generated by our proposed model. Our model learns to transform source images that reflect the style of a given reference image, following high-level semantics, such as hairstyle, makeup, beard, and age, from the reference image, while preserving the pose and identity of the source image. Note that the images in each column share a logo with a different style, and the images in each row share a style with a different identity.

Figure 10. Reference guided image synthesis in CelebA-HQ (The first column is the original image; the first row is the reference image).

Figure 11 is a qualitative comparison of the synthetic results of reference guided images on the CelebA-HQ and AFHQ datasets. Each method transforms the source image into the target domain, reflecting the style of the reference image. The first column is the source image and the second column is the reference image.

Figure 11. Comparison of reference guided syntheses.

Table 2 is a qualitative comparison with related methods, FID and LPIPS. When compared to these methods, ours performs the best; the images we generate are the best visually and have the most diversity of pictures.

Table 2. Quantitative comparison of reference-guided synthesis. (The bold number indicates the best result.)

Method	CelebA-HQ		AFHQ	
	FID	LPIPS	FID	LPIPS
MUNIT	106.8	0.178	183.6	0.197
DRIT	53.4	0.311	114.4	0.192
MSGAN	38.9	0.324	68.7	0.159
Ours	**28.1**	**0.382**	**26.6**	**0.399**

We present some of our experimental results in the Appendix A. Figure A1 is the image generated using cycle consistency, Figure A2 shows the result of using random noise to guide generated images, and Figure A3 is the guided synthesis of images with reference images.

5. Conclusions

The framework we propose mainly solves the problem of converting images from one domain to different images in the target domain and supports multiple target domains in image conversion. The results show that our model can generate pictures of diverse styles in multiple domains, including some shown previously [25,28,30]. However, the quality of the generated images can be further improved and the diversity can be richer. In our design, the number of different domains does not affect the quality of the output and model performance is not much different when using only a single-domain dataset compared to using a multi-domain dataset. In addition, the use of skip connections and CBAM attention modules can also make the generated images have higher visual quality, but we speculate that adding CBAM in the generator decoding part may affect the diversification of the generated images, so there is still much room for improvement. We hope that our work can be applied to the development of image translation programs in multiple domains.

Author Contributions: Funding acquisition, Z.Q.; Resources, Z.Q.; Supervision, Z.Q.; data curation, Z.Y.; Writing—original draft, Z.Y.; Writing—review and editing, Z.Y.; visualization, Z.Y. All authors have read and agreed to the published version of the manuscript.

Funding: This work was supported by Hainan Provincial Key Research and Development Program 361 (NO: ZDYF2020018), Hainan Provincial Natural Science Foundation of China (NO: 2019RC100), Haikou key research and development program (NO: 2020-049).

Institutional Review Board Statement: Not applicable.

Informed Consent Statement: Not applicable.

Data Availability Statement: The datasets used in this paper are public datasets. The CelebA-HQ could be found from https://drive.google.com/drive/folders/0B4qLcYyJmiz0TXY1NG02bzZVRGs (accessed on 22 March 2022). And the AFHQ could be found from https://github.com/clovaai/stargan-v2/blob/master/README.md#animal-faces-hq-dataset-afhq (accessed on 22 March 2022).

Conflicts of Interest: The authors declare no conflict of interest.

Appendix A

Figure A1. Images generated by cycle consistency (The first and second rows are real pictures).

Figure A2. Using random noise to guide generated images s (The first row is the real picture).

Figure A3. Reference guided synthesis images (The first column is the original image, the first row is the reference image).

References

1. Goodfellow, I.J.; Pouget-Abadie, J.; Mirza, M.; Xu, B.; Warde-Farley, D.; Ozair, S.; Courville, A.; Bengio, Y. Generative adversarial nets. In Proceedings of the Annual Conference on Neural Information Processing Systems 2014, Montreal, QC, Canada, 8–13 December 2014; pp. 2672–2680.
2. Mirza, M.; Osindero, S. Conditional generative adversarial nets. *arXiv* **2014**, arXiv:1411.1784.
3. Reed, S.; Akata, Z.; Yan, X.; Logeswaran, L.; Schiele, B.; Lee, H. Generative adversarial text to image synthesis. *Int. Conf. Mach. Learn.* **2016**, *48*, 1060–1069.
4. Dash, A.; Gamboa, J.C.B.; Ahmed, S.; Liwicki, M.; Afzal, M.Z. TAC-GAN—Text conditioned auxiliary classifier generative adversarial network. *arXiv* **2017**, arXiv:1703.06412.
5. Zhang, H.; Xu, T.; Li, H.; Zhang, S.; Wang, X.; Huang, X.; Metaxas, D.N. Stackgan: Text to photo-realistic image synthesis with stacked generative adversarial networks. In Proceedings of the IEEE International Conference on Computer Vision, Venice, Italy, 22–29 October 2017; pp. 5907–5915.
6. Zhu, J.Y.; Park, T.; Isola, P.; Efros, A.A. Unpaired Image-to-Image Translation Using Cycle-Consistent Adversarial Networks. *arXiv* **2017**, arXiv:1703.10593.
7. Yi, Z.; Zhang, H.; Tan, P.; Gong, M. Dualgan: Unsupervised dual learning for image-to-image translation. In Proceedings of the IEEE International Conference on Computer Vision, ICCV 2017, Venice, Italy, 22–29 October 2017; pp. 2868–2876. [CrossRef]
8. Choi, Y.; Choi, M.; Kim, M.; Ha, J.W.; Kim, S.; Choo, J. Stargan: Unified Generative Adversarial Networks for Multi-Domain Imageto-Image Translation. *arXiv* **2017**, arXiv:1711.09020.
9. Huang, X.; Belongie, S.J. Arbitrary style transfer in real-time with adaptive instance normalization. In Proceedings of the IEEE International Conference on Computer Vision, ICCV 2017, Venice, Italy, 22–29 October 2017; pp. 1510–1519.
10. Arjovsky, M.; Chintala, S.; Bottou, L. Wasserstein GAN. *arXiv* **2017**, arXiv:1701.07875.

11. Gulrajani, I.; Ahmed, F.; Arjovsky, M.; Dumoulin, V.; Courville, A.C. Improved training of wasserstein gans. In Proceedings of the Annual Conference on Neural Information Processing Systems 2017, Long Beach, CA, USA, 4–9 December 2017; pp. 5767–5777.
12. Isola, P.; Zhu, J.Y.; Zhou, T.; Efros, A.A. Image-to-image translation with conditional adversarial networks. In Proceedings of the 2017 IEEE Conference on Computer Vision and Pattern Recognition, CVPR 2017, Honolulu, HI, USA, 21–26 July 2017; pp. 5967–5976.
13. Kim, T.; Cha, M.; Kim, H.; Lee, J.K.; Kim, J. Learning to discover cross-domain relations with generative adversarial networks. In Proceedings of the 34th International Conference on Machine Learning, ICML 2017, Sydney, NSW, Australia, 6–11 August 2017; Volume 70, pp. 1857–1865.
14. Liu, M.Y.; Breuel, T.; Kautz, J. Unsupervised image-to-image translation networks. In Proceedings of the International Conference on Neural Information Processing Systems (NIPS), Long Beach, CA, USA, 4–9 December 2017.
15. Chen, R.; Huang, W.; Huang, B.; Sun, F.; Fang, B. Reusing discriminators for encoding: Towards unsupervised image-to-image translation. In Proceedings of the 2020 IEEE/CVF Conference on Computer Vision and Pattern Recognition, CVPR 2020, Seattle, WA, USA, 13–19 June 2020; pp. 8165–8174.
16. Kingma, D.P.; Welling, M. Auto-encoding variational bayes. In Proceedings of the 2nd International Conference on Learning Representations, ICLR 2014, Banff, AB, Canada, 14–16 April 2014.
17. Liu, M.; Tuzel, O. Coupled generative adversarial networks. In Proceedings of the Annual Conference on Neural Information Processing Systems 2016, Barcelona, Spain, 5–10 December 2016; pp. 469–477.
18. Yang, S.; Jiang, L.; Liu, Z.; Loy, C.C. Pastiche Master: Exemplar-Based High-Resolution Portrait Style Transfer. In Proceedings of the IEEE/CVF Conference on Computer Vision and Pattern Recognition 2022, New Orleans, LA, USA, 19–24 June 2022; pp. 7693–7702. [CrossRef]
19. Karras, T.; Laine, S.; Aila, T. A style-based generator architecture for generative adversarial networks. In Proceedings of the IEEE Conference on Computer Vision and Pattern Recognition, CVPR 2019, Long Beach, CA, USA, 16–20 June 2019; pp. 4401–4410.
20. Yuan, Q.L.; Zhang, H.L. RAMT-GAN: Realistic and accurate makeup transfer with generative adversarial network. *Image Vis. Comput.* **2022**, *120*, 104400. [CrossRef]
21. Li, T.; Qian, R.; Dong, C.; Liu, S.; Yan, Q.; Zhu, W.; Lin, L. Beautygan: Instance-level facial makeup transfer with deep generative adversarial network. In Proceedings of the 2018 ACM Multimedia Conference on Multimedia Conference, MM 2018, Seoul, Korea, 22–26 October 2018; pp. 645–653.
22. Woo, S.; Park, J.; Lee, J.Y.; Kweon, I.S. CBAM: Convolutional block attention module. In Proceedings of the Computer Vision—ECCV 2018—15th European Conference, Munich, Germany, 8–14 September 2018; Ferrari, V., Hebert, M., Sminchisescu, C., Weiss, Y., Eds.; Proceedings, Part VII; ser. Lecture Notes in Computer Science. Springer: Berlin/Heidelberg, Germany, 2018; Volume 11211, pp. 3–19.
23. Liu, M.Y.; Huang, X.; Mallya, A.; Karras, T.; Aila, T.; Lehtinen, J.; Kautz, J. Few-shot unsupervised image-to-image translation. In Proceedings of the 2019 IEEE/CVF International Conference on Computer Vision, ICCV 2019, Seoul, Korea, 27 October–2 November 2019; pp. 10550–10559.
24. Mescheder, L.; Geiger, A.; Nowozin, S. Which training methods for gans do actually converge? In Proceedings of the 35th International Conference on Machine Learning, ICML 2018, Stockholmsmassan, Stockholm, Sweden, 10–15 July 2018; Volume 80, pp. 3478–3487.
25. XHuang, X.; Liu, M.Y.; Belongie, S.; Kautz, J. Multimodal unsupervised image-to-image translation. In Proceedings of the Computer Vision—ECCV 2018—15th European Conference, Munich, Germany, 8–14 September 2018; Ferrari, V., Hebert, M., Sminchisescu, C., Weiss, Y., Eds.; Proceedings, Part III; ser. Lecture Notes in Computer Science. Springer: Berlin/Heidelberg, Germany, 2018; Volume 11207, pp. 179–196.
26. Zhu, J.; Zhang, R.; Pathak, D.; Darrell, T.; Efros, A.A.; Wang, O.; Shechtman, E. Toward multimodal image-to-image translation. In Proceedings of the Annual Conference on Neural Information Processing Systems 2017, Long Beach, CA, USA, 4–9 December 2017; pp. 465–476.
27. Donahue, J.; Simonyan, K. Large scale adversarial representation learning. In Proceedings of the Annual Conference on Neural Information Processing Systems 2019, NeurIPS 2019, Vancouver, BC, Canada, 8–14 December 2019; pp. 10541–10551.
28. Mao, Q.; Lee, H.Y.; Tseng, H.Y.; Ma, S.; Yang, M.H. Mode seeking generative adversarial networks for diverse image synthesis. In Proceedings of the IEEE Conference on Computer Vision and Pattern Recognition, CVPR 2019, Long Beach, CA, USA, 16–20 June 2019; pp. 1429–1437.
29. Yang, D.; Hong, S.; Jang, Y.; Zhao, T.; Lee, H. Diversity-sensitive conditional generative adversarial networks. In Proceedings of the 7th International Conference on Learning Representations, ICLR 2019, New Orleans, LA, USA, 6–9 May 2019. Available online: https://openreview.net/forum?id=rJliMh09F7 (accessed on 25 March 2022).
30. Lee, H.Y.; Tseng, H.Y.; Huang, J.B.; Singh, M.; Yang, M.H. Diverse imageto-image translation via disentangled representations. In Proceedings of the Computer Vision—ECCV 2018—15th European Conference, Munich, Germany, 8–14 September 2018; Ferrari, V., Hebert, M., Sminchisescu, C., Weiss, Y., Eds.; Proceedings, Part I; ser. Lecture Notes in Computer Science. Springer: Berlin/Heidelberg, Germany, 2018; Volume 11205, pp. 36–52.
31. Karras, T.; Aila, T.; Laine, S.; Lehtinen, J. Progressive growing of gans for improved quality, stability, and variation. In Proceedings of the 6th International Conference on Learning Representations, ICLR 2018, Vancouver, BC, Canada, 30 April–3 May 2018.

32. Loshchilov, I.; Hutter, F. Decoupled weight decay regularization. In Proceedings of the 7th International Conference on Learning Representations, ICLR 2019, New Orleans, LA, USA, 6–9 May 2019.
33. Heusel, M.; Ramsauer, H.; Unterthiner, T.; Nessler, B.; Hochreiter, S. Gans trained by a two time-scale update rule converge to a local nash equilibrium. In Proceedings of the Advances in Neural Information Processing Systems 30: Annual Conference on Neural Information Processing Systems 2017, Long Beach, CA, USA, 4–9 December 2017; pp. 6626–6637.
34. Zhang, R.; Isola, P.; Efros, A.A.; Shechtman, E.; Wang, O. The unreasonable effectiveness of deep features as a perceptual metric. In Proceedings of the 2018 IEEE Conference on Computer Vision and Pattern Recognition, CVPR 2018, Salt Lake City, UT, USA, 18–22 June 2018.

electronics

MDPI

Article

An Improved RandLa-Net Algorithm Incorporated with NDT for Automatic Classification and Extraction of Raw Point Cloud Data

Zhongli Ma, Jiadi Li *, Jiajia Liu *, Yuehan Zeng, Yi Wan and Jinyu Zhang

School of Automation, Chengdu University of Information Technology, Chengdu 610225, China
* Correspondence: lijiadi96@163.com (J.L.); liujj@cuit.edu.cn (J.L.)

Abstract: A high-definition map of the autonomous driving system was built with the target points of interest, which were extracted from a large amount of unordered raw point cloud data obtained by Lidar. In order to better obtain the target points of interest, this paper proposes an improved RandLa-Net algorithm incorporated with NDT registration, which can be used to automatically classify and extract large-scale raw point clouds. First, based on the NDT registration algorithm, the frame-by-frame raw point cloud data were converted into a point cloud global map; then, the RandLa-Net network combined random sampling with a local feature sampler is used to classify discrete points in the point cloud map point by point. Finally, the corresponding point cloud data were extracted for the labels of interest through numpy indexing. Experiments on public datasets senmatic3D and senmatickitti show that the method has excellent accuracy and processing speed for the classification and extraction of large-scale point cloud data acquired by Lidar.

Keywords: NDT registration; map building; RandLa-Net; random sampling; semantic segmentation

Citation: Ma, Z.; Li, J.; Liu, J.; Zeng, Y.; Wan, Y.; Zhang, J. An Improved RandLa-Net Algorithm Incorporated with NDT for Automatic Classification and Extraction of Raw Point Cloud Data. *Electronics* **2022**, *11*, 2795. https://doi.org/10.3390/electronics11172795

Academic Editors: Chunwei Tian, Wenqi Ren and Yudong Liang

Received: 23 June 2022
Accepted: 1 September 2022
Published: 5 September 2022

1. Introduction

Lidar is an application system that integrates laser technology, global positioning system technology, and inertial measurement unit (IMU) technology, and it is also an important sensor for environmental perception in autonomous driving systems. The proper processing of the collected raw point cloud data can provide important data for a high-definition map, and furthermore, as a scarce resource and just-needed product in the field of autonomous driving, high-definition road maps also play a central role in its entire field, which helps cars perceive complex road information (such as slope, curvature, heading, etc.) in advance, and to make the right decisions combine intelligent path planning [1].

Using the point cloud data obtained by the vehicle laser to build high-definition road maps mainly includes three steps: data acquisition, point cloud data processing and drawing. Among them, point cloud data processing is the key step to ensure the accuracy and quality of the point cloud. The traditional method firstly maps the road surface according to discriminative point cloud mapping features, and then manually labeled as point clouds for classification. Some scene elements and blind areas that cannot be recognized and scanned by point cloud classification algorithm still need to be accounted for by additional manual mining. In short, traditional methods have shortcomings, such as needing a long production cycle due to a large amount of control and measurement tasks, and thus they cannot meet the needs of making high-definition maps very well. Traditional methods also have difficulty in identifying path elements when dealing with some road sections [2].

An important purpose of raw point cloud data classification and extraction research is to quickly and accurately distinguish points in roads, cars, people, traffic signs and classes of interest; however, most of the defects of traditional methods come from manual point cloud classification. In response to this problem, many scholars have begun to pay attention to the research on the automatic classification and extraction of point clouds, which can greatly improve the timeliness of processing point cloud data when making high-definition maps. Deep learning has shown excellent performance in object classification, extraction and recognition in computer vision, but it cannot directly deal with such discrete and irregular point cloud data and it can only process the point cloud data frame by frame. Based on the above description, there will be some problems as follows:

(1) The quantity of point cloud data is very large. Current point cloud semantic segmentation algorithms take a long time to train on large-scale data sets.
(2) In order to quickly obtain complete road sections, trees and other components to make high-definition maps, most point cloud semantic segmentation algorithms segment data frame by frame to reduce the amount of computation. Therefore, it is necessary to construct the original data into a complete point cloud map, and then perform semantic segmentation. In this way, the complete information of the road section can be obtained.
(3) The point cloud semantic segmentation is designed to assign labels to each point and classify points, though it is impossible to obtain point cloud sets from unclassified point data. In other words, the region of interest cannot be directly extracted, it is necessary to make labels for the point clouds, and then extract the corresponding point cloud data in turn according to the corresponding index.

To solve the above problems, the aim of this paper was to find an automatic classification and extraction method for large-scale point cloud data, and our method can achieve the fast and efficient classification and extraction of a large number of original point cloud data for the construction of high-definition maps of autonomous driving systems. Furthermore, our method can also provide accurate data. The contributions of our method are as follows:

(1) Based on analyzing and comparing the existing registration and semantic segmentation methods, we chose to integrate NDT registration into the RandLa semantic segmentation algorithm to process original point cloud data.
(2) We tested the two algorithms on the public datasets KITTI [3], Semantic3D [4], and SemanticKITTI [5], respectively, and we chose to fuse the two algorithms according to the experimental results.
(3) We give the description of the basic process of the improved RandLa-Net (network structure based on random sampling and local feature aggregation) algorithm incorporated with NDT, and use the improved method to perform many experiments on public datasets. The experimental results show that the data processed by our method can be directly used for the construction of a high-definition map.

2. Related Work

2.1. Methods for Point Cloud Data Registration and Semantic Segmentation

The traditional algorithms of point cloud registration are roughly divided into two categories: coarse registration and fine registration [6]. Coarse registration refers to the registration by calculating an approximate rotation and translation matrix between two point clouds. This method is generally used when the relative positional relationship between two point clouds is unknown. The fine registration refers to making the rotation and translation matrix more accurate by calculation when the rotation and translation matrix is known. Most of the point cloud information collected by vehicle radar is coarse registration, which mainly includes:

(1) A registration method based on local feature description. The point feature histograms (PFH) methods proposed by Rusu et al. [7] use point feature histograms to characterize the local geometry of 3D points for registration;
(2) Method based on probability distribution. The normal distributions transform (NDT) algorithm proposed by Biber [8] et al. is a rough registration method that uses range scanning first, that is, the normal distribution transformation is performed after point

cloud matching. In recent years, point cloud registration methods based on deep learning have also been widely proposed and applied. Aoki [9] et al. used PointNet to map the found feature points to a high-dimensional space, and then regarded the vector formed by each feature point as an image in the high-dimensional space, and finally used the traditional image registration algorithm (Lucas-Kanada, LK) [10] for point cloud registration. Wang [11] et al. extracted the features of the point cloud to be registered; they used the improved transformer network to merge the information between the point clouds, calculated the soft matching between the point clouds, and then used the differentiable singular value decomposition module to extract the rigid body changes for point clouds. Here, cloud registration [12] was combined with keypoint detection to solve the non-convexity and local registration problems of registration.

One type of method is to improve the traditional point cloud semantic segmentation algorithms, such as random sample consensus (RANSAC), density-based spatial clustering of applications with noise (DBSCAN), and region growth algorithm (region growing) [13–15]. These algorithms have high requirements for the quality of point cloud data and have low accuracy and slow speed when processing large-scale point cloud data. The other type of method is based on deep learning network, which mainly includes:

(1) Projection-based network [16]: due to the inhomogeneity of point cloud data, it is impossible to directly use the convolutional neural network on point cloud. To make use of two-dimensional convolutional neural network, the projection-based network chooses to project a three-dimensional point cloud onto a two-dimensional image and then input it into the network [17], but the projection process may lead to the loss of geometric information, and the method lacks the ability of non-local geometric features [18];
(2) Voxelization-based network [19]: for the disorder and irregularity of the point cloud, the disordered point cloud is voxelized into an ordered voxel block, and then a three-dimensional convolutional neural network is used to process the ordered voxel block. The main limitation of the method is that the computational cost is too high, especially when dealing with large-scale point clouds. It cannot meet practical applications [20], and the volume setting of the voxel block will affect the final segmentation effect. Furthermore, due to the sparsity of the point cloud, there will be empty voxel blocks generated, wasting the computational cost [21];
(3) Network based on the neural architecture: Hu Q [22] et al. designed an efficient neural architecture network structure based on random sampling and local feature aggregation (RandLa-Net). It can directly process large-scale point cloud data without any preprocessing.

2.2. Data Format Requirements in Unmanned High-Definition Map Construction

In order to make the high-definition electric urban map, a designed unmanned vehicle (shown in Figure 1) equipped with two 32-line lidars was used to collect the point cloud data of some road sections, and the sampling frequency of this lidar is 10 Hz. In order to restore the real road, we need to extract the feature information of the road from the large amount of point cloud data collected by this lidar, such as roads, vehicles, trees, buildings and pedestrians.

Figure 1. Unmanned vehicle with two 32-line lidars.

The construction of high-definition maps requires the classification of road point features by class and the three-dimensional coordinates of each point. The coordinates are listed in $(x,\ y,\ z)$ format, and the coordinate system is based on the starting position of the driverless car as the origin.

Because there are a lot of complex targets in the point clouds and the number of data collected is large, not all the collected road section information have a unified initial position, so the registration method based on the deep learning registration method is difficult and time-consuming to train. At the same time, although the RandLa-Net algorithm can effectively segment the point cloud map, the format of the single frame data output is '.label', which cannot be directly used to build the high-definition electric urban map.

Based on the above analysis, we integrated the traditional mature NDT registration algorithm into the RandLa-Net algorithm, so that the results of the semantic segmentation are more suitable for the construction of high-definition maps.

3. Creation of Global Map of Point Cloud Based on NDT

3.1. Registration Algorithm Based on NDT

NDT is normal distribution transformation. After gridding the reference point, the normal distribution transformation is performed one by one to complete the modeling of all reconstructed points. The specific operations are as follows.

First, the point cloud space is divided into cells with the same size according to certain rules.

Then, the following actions are performed on each cell:

Step1. Collect all points contained in this box: $X_i = 1 \ldots n$;

Step2. Calculate the average:

$$q = \frac{1}{n}\sum_i X_i \tag{1}$$

Step3. Calculate the covariance matrix:

$$\Sigma = \frac{1}{n}\sum_i (X_i - q)(X_i - q)^t \tag{2}$$

Step4. The probability of measuring a sample at point x contained in this cell is now modeled by the normal distribution N (q, Σ):

$$P(x) \sim exp\left(-\frac{(x-q)^t\Sigma^{-1}(x-q)}{2}\right) \tag{3}$$

Figure 2 shows the effect after meshing the 3D point cloud, the original frame and the visualization after NDT. The visualization is created by evaluating the probability density of each point, with bright areas indicating high probability density. Then, the normal distribution transformation is used for the registration between the two frames of point clouds, and the spatial mapping T between the radar coordinate systems of the two frames of point clouds is given by:

$$T : \begin{matrix} x' \\ y' \end{matrix} = \begin{pmatrix} cos(\varphi) & -sin(\varphi) \\ sin(\varphi) & cos(\varphi) \end{pmatrix}\begin{pmatrix} x \\ y \end{pmatrix} + \begin{pmatrix} t_x \\ t_y \end{pmatrix} \tag{4}$$

where (t_x, t_y) refers to the original position of the reference frame, (x', y') is the position of the frame to be registered, T describes the translation and rotation relationship between the two, φ is the rotation angle, and x and y are the translation distances.

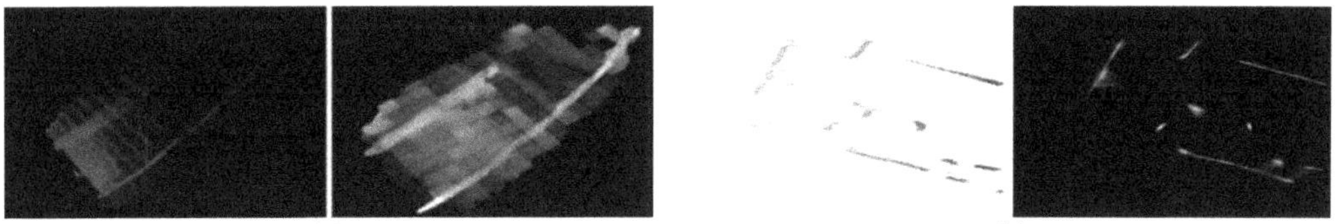

Figure 2. The original data and the resulting probability density: (**a**) mesh the point cloud; and (**b**) the original scan and the resulting probability density.

The specific operations of registration are as follows:

Step1. Build the NDT based on the first frame scan;

Step2. Estimate initialization parameters;

Step3. For each sample to be registered, map the reconstructed point into the coordinate system of the reference frame according to the parameters;

Step4. Determine the corresponding normal distribution for each mapped point;

Step5. Determine the score of the parameter by evaluating the distribution of each mapped point and sum up the results;

Step6. Return to step 3 until the convergence criteria are met and the registration is completed.

The score is calculated as follows:

$$score(p) = \Sigma_i exp\left(\frac{-(x' - q_i)^t \Sigma_i^{-1}(x_i' - q_i)}{2}\right) \quad (5)$$

where p is a vector of parameters to be estimated, x_i is the point in the second frame of point cloud data, x_i' is the point x_i mapped to the coordinate system of the first frame point cloud data according to the parameter p, that is, $x_i' = T(x_i, p)$. Σ_i and q_i are the covariance matrix of the point cloud data in the first frame and the mean value of the normal distribution corresponding to the point x_i'. A mapping according to p can be considered optimal if the sum of the normal distributions of all points x_i' evaluated using the parameters Σ_i and q_i are at maximum, that is, the sum of the scores of p is optimal.

The overall process of point cloud data registration based on NDT is shown in Figure 3.

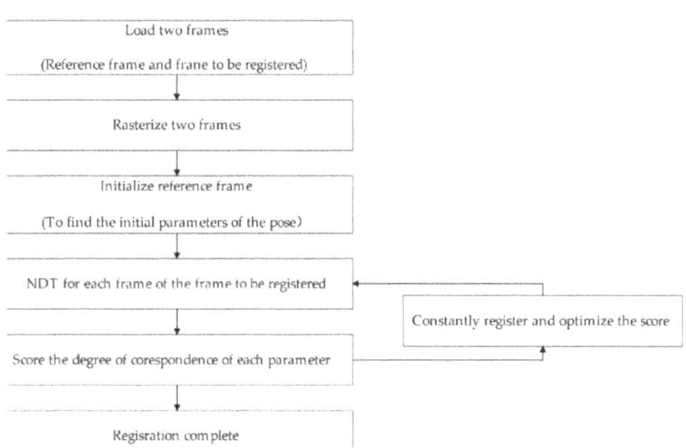

Figure 3. The flow of the NDT registration algorithm.

3.2. Point Cloud Map Creation Based on NDT

The point cloud map is created to avoid frame-by-frame processing for subsequent point cloud classification, and to improve the efficiency of the automatic classification of point cloud data. The main task of point cloud mapping is to use the collected point cloud data frame by frame to build a complete point cloud map. The algorithm flow is shown in Figure 4.

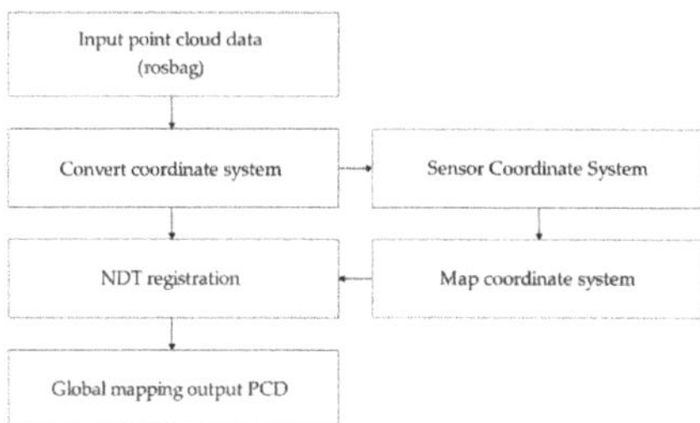

Figure 4. The process of point cloud mapping.

In Figure 4, the coordinate system transformation adopts the TF coordinate system transformation. If the coordinates of a point in the radar coordinate system are $P_L\ (X_L, Y_L)$, then the coordinates of a certain point in the map after conversion are $P_M\ (x_m, y_m)$, and the map coordinate system is a fixed coordinate system. The coordinate system is the same as the world coordinate system, then:

$$R \times P_L + t = P_M \tag{6}$$

where R is the transformation matrix, so that the poses of the two coordinate systems are consistent.

$$R = \begin{bmatrix} cos(\theta) & -sin(\theta) \\ sin(\theta) & cos(\theta) \end{bmatrix} \tag{7}$$

In Formula (6), t is the position where the origin of the sensor coordinate system is mapped to the map coordinate system. In general, the initial position of the sensor is the origin of the map, and t = (x_o, y_o) can be set directly.

In Formula (7), θ is the heading angle during the driving process. According to the right-hand rule, the counterclockwise rotation around the z axis of the map coordinate system is positive, then θ can be directly brought into the transformation matrix, and the distance from the sensor coordinates to the map coordinates is transformed to:

$$P_M = \begin{pmatrix} cos(\theta) & -sin(\theta) \\ sin(\theta) & cos(\theta) \end{pmatrix} \times P_L + \begin{pmatrix} x_o \\ y_o \end{pmatrix} \tag{8}$$

The relationship between the sensor coordinate system and the map coordinate system is shown in Figure 5.

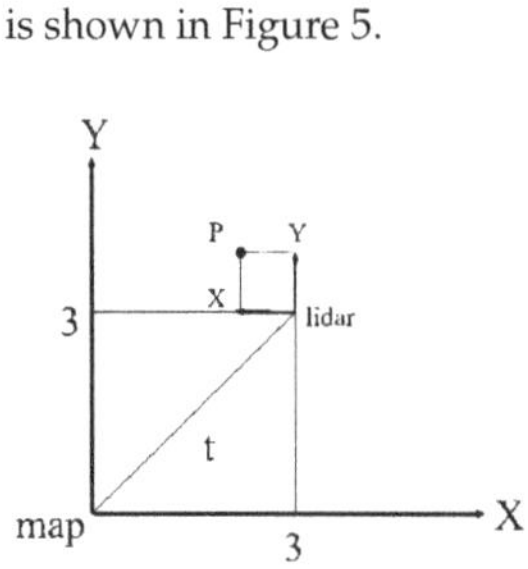

Figure 5. The relationship between the sensor coordinate and the map coordinate system.

3.3. *Test of NDT Global Mapping*

3.3.1. Dataset Selection

The KITTI dataset consists of point cloud data collected by 64-line 3D Lidar combined with two gray-scale cameras, two color cameras and four optical lenses, and the sampling frequency of Lidar is 10 Hz. The entire dataset consists of 389 pairs of stereo images and optical flow images, and more than 200 k 3D annotated objects. The KITTI dataset is often used in 3D object detection and point cloud segmentation, the part samples of dataset are shown in Figure 6.

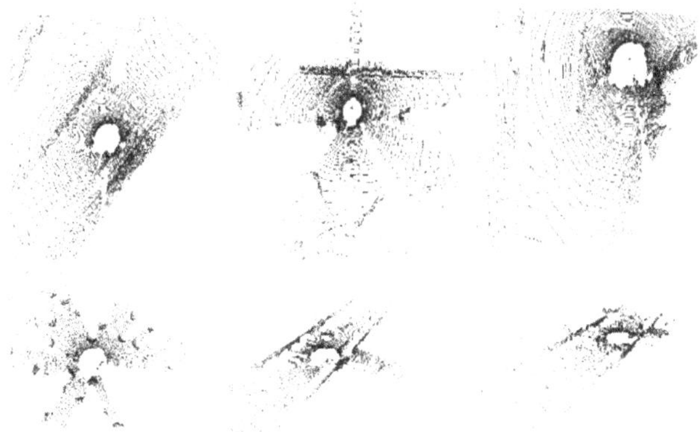

Figure 6. Partial image examples in the dataset.

3.3.2. Test Results

The CPU model of the computer is i7-10700k, the GPU model is NVIDIA RTX3090, and the memory size is 30 GB; the operating system is Ubuntu18.04, and the operating platform is Pycharm and PCL 1.10.0. The pseudocode for NDT registration is shown in Figure 7.

Algorithm 1 Register scan X to reference scan Y using NDT

Input: $X = x_i \in X$; $Y = y_i \in Y$
Output: score(p)

function M(y) Funs(X, Y, P):
    Initialisation
    allocate cell structure
    for all points Y **do**
        $findthecellb_i \in \beta$ *that contains* Y
        $store\ y_i\ in\ b_i$
        $y' = y'_1, \ldots, y'_n$
        $\sum = \frac{1}{n} \sum_i (X_i - q)(X_i - q)^t$
    end for

    while not converged **do**
        score(p) = 0
        for all points X **do**
            $score\,(P) = score + \sum_i \exp(\frac{-(x_i - q_i)^t \sum_i^{-1} (x_i - q_i)}{2})$
        end for
        update score(P)
    end while
    return score(P)

Figure 7. Pseudocode for NDT registration algorithm.

The point cloud global mapping is performed on the test sets 11–21 of KITTI and the results are shown in Figure 8, where we can see that the point cloud global mapping is very densely reconstructed by NDT algorithm, and we can clearly see the vehicles on the road and the trees and buildings on both sides, even in the unsegmented state. In addition, due to the NDT registration using a one-time initialization, the process of algorithm execution does not need to consume much computing power to calculate the nearest search matching point, and the registration only takes 0.18 s per meter distance.

Figure 8. The effect of point cloud mapping.

4. Point Cloud Semantic Segmentation Based on RandLa-Net

4.1. Point Cloud Semantic Segmentation Algorithm Based on RandLa-Net

Point cloud semantic segmentation is to add semantic labels for each point, and to classify point clouds into different point subsets, and to make sure the same point cloud set has similar features, such as vehicles, roads, or pedestrians.

The semantic segmentation of a point cloud based on RandLa-Net: first, reduce the density and computational cost of the point cloud by random sampling, and then use the local feature aggregator to collect the features of the point cloud so as to avoid losing some important feature information of the point cloud due to random sampling. Finally, these features are aggregated and the point cloud is classified so that each point has corresponding label information. The specific segmentation process is shown in Figure 9.

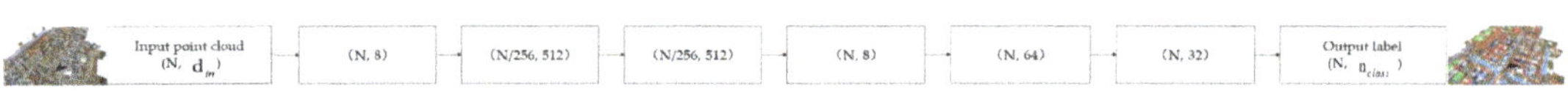

Figure 9. The process of semantic segmentation.

In Figure 9, (N, D) represent the point cloud number and feature dimension, respectively. FC represents the fully connected layer, LFA represents local feature aggregation, RS represents the random sampling, MLP represents shared multilayer perceptron and US represents upsampling.

4.1.1. Random Sampling

RS is to take n points from N points as samples, where each point has the same probability of being selected, and there is no special correlation between any two points. Its computational complexity is O (1). Compared with farthest point sampling and inverse density importance sampling, random sampling is the most computationally efficient, which only takes 0.004 s to process 106 points [22].

To evaluate the sampling efficiency of common types of samplings including farthest point sampling (FPS), inverse density importance sampling (IDIS), random sampling (RS), generator-based sampling (GS), continuous relaxation-based sampling (CRS) and policy gradient-based sampling (PGS), each of the above sampling methods is tested with point cloud data with the numbers of 103, 104, 105 and 106 in turn. Point cloud data generally need to be downsampled five times, and each downsampling on a single GPU only retains 1/4 of the original points. The time and memory consumption of each sampling method are compared, and the results are shown in Figure 10 [23], the dashed lines represent the estimated value of the memory consumption.

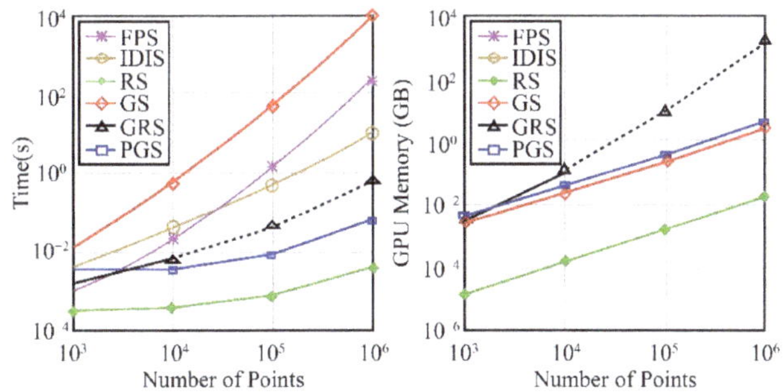

Figure 10. Efficiency comparison of different sampling methods.

4.1.2. Local Feature Aggregator

In order to retain the important feature information of the next point, the algorithm performs local feature aggregation after random sampling. This local feature aggregator is applied once at each point, which consists of three parts: (1) local spatial encoder (LocSE); (2) attention pooling layer; and (3) dilated residual block. The specific network structure is shown in Figure 11.

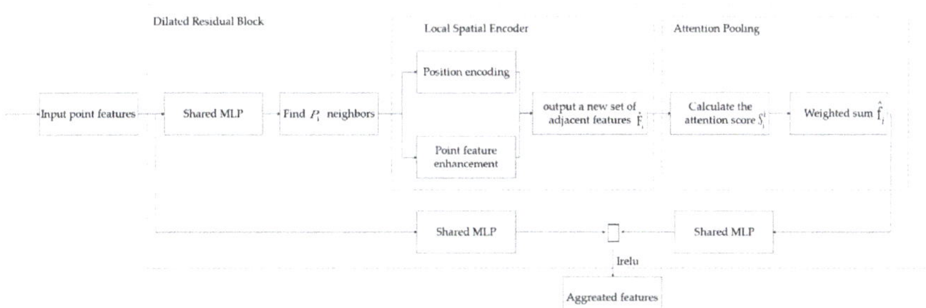

Figure 11. Local feature aggregator network structure.

(1) Local Spatial Encoder

Given a point cloud P and the features of all points, this local space encoder stores the xyz coordinates of all adjacent points so that the features of the corresponding points also have corresponding coordinate positions. The local spatial encoder can also observe geometric patterns in blocks or regions, and the entire network can efficiently learn the complex local structure of point cloud data. Specific steps are as follows:

Step1. Find neighbors.

Step2. Position encoding of relative points. For each nearest point $\{p_i^1 \dots p_i^k \dots p_i^K\}$ of center point p_i, the corresponding positions are encoded as follows:

$$r_i^k = MLP\left(p_i \oplus p_i^k \oplus \left(p_i - p_i^k\right) \oplus \|p_i - p_i^k\|\right) \tag{9}$$

where $\oplus$ is the connection operation and $\|\cdot\|$ represents the calculation of the Euclidean distance between adjacent points and a given point p .

Step3. Point feature enhancement. For each adjacent point p_i^k, concatenate the encoded relative point position r_i^k with its corresponding point feature f_i^k to obtain an enhanced feature vector $\hat{f}_i^k$.

Step4. The output is a new set of adjacent features:

$$\hat{F}_i = \left\{ \hat{f}_i^1 \ldots \hat{f}_i^k \ldots \hat{f}_i^K \right\} \tag{10}$$

(2) Attention Pooling Layer

This module is used to aggregate adjacent point features $\hat{F}_i$ and uses the attention mechanism to learn important local features spontaneously. The attention mechanism consists of the following parts:

Part1. Calculate the attention score. The formula is as follows:

$$s_i^k = g\left(\hat{f}_i^k, W\right) \tag{11}$$

where $g()$ represents a shared function to learn a unique attention score for each feature and W is the learnable weight of the shared MLP.

Part2. The formula for the weighted summation is as follows:

$$\hat{f}_i = \sum_{K=1}^{K} \left(\hat{f}_i^k \cdot s_i^k\right) \tag{12}$$

(3) Dilated Residual Block

Finally, in order to preserve an important feature information of all points before sampling as much as possible, this algorithm uses multiple local spatial encoders and expands the residual block formed in the stack of attention-eating layers and skip connections.

4.2. Point Cloud Semantic Segmentation Test Based on RandLa-Net

4.2.1. Dataset Selection

The experiment was conducted on the SemanticKITTI dataset. We selected the point cloud data from the first sequence to tenth sequence as the training set, and the point cloud data from the eleventh sequence to twenty-first sequence as the test set. Annotation categories are divided into: cars, bicycles, motorcycles, trucks, other vehicles, people, cyclists, motorcyclists, roads, parking lots, sidewalks, other surfaces, buildings, fences, vegetation, tree trunks, terrain, utility poles and traffic signs—19 categories in total, which are all important components in the autonomous driving traffic environment [5].

The Semantic3D dataset is a global point cloud image of different urban scenes obtained by static scanning with advanced equipment, and the dataset has more than 4 billion points in total. The annotation categories are artificial terrain; natural terrain; high vegetation; low vegetation; buildings; landscapes; objects; and cars—all of which are the main components of the urban environment [4].

4.2.2. Algorithm Testing

The CPU model of the test computer is i7-10700k, the GPU model is NVIDIA RTX3090 and the memory size is 30 GB; the operating system is Ubuntu18.04, the platform for deep learning uses the TensorFlow, the initial learning rate is set to 0.01 and it is reduced by 5% after each epoch. The number of closest points K is set to 16, the batch processed per iteration is set to 8 and a fixed number of points (approximately 10^5) are sampled as input. The SemanticKITTI and Semantic3D datasets are used as the training and testing sets, respectively.

Table 1 shows the results of the training with different approaches on the semantickitti dataset. The average intersection ratio (mIoU) of all categories is used as a quasi-index, and the parameter column refers to the number of network parameters in the algorithm. It can be seen that the mIoU of all categories obtained by RandLa-Net is significantly better than that of other algorithms, and RangeNet53++ has the best segmentation accuracy for small targets such as traffic signs and bicycles [24], which is because the network parameters of RangeNet53++ [24] are more than 40 times higher than those of RandLa-Net.

Table 1. Quantitative results of different approaches on semantickitti.

Methods	PointNet	PointNet++	SqueezeSeg	DarkNet21Seg	RangeNet53++	RandLa-Net
mIoU(%)	14.1	19.8	28.8	45.6	52.1	53.7
parameter (M)	3	6	1	25	50	1.24
road	61.6	71	85.3	91.4	91.8	91.7
side-walk	35.5	41.3	54.1	73	74.2	77.1
parking	15.6	18.3	26.9	56	63.9	41.2
other-ground	1.2	5.2	4.4	26.4	27.8	38.9
building	41.2	61.5	56.3	81.9	87.4	88.2
car	46.1	53.7	68.5	85.1	90.4	93.3
truck	0.1	0.9	3.3	18.1	24.7	40.1
bicycle	1.3	1.9	15	26.2	25.7	15.5
motorcycle	0.3	0.2	4.1	26.5	34.4	28.8
other-vehicle	0.7	0.2	3.5	15.6	22.9	38.5
vegetation	30	46.5	60	77.6	80.5	84.5
trunk	4.6	13.8	24.3	47.4	55.1	40.1
terrain	17.6	30	53.7	63.6	64.5	72.1
person	0.2	0.9	12.9	31.8	38.3	53.4
bicyclist	0.2	1	13.1	33.6	38.8	53.36
motorcyclist	0	0	0.9	4	4.8	7.2
fence	12.9	16.9	29.9	52.3	58.6	44.5
pole	2.4	6	17.8	36	47.9	51.3
traffic sign	3.7	8.9	24.5	50	55.9	38.6

Table 2 shows the train results of different approaches on Semantic3D, and the mean cross-over-union ratio (mIoU) and overall accuracy (OA) for all classes were used as standard metrics.

Table 2. Quantitative results of different approaches on Semantic3D.

Methods	SnapNet	ShellNet	GACNet	KPConv	RandLa-Net
mIoU(%)	59	69.1	70.7	74.6	77.4
OA(%)	88.6	91.8	94	92.9	94.8
man-made terrain	81	86.4	96.4	90.9	94.2
natural terrain	77.2	77.7	92.6	82.8	91.4
high vegetation	79.7	88.5	87.9	84.1	82.9
low vegetation	22.9	60.6	44	47.8	52
buildings	91.1	94.2	83.2	94.5	94.7
hard scape	18.4	37.3	31	40	54.9
scanning artefacts	37.3	43.5	63.5	77.3	70.9
Cars	64.4	77.8	76.2	79.7	76.9

RandLa-Net significantly outperforms other algorithms in mIoU and OA. The test accuracy of different methods for eight kinds of targets can be seen in Table 2, and the test accuracy of RandLa-Net algorithm is also better than most algorithms. The point cloud segmentation effect on different datasets is shown in Figure 12.

Figure 12. The segmentation effect on different datasets: (a) frame-by-frame segmentation effect of SemanticKITTI-11; (b) frame-by-frame segmentation effect of SemanticKITTI-12; (c) panoramic split effect of Semantic3D-1; and (d) panoramic split effect of Semantic3D-2.

Table 3 shows the total time and memory consumption of different methods. It can be seen in Table 3. RandLa-Net has the shortest processing time and the most maximum inference points. Although the number of network parameters of the SPG algorithm is the least, the processing time of point clouds is very long and the overall effect is not as good as RandLa-Net, due to the complex geometric division and hypergraph construction steps. Therefore, RandLa-Net is the most efficient network.

Table 3. Efficiency of the semantic segmentation of different methods on sequence 08.

	Total Time (s)	Parameters (Millions)	Maximum Inference Points (Millions)
PointNet	192	0.8	0.49
PointNet++	9831	0.97	0.98
PointCNN	8142	11	0.05
SPG	43584	0.25	-
KPConv	717	14.9	0.54
RandLa-Net	185	1.24	1.03

5. Comprehensive Test of Automatic Classification and Extraction of Raw Point Cloud Data

5.1. The Flow of the Algorithm

The improved RandLa-Net algorithm incorporated with NDT registration can directly process the complete point cloud map, and further obtain the data required by the high-definition map. The specific algorithm flow is shown in Figure 13.

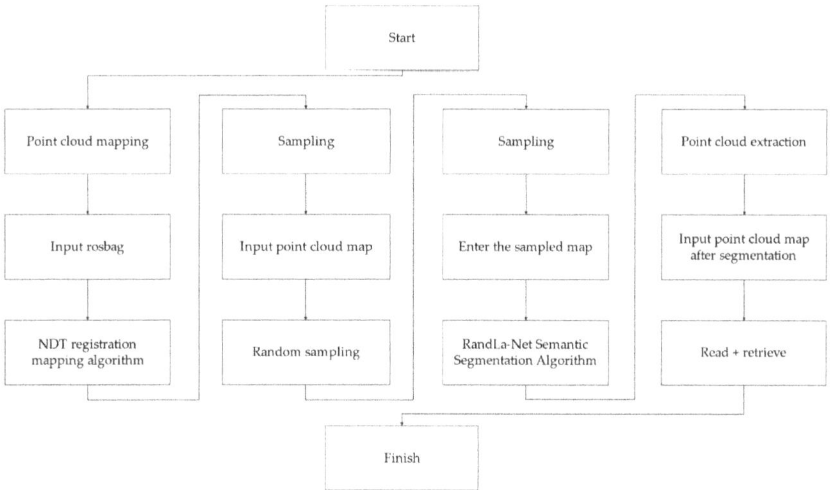

Figure 13. Process of the improved RandLa-Net algorithm.

5.2. Test Data

Take the SemanticKITTI Dataset 03 as an example for testing. Figure 14 shows the original frame-by-frame data of the 03 point cloud.

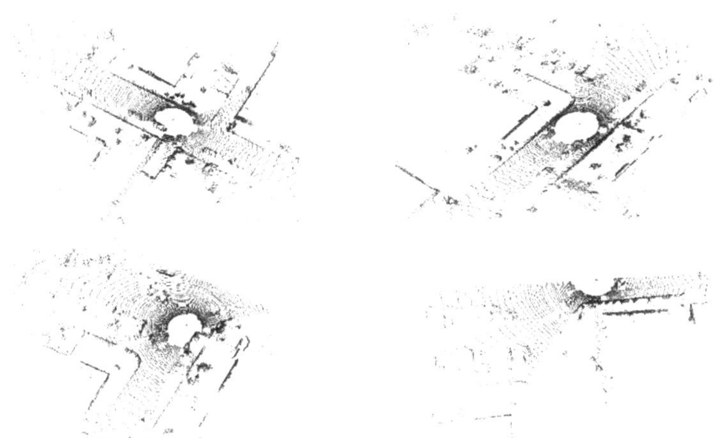

Figure 14. Original dataset.

5.3. Point Cloud Mapping

Firstly, load the rosbag of the original dataset, then use the NDT algorithm to perform the coordinate transformation and registration on the frame-by-frame point cloud data, and build a map globally. The output effect is shown in Figure 15.

Figure 15. Point cloud mapping.

5.4. Random Sampling

In order to reduce the amount of computation, the random sampling of data is required before semantic segmentation, and the sampling effect is shown in Figure 16. We can see that the number of point clouds is significantly reduced and the features become blurred. However, the local feature aggregator of the subsequent RandLa-Net algorithm will solve this problem.

Figure 16. Random sampling.

5.5. Point Cloud Semantic Segmentation

Load the bin format file of the global point cloud image, and use the RandLa-Net algorithm to semantically segment the point cloud image. Figure 17a is the visualization of semantic segmentation on the dataset using our improved RandLa-Net algorithm, and Figure 17b is the visualization of semantic segmentation on the same dataset using the original RandLa-Net algorithm. Table 4 shows the specific effect of the test, which is reflected from the indicator mIoU. We can see that the segmentation effect does not decrease the accuracy due to the change from frame-by-frame segmentation to global segmentation.

Figure 17a is the visualization effect of the global point cloud map classification using RandLa-Net fused with NDT. Figure 17b is the visualization effect of the frame-by-frame point cloud classification using the original RandLa-Net network. Figure 17c is the label information corresponding to each color in the visualization. It can be seen that the algorithm in this paper can be used to classify the point cloud map at one time, and then the required area can be directly extracted. However, since the point cloud data processed by the original RandLa-Net algorithm was not registered and then there are a large number of redundancy points so that the point cloud of each road element cannot be directly extracted for making high-definition maps.

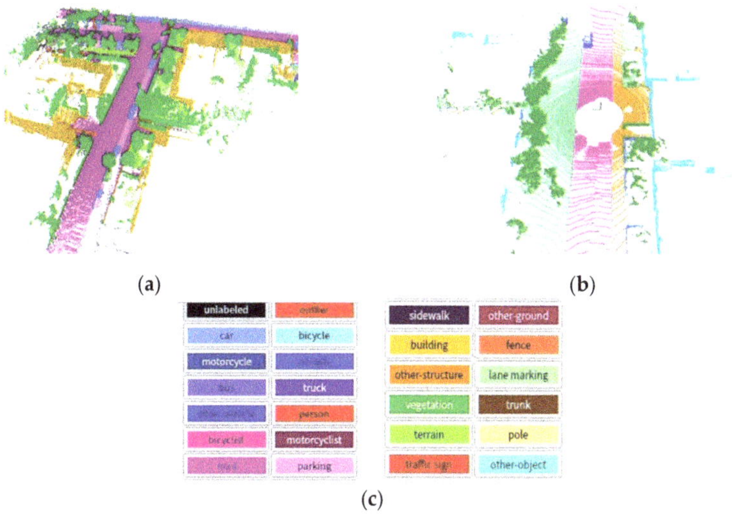

Figure 17. The comparison of point cloud semantic segmentation results: (**a**) segmentation visualization based on the improved RandLa-Net; (**b**) segmentation visualization based on the original RandLa-Net; and (**c**) the label information corresponding to the color.

Table 4. Test accuracy on semantickitti 03.

Methods	RandLa-Net	Ours
mIoU(%)	60.6	61.5
parameter(M)	1.24	1.24
road	92.1	92.3
side-walk	77.3	78.1
parking	42.6	43.9
other-ground	40.8	41.2
building	89.4	90.2
car	92.1	92.3
truck	50.8	51.3
bicycle	15.6	16.5
motorcycle	30.9	31.2
other-vehicle	41.7	41.5
vegetation	85.9	85.6
trunk	42.3	43.1
terrain	72.5	73.1
person	65.9	65.4
bicyclist	55.8	55.4
motorcyclist	8.9	9.7
fence	47.2	48.3
pole	53.1	53.3
traffic sign	40.4	41.7

From Table 4, it can be seen that the mIoU of all categories are improved after using the new method. This is because the shape features of each element on the global point cloud map are more complete, which is more conducive to be identified and extracted. However, in the classification of some small targets, the accuracy of the original algorithm is not as high as the frame-by-frame segmentation. This is because the registration of small targets in the registration time is not complete, which leads to some deviation in the subsequent classification.

5.6. Point Cloud Extraction

Since the file is in bin format, in order to find the point cloud data corresponding to the desired label from the classified point cloud map, the point cloud can be extracted according to the label. Figure 18a is the point extracted on label 15 cloud data, and Figure 18b is the point cloud data extracted on label 9.

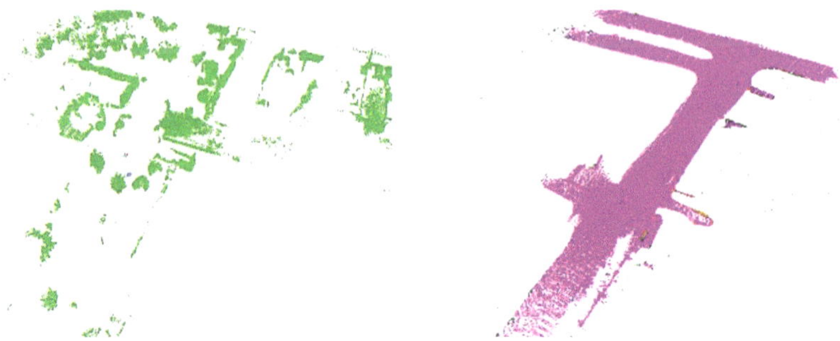

(**a**) (**b**)

Figure 18. Point cloud extraction: (**a**) vegetation; and (**b**) road.

Figure 18 is the visual effect of extracting all the points marked vegetation and road in the SemanticKITTI dataset 03. It can also be converted into txt type and data in (x, y, z, label) format can be obtained for further operation. In addition to the two extraction examples of Figure 18, specific extractions can be performed based on 19 labels. The extracted data can be directly used to create high-definition maps after simple sorting. The sorted part of the data is shown in Table 5, where the category corresponding to the number is represented by a label.

Table 5. The form after data output.

Numb	Object Type	Position Coordinates (x, y, z)		
1	car	20.354 20.356 20.359	40.375 40.374 40.374	−2.404 −2.404 −2.399
…	…	…	…	…
6	person	−0.847 −0.852 −0.852	−34.686 −34.683 −34.683	3.215 3.212 3.215
…	…	…	…	…
9	road	−0.003 −0.001 −0.002	−31.752 −31.756 −31.759	0.002 0.002 0.002
…	…	…	…	…
15	vegetation	−8.033 −8.053 −8.062	−0.995 −0.982 −0.975	−1.201 −1.200 −1.201

The data are labeled in the following order: 1. car; 2. bicycle; 3. motorcycle; 4. truck; 5. other-vehicle; 6. person; 7. bicyclist; 8. motorcyclist; 9. road; 10. parking; 11. sidewalk; 12. other-ground; 13. building; 14. fence; 15. vegetation; 16. trunk; 17. terrain; 18. pole; and 19. traffic sign.

6. Conclusions

This paper proposes a fast automatic classification and extraction method for large-scale point cloud data. The original point cloud data are globally mapped by NDT registration. In order to reduce the amount of computation, the point cloud map is randomly sampled. The cloud global map is semantically segmented, with each point assigned to a corresponding label, and then the numpy index is used to extract the point cloud data corresponding to the label of interest. The training results show that the point cloud data classification reaches 53.7% on the public dataset SemanticKITTI and 77.4% on Semantic3D. The test on the SemanticKITTI-03 dataset reflects that our method is more efficient than traditional manual annotation on large-scale point cloud datasets. However, in order to save computing power, the algorithm network parameters in the point cloud classification part are too few, resulting in an unsatisfactory classification effect on small target objects. In the follow-up research, the classification of small target objects will be further optimized and tested.

Author Contributions: Conceptualization, J.L. (Jiadi Li); methodology, J.L. (Jiadi Li); validation, J.L. (Jiadi Li); formal analysis, Z.M. and J.L. (Jiajia Liu); investigation, Y.Z., Y.W. and J.Z.; project administration, Z.M.; writing—original draft, J.L. (Jiadi Li); writing—review and editing, Z.M. and J.L. (Jiajia Liu) All authors have read and agreed to the published version of the manuscript.

Funding: This paper is supported by the International Cooperation Project of Science and Technology Bureau of Chengdu (no. 2019-GH02-00051-HZ), Sichuan Unmanned System-Intelligent Perception, Engineering Laboratory Open Fund, and the research fund of the Chengdu University of Information Engineering, under grant nos. WRXT2020-001, WRXT2020-002, WRXT2021-002 and KYTZ202142. This paper is also supported by the Sichuan Science and Technology program of China, grant no. 2022YFS0565.

Institutional Review Board Statement: Not applicable.

Informed Consent Statement: Not applicable.

Data Availability Statement: Not applicable.

Acknowledgments: The authors are grateful to the College of Automation, Chengdu University of Information Technology.

Conflicts of Interest: The authors declare that they have no conflict of interest to report regarding the present study.

References

1. Xu, J.; Hou, F.; Cao, G.H. High-definition road map production method and key technology. *Surv. Mapp. Bull.* **2022**, *1*, 155–158. [CrossRef]
2. Wang, Y.; He, W.; Zhou, L.; Peng, X.T.; Li, W. High-definition road map production based on vehicle LiDAR data. *Geospat. Inf.* **2022**, *20*, 92–95.
3. Andreas, G.; Philip, L.; Raquel, U. Are we ready for autonomous driving? The KITTI vision benchmark suite. In Proceedings of the IEEE Conference on Computer Vision and Pattern Recognition (CVPR), Providence, RI, USA, 16–21 June 2012; pp. 3354–3361.
4. Hackel, T.; Savinov, N.; Ladicky, L.; Wegner, J.D.; Schindler, K.; Pollefeys, M. Semantic3d. net: A new large-scale point cloud classification benchmark. *arXiv* **2017**, arXiv:1704.03847.
5. Behley, J.; Garbade, M.; Milioto, A.; Quenzel, J.; Behnke, S.; Stachniss, C.; Gall, J. SemanticKITTI: A dataset for semantic scene understanding of lidar sequences. In Proceedings of the IEEE/CVF International Conference on Computer Vision (ICCV), Seoul, Korea, 27 October–2 November, 2019.
6. Yuan, M.; Li, X.; Cheng, L.; Li, X.; Tan, H. A coarse-to-fine registration approach for point cloud data with bipartite graph structure. *Electronics* **2022**, *11*, 263. [CrossRef]
7. Rusu, R.B.; Marton, Z.C.; Blodow, N.; Beetz, M. Persistent point feature histograms for 3Dpoint cloud. In Proceedings of the 10th International Conference on Intelligent Autonomous Systems (IAS-10), Baden-Baden, Germany, 2008; Volume 1, pp. 119–128.
8. Biber, P.; Strasser, W. The normal distributions transform: A new approach to laser scan matching. In Proceedings of the 2003 IEEE/RSJ International Conference on Intelligent Robots and Systems (IROS 2003) (Cat. No. 03CH37453), Las Vegas, NV, USA, 27–31 October 2003; pp. 2743–2748.
9. Aoki, Y.; Goforth, H.; Srivatsan, R.A.; Lucey, S. PointNetLK: Robust & efficient point cloud registration using pointnet. In Proceedings of the IEEE/CVF Conference on Computer Vision and Pattern Recognition, Long Beach, CA, USA, 16–20 June 2019; pp. 7163–7172.
10. Lucas, D.D. An iterrative image registration technique with an application to stereo vision. In Proceedings of the 1981 International Conference on Imaging Understanding Workshop, Piscataway, NJ, USA, 24–28 August 1981; Volume 4, pp. 121–130.

11. Wang, Y.; Solomon, J. Deep closest point: Learning representations for point cloud registration. In Proceedings of the IEEE/CVF International Conference on Computer Vision, Seoul, Korea, 27 October–2 November 2019; pp. 3523–3532.
12. Wang, Y.; Solomon, J. PRNet: Self-supervised learning for partial-to-partial registration. In Proceedings of the Annual Conference on Neural Information Processing Systems 2019, NeurIPS 2019, Vancouver, BC, Canada, 8–14 December 2019; pp. 8814–8826.
13. Shan, J.C.; Li, X.Z.; Zhang, X.Y.; Jia, S.M. Real-time 3D semantic map construction in indoor scenes. *J. Instrum.* **2019**, *40*, 240–248.
14. Qiu, J.Y.; Lai, J.Z.; Li, Z.M.; Huang, K.; Liu, J.Y. LiDAR ground segmentation method for complex scenes. *J. Instrum.* **2020**, *41*, 244–251.
15. Qian, Y.L.; Gai, S.Y.; Zheng, D.L.; Da, F.P. Fast 3D human ear recognition based on local and global information. *J. Instrum.* **2019**, *40*, 99–106.
16. Su, H.; Maji, S.; Kalogerakis, E.; Learned-Miller, E. Multi-view convolutional neural networks for 3d shape recognition. In Proceedings of the 2015 IEEE International Conference on Computer Vision, Santiago, Chile, 13–16 December 2015; pp. 945–953.
17. Chen, X.; Ma, H.; Wan, J.; Li, B.; Xia, T. Multi-view 3d object detection network for autonomous driving. In Proceedings of the 2017 IEEE Conference on Computer Vision and Pattern Recognition, Honolulu, HI, USA, 21–26 July 2017; pp. 1907–1915.
18. Lang, A.H.; Vora, S.; Caesar, H.; Zhou, L.; Yang, J.; Beijbom, O. Pointpillars: Fast encoders for object detection from point clouds. In Proceedings of the 2019 IEEE/CVF Conference on Computer Vision and Pattern Recognition, Long Beach, CA, USA, 15–21 June 2019; pp. 12697–12705.
19. Truc, L.; Ye, D. Pointgrid: A deep network for 3d shape understanding. In Proceedings of the 2018 IEEE Conference on Computer Vision and Pattern Recognition, Salt Lake City, UT, USA, 18–21 June 2018; pp. 9204–9214.
20. Meng, H.Y.; Gao, L.; Lai, Y.K.; Manocha, D. Vv-net: Voxel vae net with group convolutions for point cloud segmentation. In Proceedings of the 2019 IEEE/CVF International Conference on Computer Vision, Seoul, Korea, 27 October–3 November 2019; pp. 8500–8508.
21. Wang, C.Y.; Liao, H.Y.M.; Wu, Y.H.; Chen, P.Y.; Hsieh, J.W.; Yeh, I.H. Fast point r-cnn. In Proceedings of the 2019 IEEE/CVF International Conferenceon Computer Vision, Seoul, Korea, 27 October–3 November 2019; pp. 9775–9784.
22. Hu, Q.; Yang, B.; Xie, L.; Rosa, S.; Guo, Y.; Wang, Z.; Markham, A. RandLa-Net: Efficient semantic segmentation of large-scale point clouds. In Proceedings of the 2020 IEEE/CVF Conference on Computer Vision and Pattern Recognition (CVPR), Seattle, WA, USA, 13–19 June 2020.
23. Loic, L.; Martin, S. Large-scale point cloud semantic segmentation with superpoint graphs. In Proceedings of the IEEE Conference on Computer Vision and Pattern Recognition, Salt Lake City, UT, USA, 18–23 June 2018.
24. Milioto, A.; Vizzo, I.; Behley, J.; Stachniss, C. RangeNet++: Fast and accurate lidar semantic segmentation. In Proceedings of the 2019 IEEE/RSJ International Conference on Intelligent Robots and Systems (IROS), The Venetian Macao, Macau, China, 4–8 November 2019.

electronics

MDPI

Article

An Improved YOLOv5s Algorithm for Object Detection with an Attention Mechanism

Tingyao Jiang, Cheng Li *, Ming Yang and Zilong Wang

College of Computer and Information, China Three Gorges University, Yichang 443002, China
* Correspondence: lc@ctgu.edu.cn

Abstract: To improve the accuracy of the You Only Look Once v5s (YOLOv5s) algorithm for object detection, this paper proposes an improved YOLOv5s algorithm, CBAM-YOLOv5s, which introduces an attention mechanism. A convolutional block attention module (CBAM) is incorporated into the YOLOv5s backbone network to improve its feature extraction ability. Furthermore, the complete intersection-over-union (CIoU) loss is used as the object bounding-box regression loss function to accelerate the speed of the regression process. Experiments are carried out on the Pascal Visual Object Classes 2007 (VOC2007) dataset and the Microsoft Common Objects in Context (COCO2014) dataset, which are widely used for object detection evaluations. On the VOC2007 dataset, the experimental results show that compared with those of the original YOLOv5s algorithm, the precision, recall and mean average precision (mAP) of the CBAM-YOLOv5s algorithm are improved by 4.52%, 1.18% and 3.09%, respectively. On the COCO2014 dataset, compared with the original YOLOv5s algorithm, the precision, recall and mAP of the CBAM-YOLOv5s algorithm are increased by 2.21%, 0.88% and 1.39%, respectively.

Keywords: object detection; YOLOv5s; attention mechanism; deep learning

Citation: Jiang, T.; Li, C.; Yang, M.; Wang, Z. An Improved YOLOv5s Algorithm for Object Detection with an Attention Mechanism. *Electronics* **2022**, *11*, 2494. https://doi.org/10.3390/electronics11162494

Academic Editor: Stefanos Kollias

Received: 1 July 2022
Accepted: 9 August 2022
Published: 10 August 2022

1. Introduction

In recent years, due to the advent of the era of big data and the rapid development of computer graphics cards, the computing power of computers has also improved, which has accelerated the development of artificial intelligence in the computer field. There are more and more studies related to artificial intelligence, for example, the research in [1–4] has good application value, and the research of object detection has also developed accordingly.

Object detection has a wide range of applications in many areas of artificial intelligence, including robot navigation [5], autonomous driving [6], medical imaging [7] and human–object interaction [8]. Current object detection algorithms are mainly divided into single-stage detection algorithms and two-stage detection algorithms. Single-stage detection algorithms are represented by the You Only Look Once (YOLO) series [9–13], single-shot multibox detector (SSD) series [14–17], etc. Two-stage detection algorithms are represented by the region-based convolutional neural network (R-CNN) series [18–20]. A single-stage detection algorithm simultaneously classifies and locates the object of interest during object detection, while a two-stage detection algorithm performs these tasks separately. The characteristics of single-stage detection algorithms include that their detection speeds are very fast, but their accuracies are low. A two-stage detection algorithm is the opposite of a single-stage detection algorithm, with high accuracy but a slow detection speed. At present, most object detection tasks are real-time detection problems based on video, which require high detection speed, so a single-stage object detection algorithm is more suitable.

The latest single-stage object detection algorithm is the YOLOv5 algorithm. Compared with other single-stage object detection algorithms, the YOLOv5 algorithm has a faster detection speed and a smaller model. YOLOv5 is divided into four different algorithms: YOLOv5s, YOLOv5m, YOLOv5l and YOLOv5x. The network structures of these four different algorithms are roughly the same, but their differences lie in the depths and widths of the networks. Among them, YOLOv5s has a faster detection speed and a smaller model than the other three algorithms, but the disadvantage is that its accuracy is low. In response to this problem, this paper proposes an improved YOLOv5s algorithm, CBAM-YOLOv5s, which introduces an attention mechanism.

In recent years, the attention mechanism has been widely used in various fields of deep learning [21–24], including image processing, speech recognition and natural language processing. There are many attention mechanism modules in the field of computer vision, among which the most classic ones are the squeeze-and-excitation network (SENet) [23] and the convolutional block attention module (CBAM) [24]. SENet is the champion of the ImageNet2017 image recognition competition, and CBAM is the champion of the 2018 classification competition. In this paper, these two classical modules are introduced respectively for comparative experiments.

At present, there are many improved object detection models based on the attention mechanism, which have good results, but most of the model parameters are large, and the detection speed is not fast enough. The improved method in this paper has good performance in both detection effect and detection speed.

2. CBAM-YOLOv5s

In this section, the YOLOv5s algorithm is first introduced, followed by detailed descriptions of the improvements made to the YOLOv5s network structure and the object bounding-box regression loss function used by the algorithm proposed in this paper.

2.1. YOLOv5s Algorithm

The YOLOv5s algorithm includes three parts: a feature extraction backbone network, a feature fusion neck network and a detection head. The network structure is shown in Figure 1. The detection process of the YOLOv5s algorithm is roughly divided into three steps. The first step is to extract features, adjust the scale of the input image to 640 × 640, and input the adjusted image into the backbone network. The BottleneckCSP-2 module, the BottleneckCSP-3 module, and the BottleneckCSP-4 module output three different scales of feature maps with sizes of 80 × 80, 40 × 40 and 20 × 20, respectively; these three feature maps contain different feature information. The second step is feature fusion. The three different scales of feature maps obtained through the backbone network are transmitted to the neck network, and the neck network performs a series of upsampling, convolution, channel concatenation and other operations to fully integrate the information provided by the feature maps. The third step is to output the detection heads. After the neck network fully integrates the features, three detection heads with sizes of 80 × 80, 40 × 40 and 20 × 20 are output. These three detection heads with different scales are used to detect small objects, medium objects and large objects.

Compared with YOLOv4, YOLOv5s adds a focus module to the backbone network. The main function of this module is to periodically extract pixels from high-resolution images and reconstruct them into low-resolution images to improve the receptive field of each pixel while retaining relatively complete original information. The design of the module is mainly used to reduce the number of calculations and speed up the algorithm. YOLOv4 only uses a cross-stage partial network (CSP) [25] structure in the backbone network, while YOLOv5s uses CSP structures in both the backbone network and the neck network. A CSP structure is used for local cross-layer network fusion, which reduces the number of calculations while simultaneously ensuring accuracy.

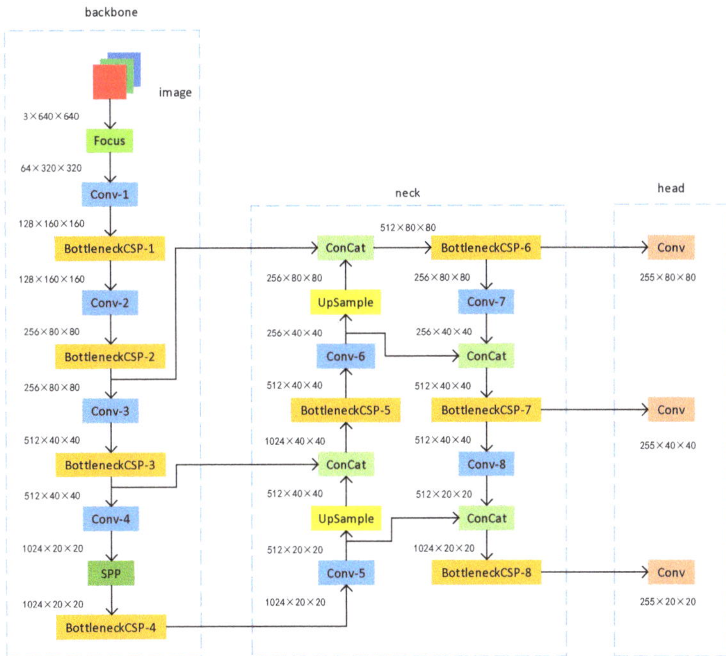

Figure 1. Structure of the YOLOv5s network.

2.2. Improved YOLOv5s with an Attention Mechanism

An attention mechanism is a data processing method in that is widely used in various types of machine learning tasks, such as natural language processing, image recognition and speech recognition. An attention mechanism is essentially similar to the mechanism by which humans observe external objects; when humans observe external objects, they are first inclined to observe some important local information about these objects and then combine the information derived from different regions to form an overall impression of the observed objects.

2.2.1. CBAM

The CBAM is a lightweight module that includes a channel attention submodule and a spatial attention submodule. The channel attention submodule focuses on important feature information, and the spatial attention submodule focuses on object location information. The structure of the CBAM is shown in Figure 2.

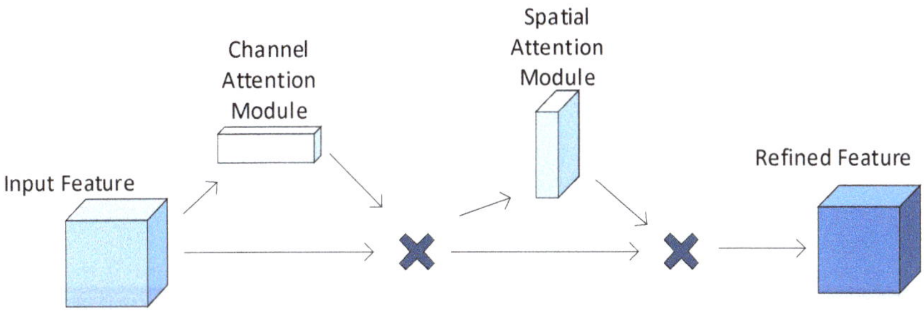

Figure 2. Structure of the CBAM.

The operation process of the channel attention submodule: The input feature map uses a global average pooling operation and a global maximum pooling operation to aggregate the spatial information of the input feature map to obtain a one-dimensional channel attention vector, sends it to a shared network, passes the added elements through a sigmoid activation function to obtain the resulting channel attention vector and finally multiplies the channel attention vector with the initial input to obtain the output of the channel attention submodule.

The operation process of the spatial attention submodule: The output of the channel attention submodule is subjected to an average pooling operation and a maximum pooling operation to obtain a spatial attention tensor; this is followed by channel concatenation. Then, the spatial attention tensor is obtained through a convolution operation and the sigmoid activation function; finally, the spatial attention tensor is multiplied with the output of the channel attention submodule to obtain the output of the spatial attention submodule.

2.2.2. YOLOv5s Introduces the CBAM

The CBAM is incorporated into the backbone network of YOLOv5s, and the network structure is shown in Figure 3. The function of the module is to let the network know which part to focus on and to accordingly achieve prominent representations of important features while suppressing the less important features; this module can adjust the attention weight of the feature map and improve the feature extraction ability of the network.

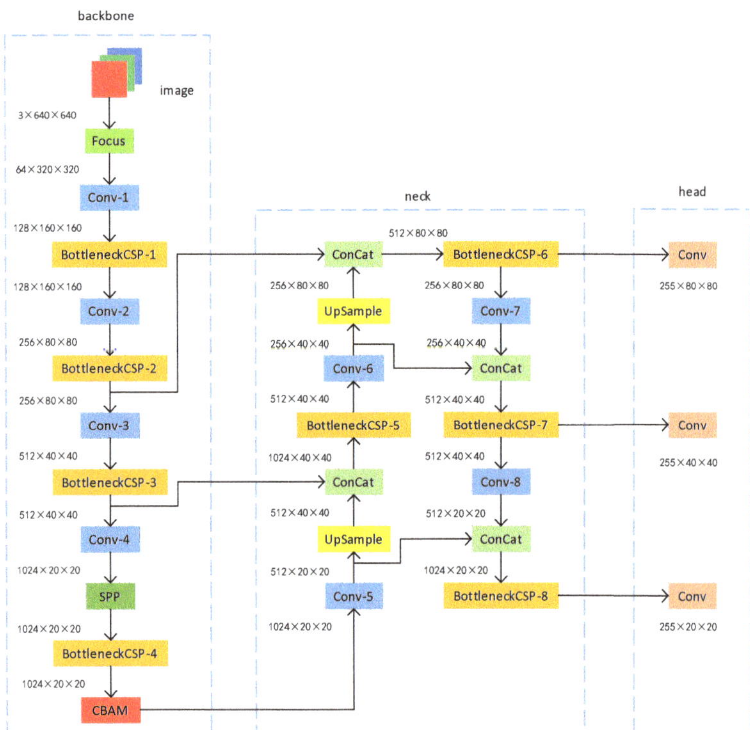

Figure 3. Structure of the CBAM-YOLOv5s network.

The specific operation of the CBAM is mainly divided into two steps, as shown in Figure 4.

In the first step, the channel attention operation is performed on the input feature map. The input 1024 × 20 × 20 feature map is processed through a maximum pooling operation and an average pooling operation to obtain two 1024 × 1 × 1 feature maps, and then these two feature maps are each compressed by the first fully connected layer to compress the number of channels to 64, thereby reducing the computational cost. This is followed by an expansion operation performed through the second fully connected layer to output two 1024 × 1 × 1 feature maps. Then, the feature information of the two feature maps is added and passed through the sigmoid activation function to obtain a 1024 × 1 × 1 feature map, and finally, the feature map is multiplied by the initial input to obtain an output of size 1024 × 20 × 20 with constant dimensions.

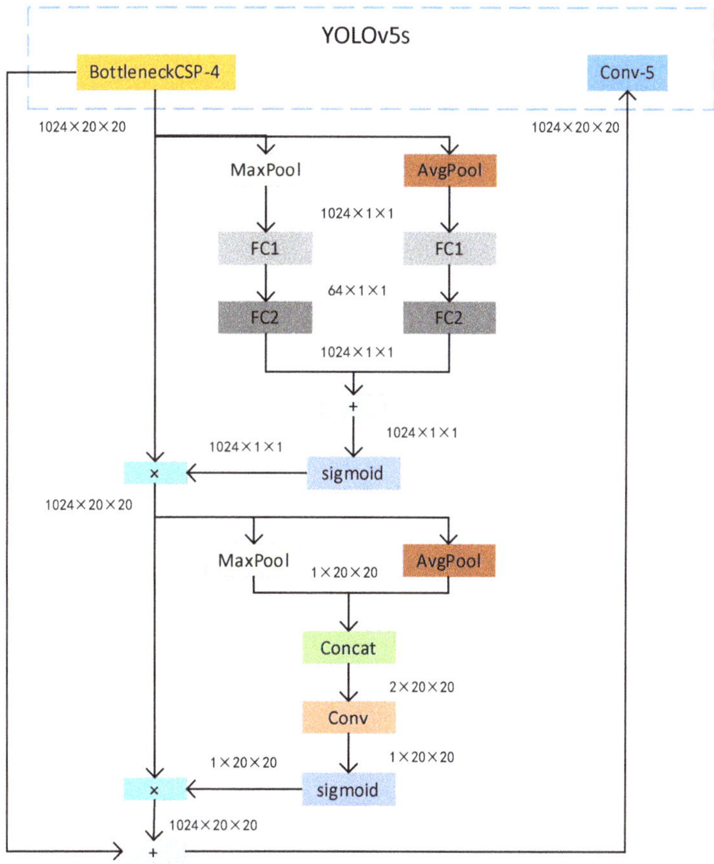

Figure 4. CBAM operation.

In the second step, the spatial attention operation is performed. The 1024 × 20 × 20 feature map obtained through the channel attention operation is subjected to a maximum pooling operation and an average pooling operation to output two 1 × 20 × 20 feature maps, and then 2 × 20 × 20 feature maps are output through the channel concatenation operation. Next, the dimensions of the feature map are restored to 1 × 20 × 20 through a convolution operation, and this is followed by the sigmoid activation function, which outputs a 1 × 20 × 20 feature map. Then, the feature map is multiplied by the initial input to obtain a 1024 × 20 × 20 feature map, and this feature map is added to the input of the BottleneckCSP-4 module to obtain the final output: a 1024 × 20 × 20 feature map. Finally, the extracted 1024 × 20 × 20 feature map is input back into the Conv-5 module in the neck network.

2.3. The Object Bounding-Box Regression Loss Function

The object bounding-box regression loss functions of most object detection algorithms use generalized intersection-over-union (*GIoU*) loss [26] to calculate the deviation between each prediction box and the corresponding ground truth; this loss is defined as

$$L_{GIoU} = 1 - IoU + \frac{|Ac - U|}{|Ac|} \quad (1)$$

where Ac represents the area of the smallest box that contains both the ground truth and the prediction box, IoU represents the intersection over union of two bounding boxes, U represents the union of the two bounding boxes and L_{GIoU} represents the *GIoU* loss.

The advantage of the *GIoU* loss is that it not only focuses on the overlapping area between the prediction box and the ground truth but also focuses on other nonoverlapping areas, so it can better reflect the degree of overlap between the prediction box and the ground truth. However, the disadvantage of the *GIoU* loss is that when the ground truth or the prediction box surrounds the other, the *GIoU* loss function deteriorates, causing slow convergence and a large localization bias during the training process. The complete *IoU* (*CIoU*) loss [27] was developed in view of this problem; in addition to considering the overlapping area between the prediction box and the corresponding ground truth, the distance between the center points and the aspect ratio of the two bounding boxes are also considered. The *CIoU* loss is given as

$$L_{CIoU} = 1 - IoU + \frac{\rho^2(b, b^{gt})}{c^2} + \alpha v, \quad (2)$$

$$\alpha = \frac{v}{(1 - IoU) + v}, \quad (3)$$

$$v = \frac{4}{\pi^2}\left(arctan\frac{w^{gt}}{h^{gt}} - arctan\frac{w}{h}\right)^2, \quad (4)$$

where $\rho^2(b, b^{gt})$ denotes the Euclidean distance between the prediction box and the ground truth, c denotes the shortest diagonal length of the smallest box containing both the ground truth and the prediction box, α is the weight parameter, v denotes the similarity between the aspect ratios of the two bounding boxes, w^{gt} and h^{gt} denote the width and height of the ground truth, w and h denote the width and height of the prediction box, respectively, and L_{CIoU} denotes the *CIoU* loss.

Compared with the *GIoU* loss, the *CIoU* loss adds loss terms for the center distance and the aspect ratio between the prediction box and the ground truth to the loss function, which makes the prediction box converge faster and the regression localization more accurate, so the algorithm in this paper uses the *CIoU* loss as the object bounding-box regression loss function.

3. Experiments

To evaluate the improvement achieved by the CBAM introduced to YOLOv5s, the CBAM incorporated into the backbone network of YOLOv5s is replaced by another attention mechanism module called the SENet for an ablation experiment. In this section, the experimental equipment, dataset, evaluation metrics, experimental results and comparative analysis are introduced. We have put the core code of the algorithm on GitHub. Interested readers can download it, and the access link is https://github.com/2530525322/object-model (accessed on 9 August 2022).

The SENet mainly includes squeeze and excitation operations. The module structure is shown in Figure 5.

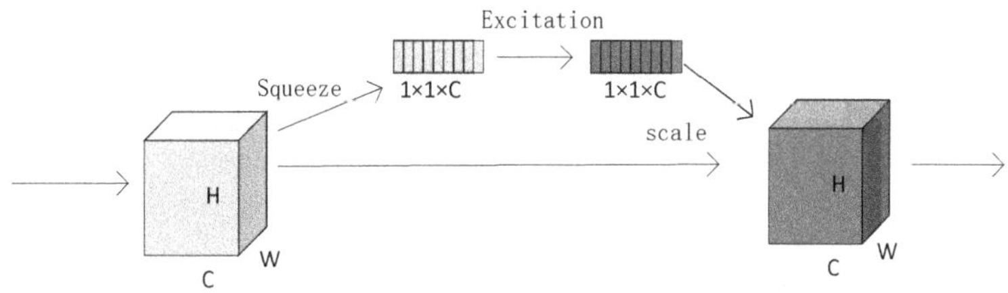

Figure 5. Structure of the SENet module.

The SENet mainly focuses the network's attention on specific channels by learning the connections between channels, thereby achieving improved accuracy. The general processing flow of the SENet is roughly divided into three steps.

Squeeze operation: The input W × H × C feature map is subjected to the global average pooling operation to obtain a 1 × 1 × C feature map.

Excitation operation: The result of the squeeze operation is transformed nonlinearly by using a fully connected layer.

Scale operation: The output obtained by the excitation operation is used as the weight and multiplied by the initial W × H × C input for the channel weights to obtain the final output.

3.1. Experimental Equipment and Training Parameters

The equipment used in the experiment is a Dell desktop computer, and its specific configuration is shown in Table 1.

Table 1. Computation system.

Name	Configuration
Processor	Intel(R) Core(TM) i9-10900X CPU @ 3.70GHz
Running Memory	64 GB
Operating System	Linux
GPU	NVIDIA GeForce RTX 3080
GPU Memory	10 GB
Programming Tool	PyCharm
Programming Language	Python
Deep Learning Framework	PyTorch

Some of the training parameters in the experiment are shown in Table 2.

Table 2. Training parameters.

Parameter	Value
Learning Rate	0.01
Batch Size	32
Weight Decay	0.0005
Momentum	0.937
Epochs	300

3.2. Dataset

The datasets used in this experiment are the Pascal Visual Object Classes 2007 (VOC2007) dataset [28] and the Microsoft Common Objects in Context (COCO2014) dataset. The

COCO2014 dataset has a total of 123,287 images with 80 categories. The VOC2007 dataset contains a total of 9963 images. Twenty classes are included in the dataset, as shown in Figure 6; these classes include the airplane, bicycle, bird, boat, bottle, bus, car, cat, chair, cow, dining table, dog, horse, motorbike, person, potted plant, sheep, sofa, train and TV monitor categories, and the associated XML file provides the object class of the input image and the coordinates of the corresponding ground truth.

Figure 6. VOC2007 dataset.

3.3. Evaluation Metrics

To evaluate the performance of the proposed algorithm, the evaluation metrics in this paper are precision (P), recall (R), mean average precision (mAP), F-score and frames per second (FPS).

Precision calculates the proportion of the number of correctly predicted positive samples to the total number of samples predicted as positive samples, that is, the accuracy of the prediction for the evaluation object. Precision is defined as follows:

$$P = \frac{TP}{TP + FP}, \tag{5}$$

where TP represents true positives, that is, the number of positive samples predicted as positive samples; FP represents false positives, that is, the number of negative samples predicted as positive samples.

Recall calculates the proportion of the number of correctly predicted positive samples to the total number of actual positive samples, that is, whether the evaluation object is completely found or not. Recall is defined as follows:

$$R = \frac{TP}{TP + FN}, \tag{6}$$

where FN represents false negatives, that is, the number of positive samples predicted as negative samples.

The mAP calculates the mean of the average precision (AP) values of all classes and is used to evaluate the overall performance of the algorithm. The mAP is given as

$$AP = \int_0^1 P\,dR, \tag{7}$$

$$mAP = \frac{\sum_{j=1}^{c} AP_j}{c}. \tag{8}$$

F-score calculates the harmonic value of precision and recall, which can comprehensively measure these two indicators. The *F-score* is defined as follows:

$$F\text{-}score = \left(1+\beta^2\right)\frac{P \cdot R}{\beta^2 \cdot P + R}, \tag{9}$$

where β is used to balance the weight of precision and recall in the *F-score*, and there are three values. When β is equal to 1, precision is as important as recall; when β is less than 1, precision is more important than recall; when β is greater than 1, recall is more important than precision.

3.4. Experimental Results and Comparative Analysis

During the experimental training process, the stochastic gradient descent (SGD) [29] optimization algorithm is used to update the model parameters. Table 3 shows the experimental results obtained on the VOC2007 dataset.

Table 3. Ablation experiment.

Dataset	Attention Mechanism		Precision	Recall	mAP@0.5	mAP@0.95	F1-Score	FPS
	SENet	CBAM						
VOC2007	×	×	75.68%	60.87%	66.35%	41.14%	67.47%	76
	✓	×	76.27%	62.09%	67.33%	42.03%	68.45%	57
	×	✓	80.20%	62.05%	69.44%	45.99%	69.97%	60

As seen in Table 3, compared with those of the original YOLOv5s algorithm that does not introduce an attention mechanism, the precision, recall and mAP of the proposed algorithm that introduces an attention mechanism are improved. Compared with the original YOLOv5s, the YOLOv5s version with the SENet module achieves a 0.59% improvement in precision, a 1.22% improvement in recall and a 0.98% improvement in mAP, while the YOLOv5s version with the CBAM yields larger improvements, with a 4.52% improvement in precision, a 1.18% improvement in recall and a 3.09% improvement in mAP. By conducting a comparative analysis on the experimental results, it can be concluded that the algorithm in this paper has better performance than the original algorithm and the algorithm with the SENet module. SENet only includes channel attention and can only obtain important feature information on the channel, while CBAM includes not only channel attention but also spatial attention. It can obtain important feature information in both channel and space, so that the network can better learn important features in the image. The more picture features the network learns, the better it can recognize the object, which will make the network's recognition accuracy higher.

The experimental comparison results of the object bounding-box regression loss function are shown in Figure 7, where the horizontal axis is the number of epochs and the vertical axis is the value of the bounding-box loss. The experimental results show that the use of the CIoU loss as the bounding-box regression loss function results in faster convergence than the GIoU loss.

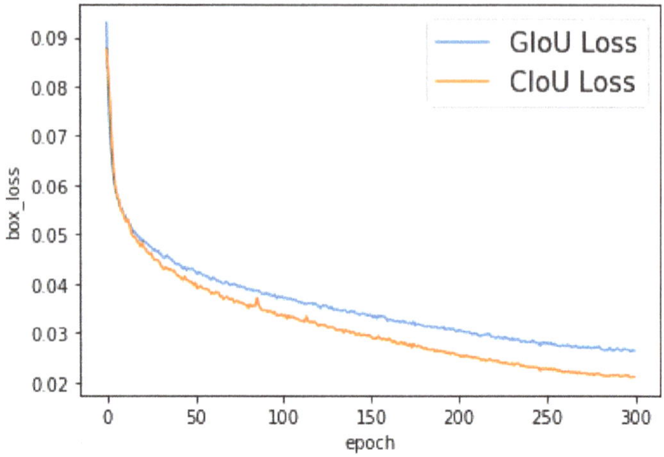

Figure 7. Variation in the two loss functions.

In order to further verify the effectiveness of the improved algorithm, this study includes comparative experiments on the COCO2014 dataset. The experimental results are shown in Table 4.

Table 4. Comparative experiment.

Dataset	Algorithm	Precision	Recall	mAP@0.5	mAP@0.95	F1-Score	FPS
COCO2014	YOLOv5s	64.48%	48.22%	52.72%	33.22%	55.18%	60
	CBAM-YOLOv5s	66.69%	49.10%	54.11%	33.98%	56.56%	58

As can be seen from Table 4, compared with the original YOLOv5s algorithm, the precision, recall and mAP of the CBAM-YOLOv5s algorithm are increased by 2.21%, 0.88% and 1.39%, respectively. Based on the experimental results in Tables 3 and 4, it can be concluded that the improved CBAM-YOLOv5s algorithm is better than the original YOLOv5s algorithm on the VOC2007 dataset and the COCO2014 dataset.

Figure 8 shows the detection effect of the CBAM-YOLOv5s algorithm on the VOC2007 dataset. It can detect different targets in the picture and frame them.

Figure 8. The detection effect of the algorithm.

To verify the effect of the algorithm proposed in this paper, this paper also compares it with other object detection algorithms, as shown in Table 5. The precision and the FPS are used as measurement indicators.

Table 5. Comparison with other algorithms.

Dataset	Algorithm	Backbone	Precision	FPS
VOC2007	SSD	VGG-16	77.5%	46
	ESSD	VGG-16	79.4%	25
	MDSSD	VGG-16	78.6%	28
	YOLOv3	Darknet-53	74.5%	36
	YOLOv4	CSPDarknet53	78.1%	35
	YOLOv5s	CSPDarknet53	75.6%	76
	CBAM-YOLOv5s	CSPDarknet53	80.2%	60

It can be seen from the results in Table 5 that the improved YOLOv5s performs better than the other detection algorithms in terms of precision and FPS. For example, it outperforms the YOLOv4 by 2.1% on the VOC2007 dataset with faster detection.

4. Conclusions

In this paper, a CBAM is incorporated into the backbone network of YOLOv5s to optimize its network structure, and the CIoU loss is used as the object bounding-box regression loss function to accelerate the speed of the regression process. To verify the performance of the proposed algorithm, extensive experiments are conducted on the VOC2007 dataset. The experimental results show that compared with those of the original YOLOv5s, the precision, recall and mAP of the proposed algorithm are significantly improved; furthermore, the CIoU loss is used because the bounding-box regression loss function is faster than the GIoU loss in terms of convergence. The algorithm in this paper solves the problem regarding the low detection accuracy of the original YOLOv5s algorithm to a certain extent, but the algorithm still exhibits certain detection errors and missed detection problems for complex images with dense objects. Future research will involve continuously optimizing the network structure of the proposed algorithm to further improve its detection accuracy.

Author Contributions: Conceptualization, T.J., C.L., M.Y. and Z.W.; data curation, T.J., C.L., M.Y. and Z.W.; methodology, T.J. and C.L.; writing—original draft, T.J. and C.L. All authors have read and agreed to the published version of the manuscript.

Funding: This research was funded by the National Natural Science Foundation of China, grant number 61871258.

Conflicts of Interest: The authors declare no conflict of interest.

References

1. Wang, W.; Zhang, Y.; Ge, G.; Jiang, Q.; Wang, Y.; Hu, L. A Hybrid Spatial Indexing Structure of Massive Point Cloud Based on Octree and 3D R*-Tree. *Appl. Sci.* **2021**, *11*, 9581. [CrossRef]
2. Liu, K.; Mulky, R. Enabling autonomous navigation for affordable scooters. *Sensors* **2018**, *18*, 1829. [CrossRef] [PubMed]
3. Conte, G.; Scaradozzi, D.; Mannocchi, D.; Raspa, P.; Panebianco, L.; Screpanti, L. Experimental testing of a cooperative ASV-ROV multi-agent system. *IFAC-PapersOnLine* **2016**, *49*, 347–354. [CrossRef]
4. Kang, T.; Yi, J.B.; Song, D.; Yi, S.J. High-speed autonomous robotic assembly using in-hand manipulation and re-grasping. *Appl. Sci.* **2020**, *11*, 37. [CrossRef]
5. Garcia, A.; Mittal, S.S.; Kiewra, E.; Ghose, K. A convolutional neural network feature detection approach to autonomous quadrotor indoor navigation. In Proceedings of the 2019 IEEE/RSJ International Conference on Intelligent Robots and Systems (IROS), Macau, China, 4–8 November 2019; pp. 74–81.
6. Levinson, J.; Askeland, J.; Becker, J.; Dolson, J.; Heldet, D.; Kammel, S.; Kolter, J.Z.; Langer, D.; Pink, O.; Pratt, V.; et al. Towards fully autonomous driving: Systems and algorithms. In Proceedings of the 2011 IEEE Intelligent Vehicles Symposium (IV), Baden-Baden, Germany, 5–9 June 2011; pp. 163–168.

7. Behrens, T.; Rohr, K.; Stiehl, H.S. Robust segmentation of tubular structures in 3-D medical images by parametric object detection and tracking. *IEEE Trans. Syst. Man Cybern. Part B Cybern.* **2003**, *33*, 554–561. [CrossRef] [PubMed]
8. Zhou, T.; Qi, S.; Wang, W.; Shen, J.; Zhu, S.C. Cascaded parsing of human-object interaction recognition. *IEEE Trans. Pattern Anal. Mach. Intell.* **2021**, *44*, 2827–2840. [CrossRef] [PubMed]
9. Redmon, J.; Divvala, S.; Girshick, R.; Farhadi, A. You only look once: Unified, real-time object detection. In Proceedings of the IEEE Conference on Computer Vision and Pattern Recognition, Las Vegas, NV, USA, 27–30 June 2016; pp. 779–788.
10. Redmon, J.; Farhadi, A. YOLO9000: Better, faster, stronger. In Proceedings of the IEEE Conference on Computer Vision and Pattern Recognition, Honolulu, HI, USA, 21–26 July 2017; pp. 7263–7271.
11. Redmon, J.; Farhadi, A. Yolov3: An incremental improvement. *arXiv* **2018**, arXiv:1804.02767.
12. Bochkovskiy, A.; Wang, C.Y.; Liao, H.Y.M. Yolov4: Optimal speed and accuracy of object detection. *arXiv* **2020**, arXiv:2004.10934.
13. GitHub. YOLOV5-Master. 2021. Available online: https://github.com/ultralytics/yolov5.git/ (accessed on 1 March 2021).
14. Liu, W.; Anguelov, D.; Erhan, D.; Szegedy, C.; Reed, S.; Fu, C.Y.; Berg, A.C. Ssd: Single shot multibox detector. In Proceedings of the European Conference on Computer Vision, Amsterdam, The Netherlands, 8–16 October 2016; Springer: Cham, Switzerland, 2016; pp. 21–37.
15. Fu, C.Y.; Liu, W.; Ranga, A.; Tyagi, A.; Berg, A.C. Dssd: Deconvolutional single shot detector. *arXiv* **2017**, arXiv:1701.06659.
16. Zheng, L.; Fu, C.; Zhao, Y. Extend the shallow part of single shot multibox detector via convolutional neural network. In Proceedings of the Tenth International Conference on Digital Image Processing (ICDIP 2018), Shanghai, China, 11–14 May 2018; International Society for Optics and Photonics: Shanghai, China, 2018; Volume 10806, p. 1080613.
17. Cui, L.; Ma, R.; Lv, P.; Jiang, X.; Gao, Z.; Zhou, B.; Xu, M. MDSSD: Multi-scale deconvolutional single shot detector for small objects. *arXiv* **2018**, arXiv:1805.07009. [CrossRef]
18. Girshick, R.; Donahue, J.; Darrell, T.; Malik, J. Rich feature hierarchies for accurate object detection and semantic segmentation. In Proceedings of the IEEE Conference on Computer Vision and Pattern Recognition, Columbus, OH, USA, 23–28 June 2014; pp. 580–587.
19. Girshick, R. Fast r-cnn. In Proceedings of the IEEE International Conference on Computer Vision, Santiago, Chile, 7–13 December 2015; pp. 1440–1448.
20. Ren, S.; He, K.; Girshick, R.; Sun, J. Faster r-cnn: Towards real-time object detection with region proposal networks. *Adv. Neural Inf. Process. Syst.* **2015**, *28*, 91–99. [CrossRef] [PubMed]
21. Zhou, T.; Li, J.; Wang, S.; Tao, R.; Shen, J. Matnet: Motion-attentive transition network for zero-shot video object segmentation. *IEEE Trans. Image Process.* **2020**, *29*, 8326–8338. [CrossRef] [PubMed]
22. Vaswani, A.; Shazeer, N.; Parmar, N.; Uszkoreit, J.; Jones, L.; Gomez, A.N.; Kaiser, Ł.; Polosukhin, I. Attention is all you need. In Proceedings of the Advances in Neural Information Processing Systems, Long Beach, CA, USA, 4–9 December 2017; pp. 5998–6008.
23. Hu, J.; Shen, L.; Sun, G. Squeeze-and-excitation networks. In Proceedings of the 2018 IEEE conference on Computer Vision and Pattern Recognition, Salt Lake City, UT, USA, 18–22 June 2018; pp. 7132–7141.
24. Woo, S.; Park, J.; Lee, J.Y.; Kweon, I.S. Cbam: Convolutional block attention module. In Proceedings of the European Conference on Computer Vision (ECCV), Munich, Germany, 8–14 September 2018; pp. 3–19.
25. Kim, D.; Park, S.; Kang, D.; Paik, J. Improved Center and Scale Prediction-Based Pedestrian Detection Using Convolutional Block. In Proceedings of the 2019 IEEE 9th International Conference on Consumer Electronics, Berlin, Germany, 8–11 September 2019; pp. 418–419.
26. Rezatofighi, H.; Tsoi, N.; Gwak, J.; Sadeghian, A.; Reid, I.; Savarese, S. Generalized intersection over union: A metric and a loss for bounding box regression. In Proceedings of the IEEE/CVF Conference on Computer Vision and Pattern Recognition, Long Beach, CA, USA, 15–20 June 2019; pp. 658–666.
27. Zheng, Z.; Wang, P.; Liu, W.; Li, J.; Ye, R.; Ren, D. Distance-IoU loss: Faster and better learning for bounding box regression. In Proceedings of the AAAI Conference on Artificial Intelligence, New York, NY, USA, 7–12 February 2020; Volume 34, pp. 12993–13000.
28. Everingham, M.; Van Gool, L.; Williams, C.K.; Winn, J.; Zisserman, A. The pascal visual object classes (voc) challenge. *Int. J. Comput. Vis.* **2010**, *88*, 303–338. [CrossRef]
29. Bottou, L. Stochastic Gradient Descent Tricks. In *Neural Networks: Tricks of the Trade*; Springer: Berlin/Heidelberg, Germany, 2012; pp. 421–436.

electronics

MDPI

Article

Recognition of Dorsal Hand Vein in Small-Scale Sample Database Based on Fusion of ResNet and HOG Feature

Jindi Li, Kefeng Li *, Guangyuan Zhang *, Jiaqi Wang, Keming Li and Yumin Yang

School of Information Science and Electric Engineering, Shandong Jiaotong University, Jinan 250000, China
* Correspondence: 205073@sdjtu.edu.cn (K.L.); zhanggy@sdjtu.edu.cn (G.Z.)

Abstract: As artificial intelligence develops, deep learning algorithms are increasingly being used in the field of dorsal hand vein (DHV) recognition. However, deep learning has high requirements regarding the number of samples, and current DHV datasets have few images. To solve the above problems, we propose a method based on the fusion of ResNet and Histograms of Oriented Gradients (HOG) features, in which the shallow semantic information extracted by primary convolution and HOG features are fed into the residual structure of ResNet for full fusion and, finally, classification. By adding Gaussian noise, the North China University of Technology dataset, the Shandong University of Science and Technology dataset, and the Eastern Mediterranean University dataset are extended and fused to from a fused dataset. Our proposed method is applied to the above datasets, and the experimental results show that our proposed method achieves good recognition rates on each of the datasets. Importantly, we achieved a 93.47% recognition rate on the fused dataset, which was 2.31% and 26.08% higher than using ResNet and HOG alone.

Keywords: ResNet; HOG; feature fusion; DHV recognition

Citation: Li, J.; Li, K.; Zhang, G.; Wang, J.; Li, K.; Yang, Y. Recognition of Dorsal Hand Vein in Small-Scale Sample Database Based on Fusion of ResNet and HOG Feature. *Electronics* **2022**, *11*, 2698. https://doi.org/10.3390/electronics11172698

Academic Editor: Dah-Jye Lee

Received: 27 July 2022
Accepted: 25 August 2022
Published: 28 August 2022

1. Introduction

Biometric identification refers to a technology that uses the physiological or behavioral features of the human body (such as fingerprint features [1], face features [2], gait features [3], signature handwriting [4], etc.) to achieve identity authentication. Compared with the traditional authentication systems based on passwords, tokens, and certificates, biometric authentication systems have many advantages [5–7], so more and more people begin to focus on the research of biometric identification. As a kind of biometric identification method, dorsal hand vein recognition is different from other biometric identification methods. It mainly uses infrared light to collect images of the back of the hand and uses this method to show the outline structure of veins [8] in order to realize the identification of an individuals' identity. An anatomy article [9] has demonstrated that the DHV has a unique structure during growth and development, which can characterize the individual to a certain extent. Therefore, the research on DHV recognition is of great significance in individual recognition.

Currently, DHV recognition research is primarily focused on a single database, which makes it easier to achieve better recognition results due to the use of similar acquisition equipment, subjects, and collection environment. DHV identification on a single database includes two methods, namely traditional features and deep learning methods. In 2019, Vairavel et al. [10] studied the recognition performance of three classical dense descriptors for DHV recognition, including Local Binary Pattern(LBP), HOG, and Weber local descriptor (WLD), and achieved good results on the Northern University of Technology(NCUT) [11] database. Liu et al. [12] proposed an improved biometric map matching method in 2020, which achieved a 98.09% recognition rate on the Xi'an Jiaotong University (XJTU) database. With the rise of artificial intelligence, some researchers have begun to use deep learning methods for DHV identification. In 2019, Wang et al. [13] used the selective convolution feature (SCF) model and spatial pyramid pooling (SPP) to obtain a more robust feature representation of images and they achieved excellent recognition rates on the China University of Mining and Technology (CUMT) databases. In 2019, Zhong et al. [14] designed a Deep Hashing Network (DHN) for DHV identification and achieved good results.

We can see from the above research that the DHV recognition of a single database has achieved good results, whether using traditional methods or deep learning methods. In recent years, researchers have begun to focus on cross-device DHV identification. In 2019, Wang et al. [15] proposed an improved scale-invariant feature transform (SIFT) algorithm, which achieved a recognition rate of 88.5% on datasets acquired by different devices by improving the scale factor α, extremum search neighborhood structure, and matching threshold R. In 2021, Wang et al. [16] proposed a two-stage coarse-to-fine matching method. First, the vein images to be matched are roughly matched in each category of the database, and then the SIFT method is used to extract the feature points of the vein images for fine matching. Such a method achieves good results on the cross-device DHV database.

Through the investigation and research on the cross-device DHV, most of the research on the cross-device DHV is based on the database of the same group of subjects and does not consider that the different subjects may have a certain impact on the experimental results. For different databases, there is not only the problem of different sampling equipment but also the diversity of subjects. Therefore, taking into account the differences between equipment and subjects is a major challenge in this field. In addition, in the cross-device DHV research, most researchers use traditional methods, but traditional methods are not robust to noise; if deep learning methods are used, they will face the problem of data volume. Given the above problems, this paper makes the following contributions:

(1) We designed a network framework that fused ResNet and HOG features, tested them on three different small-sample datasets and achieved good results.
(2) Aiming at the less researched cross-database DHV recognition, a fusion database containing three different datasets was established, and the proposed feature fusion method was applied to this database, achieving a high recognition rate and strong robustness.

2. Materials and Methods

2.1. Data Processing

2.1.1. Dataset

The databases used in this paper are the dataset of the Shandong University of Science and Technology (SDUST) [17], the dataset of the Eastern Mediterranean University of Turkey (FYO) [18], the dataset of NCUT, and the fusion dataset (Fusion Dataset).

(1) SDUST Dataset

The dataset is a database of DHVs collected from the left and right hands of 63 males and 47 females using a commercial infrared device DF-300. The dataset contains 40 images of each subject, 20 for the left and right hands, for a total of 220 categories. The pictures in each category achieve image enhancement and data enhancement by changing brightness and random rotation, the size is 640 × 480 pixels, the horizontal and vertical resolutions are 96 dpi, and the format is jpg, as shown in Figure 1a.

(2) FYO Dataset

The FYO dataset was collected by a team from the Eastern Mediterranean University, which uses homemade equipment to collect data. The dataset collected images of the DHVs, palm veins, and wrist veins of the left and right hands of 160 volunteers (111 males and 49 females) twice, with a 10-min interval between the two acquisitions. The original data of the dataset contains data collected twice, each time there were 320 images of DHVs, 320 images of palm veins, and 320 images of wrist veins, and the images were all 800 × 600 color images, as shown in Figure 1b.

(3) NCUT Dataset

This dataset builds a database of images of the backs of the hands of 102 people, including 50 males and 52 females. During the collection, the left and right hands are alternately collected, that is, after collecting a vein picture with the left hand, the right hand is placed at the collection site to collect one image and then an image of the left hand is collected. This alternating method ensures the difference between the same type of samples. Due to the differences in the distribution of veins in the left and right hands of each individual, the database can be considered a back-of-hand image library composed of 204 types of samples. When collecting, 10 pictures are taken from the back of each hand, the size is 640 × 480 pixels, the horizontal and vertical resolutions are 96 dpi, the grayscale is 256 levels, and the format is bmp, as shown in Figure 1c.

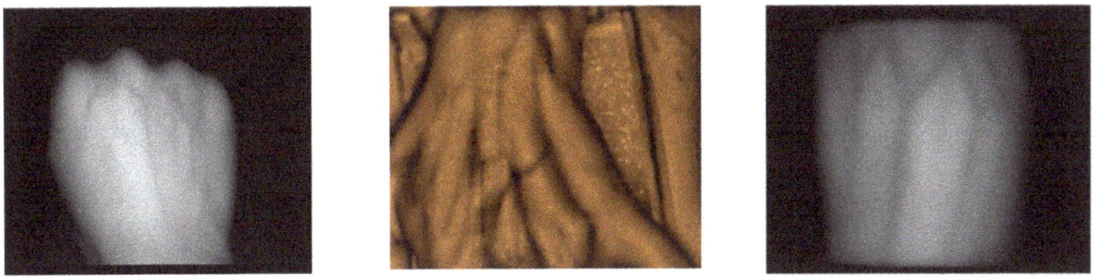

(a) (b) (c)

Figure 1. Datasets. (**a**) SDUST dataset. (**b**) FYO dataset. (**c**) NCUT dataset.

(4) Fusion Dataset

The fusion dataset includes the Shandong University of Science and Technology dataset, Northern University of Technology dataset, and Turkey Eastern Mediterranean University dataset, with a total of 372 volunteers and 744 sample data. Since the size and format of the images in different datasets are different, the images must be preprocessed first. First the images in the FYO dataset were transformed to grayscale and then their size was normalized, and the normalized size is 640 × 480 pixels. In this way, the images of the fusion dataset are all grayscale images with a size of 640 × 480 pixels, as shown in Figure 2.

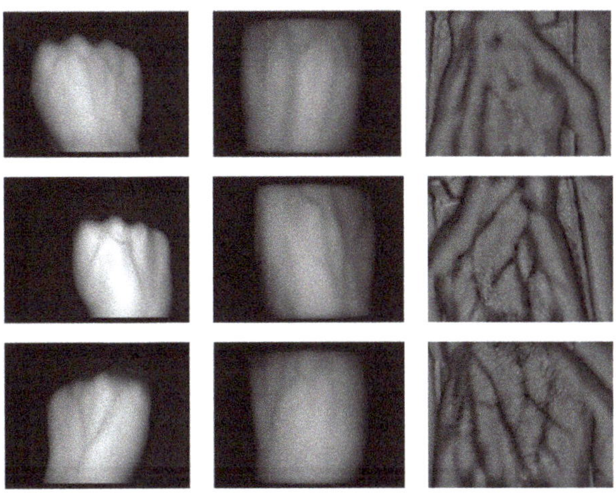

Figure 2. Fusion Dataset.

2.1.2. Extract Image ROI

As can be seen from Figures 1 and 2, there are large differences in the images of different DHV databases, including changes in rotation angle, size, brightness, and noise. This is mainly due to the differences in parameters, such as contrast, brightness, focal length, and optical performance of the lens between different acquisition devices, as well as the state of the collector's hand. These factors have a significant impact on the recognition results, and simple scale normalization is not conducive to extracting the texture features of the samples. Therefore, we need to extract the ROI of the image, and the method of extracting ROI is studied in [19,20]. In this paper, the centroid (x_0, y_0) adaptive method is used to determine the ROI area of the DHV image. The centroid of the vein image expressed by $G(x, y)$ can be calculated as:

$$x_0 = \frac{\sum\limits_{i,j} i \times g(i,j)}{\sum\limits_{i,j} g(i,j)}; y_0 = \frac{\sum\limits_{i,j} j \times g(i,j)}{\sum\limits_{i,j} g(i,j)} \tag{1}$$

where $g(i, j)$ is the grayscale value of pixel (i, j).

A square area with the size of R × R pixels is extracted and centered as the vein image to be processed. The experiment [21] verifies that when the ROI of the vein image is 380 × 380, the recognition rate can achieve the best effect, as shown in Figure 3.

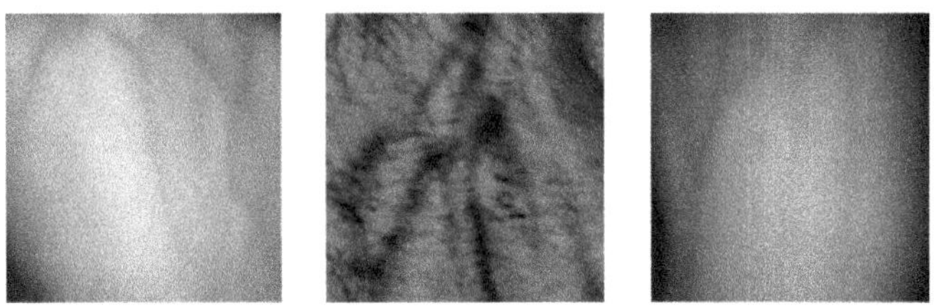

(a) (b) (c)

Figure 3. ROI images from different datasets. (**a**) ROI of SDUST dataset. (**b**) ROI of FYO dataset. (**c**) ROI of NCUT dataset.

2.1.3. Add Gaussian Noise

Gaussian noise is a kind of noise whose probability density function obeys normal distribution. The main function of Gaussian noise injection, as a data enhancement technique, is to add random Gaussian noise to samples to reduce overfitting during model training. Since there is only one image for each person in each dataset, it is not enough to prove the performance of the proposed method. Therefore, this method is adopted in this paper to expand the dorsal vein dataset. The dataset after adding Gaussian noise is shown in Figure 4.

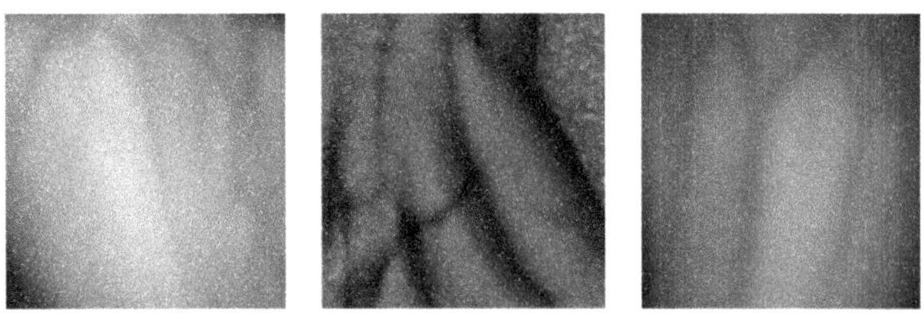

(a) (b) (c)

Figure 4. Images after adding Gaussian noise. (**a**) SDUST dataset after adding Gaussian noise. (**b**) FYO dataset after adding Gaussian noise. (**c**) NCUT dataset after adding Gaussian noise.

2.2. Related Algorithms

2.2.1. HOG

HOG [22] feature is a feature descriptor used in computer vision and image processing for object detection, which constitutes a feature by computing and counting the gradient direction histograms of local regions of an image.

The acquisition of HOG features is divided into four steps:

The first step is to normalize the color space of the DHV image, which consists of two aspects, image grayscale, and Gamma correction. Because our image is already a grayscale map, only Gamma correction is performed, and the Gamma correction formula is as shown in Formula (2).

$$I(x,y) = I(x,y)^{\gamma}, (\gamma = 0.5) \tag{2}$$

The gradient is calculated in the horizontal and vertical directions in the second step, and the gradient calculation formula are shown in Formulas (3) and (4).

$$G_x(x,y) = H(x+1,y) - H(x-1,y) \tag{3}$$

$$G_y(x,y) = H(x,y+1) - H(x,y-1) \tag{4}$$

where $G_x(x,y)$,$G_y(x,y)$, and $H(x,y)$ denote the horizontal gradient, vertical gradient, and pixel value at pixel point (x,y) in the input image, respectively. The amplitude and direction of the gradient at pixel (x,y) are shown in Formulas (5) and (6).

$$G(x,y) = \sqrt{G_x(x,y)^2 + G_y(x,y)^2} \tag{5}$$

$$\alpha(x,y) = \tan^{-1}\left(\frac{G_y(x,y)}{G_x(x,y)}\right) \tag{6}$$

The third step is to divide the image into 8 × 8 pixel cells. As shown in the red grid in Figure 5, a total of 784 cells are included, and the feature descriptors of each cell are counted. There are 9 descriptors for each cell, representing from 0° to 160°. Every 4 cells is a block, which is represented by the yellow grid in Figure 5. It contains a total of 729 blocks. The descriptors of all cells in each block are the HOG features of the block.

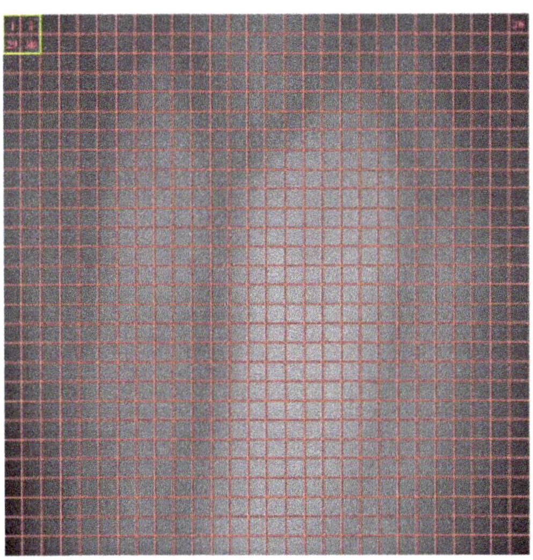

Figure 5. Divided into 28 × 28 cell images of veins.

The fourth step is to concatenate the HOG feature descriptors of all blocks in the image to represent the HOG feature of the image, which is a $1 \times 26{,}244$ vector.

2.2.2. ResNet Network

Before the idea of residual learning was proposed, traditional convolutional networks or fully connected networks had more or fewer problems, such as information loss and loss when information was transmitted. In addition, deep networks cannot be trained when gradients are small or exploding. ResNet [23] is a deep learning network that solves the problem of network degradation by introducing a deep residual learning framework. The network uses a residual unit structure, as shown in Figure 6. Assuming that the input feature is x, the learned feature is $H(x)$ and the residual unit of the learned feature can be represented as $F(x) = H(x) - x$. The equation of $F(x) + x$ can be implemented by a feedforward neural network with shortcut connections, and the residual unit structure can avoid the feature loss of the convolutional layer during the information transmission process.

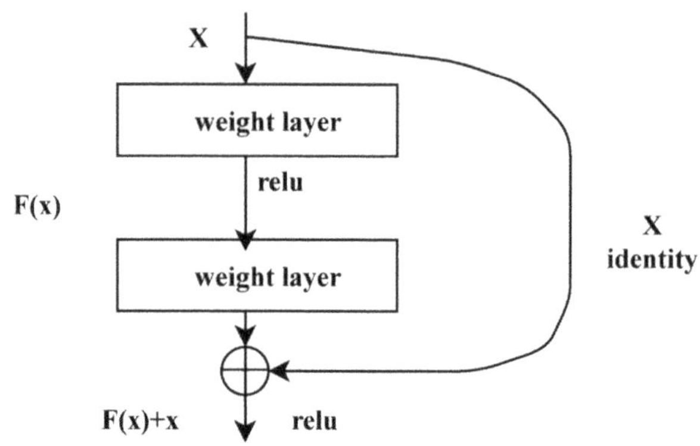

Figure 6. Residual structure.

Figure 7 shows the residual structure of ResNet34, Table 1 shows the network parameters for the ResNet34. The main branch of the residual structure is composed of two layers of 3×3 convolutional layers, and the connecting line on the right side of the residual structure is the shortcut branch, that is, the identity branch. Such branches are designed to reduce the amount of computation and parameters.

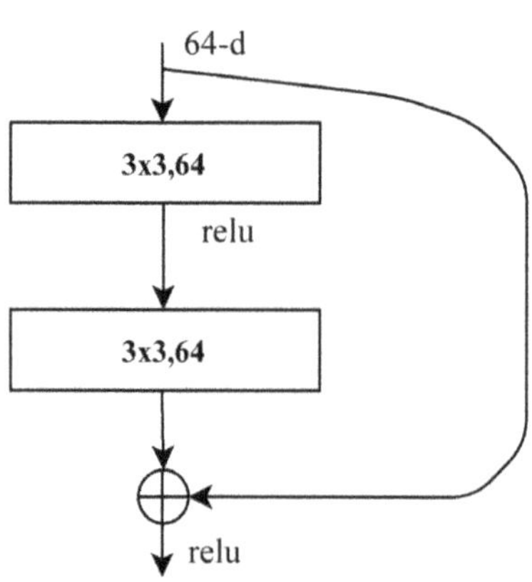

Figure 7. ResNet34 residual structure.

Table 1. The ResNet34 network parameters in this paper. (* represents the number of categories classified.).

Layer Name	Network Parameters	Input Size	Output Size
conv1	$7 \times 7, 64$, stride2	$224 \times 224 \times 3$	$56 \times 56 \times 3$
Conv_block1	$\begin{bmatrix} 3 \times 3, & 64 \\ 3 \times 3 & 64 \end{bmatrix} \times 3$, stride2	$56 \times 56 \times 64$	$56 \times 56 \times 64$
Conv_block2	$\begin{bmatrix} 3 \times 3, & 128 \\ 3 \times 3 & 128 \end{bmatrix} \times 4$, stride2	$56 \times 56 \times 64$	$28 \times 28 \times 128$
Conv_block3	$\begin{bmatrix} 3 \times 3, & 256 \\ 3 \times 3 & 256 \end{bmatrix} \times 6$, stride2	$28 \times 28 \times 128$	$14 \times 14 \times 256$
Conv_block4	$\begin{bmatrix} 3 \times 3, & 512 \\ 3 \times 3 & 512 \end{bmatrix} \times 3$, stride2	$14 \times 14 \times 256$	$7 \times 7 \times 512$
AdaptiveAvgPool2d (H, W)	H = 1, W = 1	$7 \times 7 \times 512$	$1 \times 1 \times 512$
FC	\	512	*

2.3. Proposed Methods

2.3.1. Fusion of ResNet and HOG Feature

The framework based on the fusion of ResNet and HOG features proposed in this paper is shown in Figure 8.

First, we do two-way processing on the image input to the neural network, in which we perform a convolution operation on it to extract the low-level semantic information of the image as a Feature Map. The other way is to input into the HOG function to extract the gradient information of the image. When performing HOG feature extraction on images, we make some changes to the features. First, we obtain the HOG feature of the entire image according to the general process, and the extracted feature vector is a one-dimensional vector of $1 \times 26{,}244$, as shown in Equation (7).

$$V = [l_1, l_2, \cdots, l_{26244}] \tag{7}$$

Since the acquired image HOG feature dimension is large, the feature vector needs to be normalized. Otherwise, the image features are jerkier in gradient descent and the model has difficulty converging when the neural network is learning. Therefore, the obtained one-dimensional vector is first normalized, and the normalization formula is shown in Formula (8).

$$f_{out} = \frac{x_i - \min(x)}{\max(x) - \min(x)} \tag{8}$$

After the normalization is completed, the normalized feature vector is reshaped into a feature map of 162×162, as shown in Formula (9), to obtain the HOG feature.

$$Feature\_HOG = \begin{bmatrix} l_1 & \cdots & l_{162} \\ \vdots & \ddots & \vdots \\ l_{26082} & \cdots & l_{26244} \end{bmatrix} \tag{9}$$

Since the size of HOG_Map is different from that of Feature_Map, a convolution operation is required. The convolved HOG feature is HOG_Feature, and then spatial feature fusion is performed with Feature_Map. The fusion method is shown in Equation (10), and the fusion method is shown in Formula (11). Then input the fused features into the ResNet residual block, and finally reduce the dimension of the feature map output by the ResNet residual block and input it into the fully connected layer for classification.

$$y^{sum}_{c,h,w} = \alpha x^a_{c,h,w} + (1 - \alpha) x^b_{c,h,w} \tag{10}$$

$$Feature\_Fusion\_Map = \alpha \times HOG\_Feature + (1 - \alpha) \times Feature\_Map \quad (11)$$

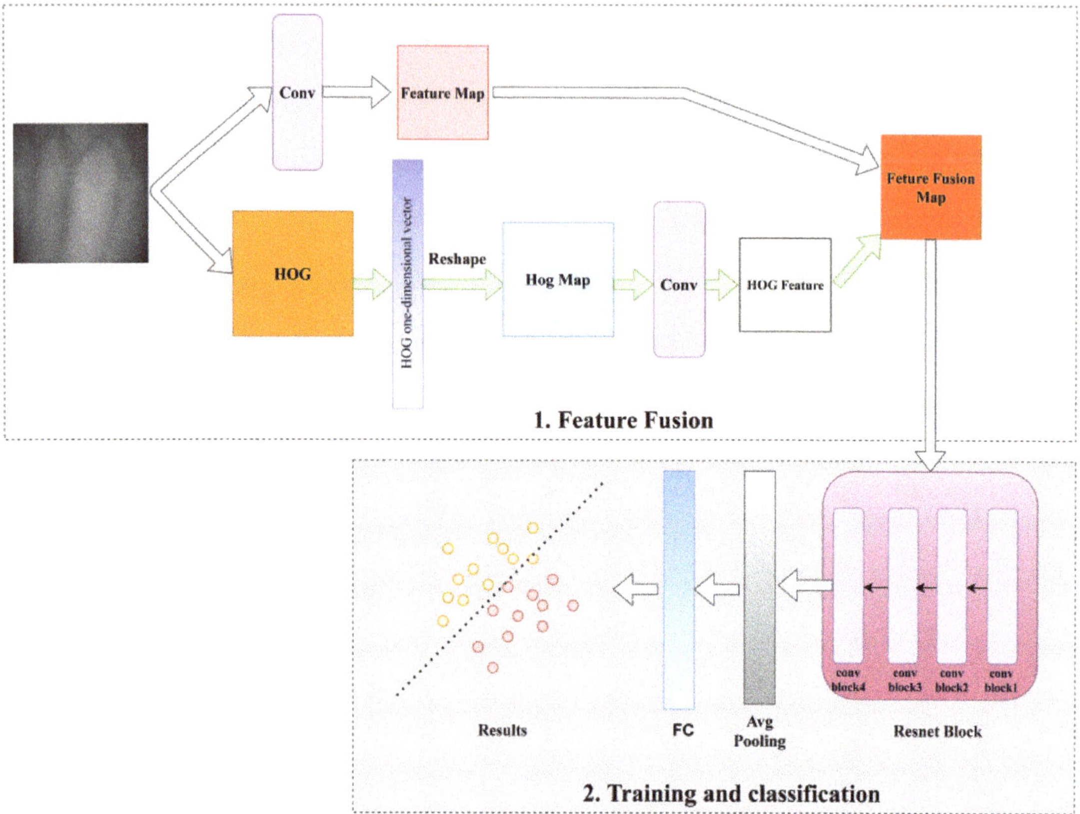

Figure 8. ResNet and HOG feature fusion methods. (1.) Feature Fusion. This part obtains the HOG feature and shallow semantic information of the image, respectively, and then performs spatial feature fusion. (2.) Training and classification. The features after feature fusion are input into the residual architecture of ResNet for training, then input into the average pooling layer for dimensionality reduction, and finally input into the fully connected layer for classification.

2.3.2. Feature Fusion Parameter Selection

When dividing the image into cells, we performed three sets of experiments on the Twenty dataset of NCUT to verify the effect of the number of cells on feature fusion. The experimental results are shown in Table 2.

Table 2. Different Cell identification results.

Number of Cells	Evaluation Methods	
	Recognition Rate (%)	Train Time (s)
196	85.94	250
256	86.12	267
784	**86.53**	**330**
3136	86.46	694

As can be seen from Table 2, as the number of cells increases from 196 to 784, the feature fusion recognition rate gradually increases. However, as the number of cells increases, when reaching 3136 cells, the recognition rate not only does not increase but instead decreases. We deduce that the reason is that when the number of cells increases to a certain level, if the number of cells is increased, the gradient information of each cell will be lost to a certain extent, resulting in a decrease in the recognition effect. It can also be seen from Table 2 that as the number of cells increases, the model training time also increases. Considering the above two factors, we chose the number of cells to be 784.

3. Experiments and Analysis

3.1. Feature Fusion Validity Experiments

In this paper, the dataset is divided into a training set, validation set, and test set according to 8:1:1 by random division. In addition, to ensure the persuasiveness of the experimental results, we conducted each experiment three times on the test set and took the average value. We used the Pytorch deep learning framework. The graphics card was NVIDIA GeForce RTX 2080 Ti 16 GB, the batch size was 16, the learning rate was 0.001, the loss function is the cross entropy loss function, and the epochs were 50.

Before feature fusion, we performed nine experiments on NCUT's Twenty dataset to find the best fusion factor α, and the experimental results are shown in Figure 9. When the fusion factor is $\alpha = 0.3$, the recognition rate can achieve the best effect. Therefore, the fusion factors of ResNet and HOG in the following experiments are both set to 0.3.

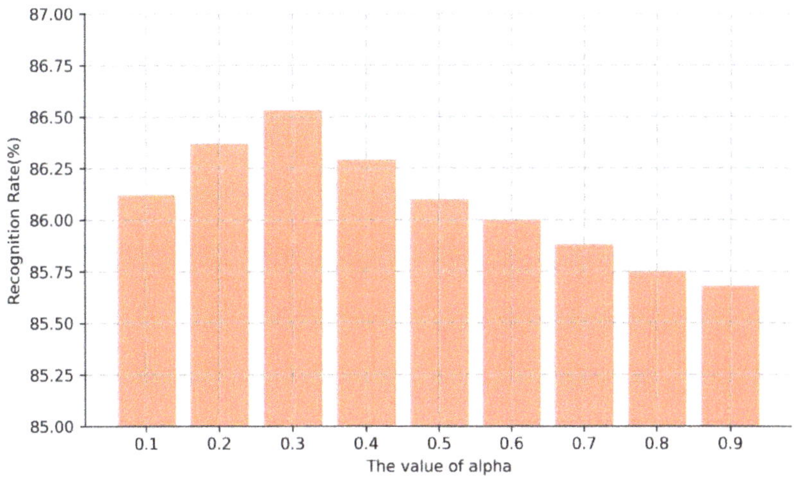

Figure 9. ResNet and HOG fusion factor take values.

Table 3 shows the recognition rates on a single dataset using ResNet, HOG, and ResNet_HOG methods. First of all, it can be seen from Table 3 that using the ResNet network can achieve a recognition rate of more than 90% on the Fifty dataset of a single database, but it does not achieve such a high effect on the Twenty dataset. This is because the larger the number of samples, the better the trained model will be, and the stronger the generalization ability of the model. Secondly, we can see from Table 3 that as the number of data increases, using the HOG algorithm cannot effectively improve the recognition rate, because the traditional method has no dependence on the amount of data in the dataset.

Table 3. ResNet and HOG feature fusion recognition rate.

Methods	Recognition Rate (%)											
	SDUST				FYO				NCUT			
	Twenty	Thirty	Forty	Fifty	Twenty	Thirty	Forty	Fifty	Twenty	Thirty	Forty	Fifty
HOG	83.06	83.13	83.15	83.16	82.01	82.04	82.08	82.10	81.30	81.32	81.34	81.35
ResNet	84.97	90.47	90.86	92.30	86.60	89.59	91.93	93.43	83.93	87.70	89.13	90.06
ResNet_HOG (ours)	**86.57**	**91.03**	**92.70**	**93.27**	**90.46**	**92.40**	**94.60**	**95.36**	**86.53**	**90.10**	**91.67**	**93.40**

In addition, it can be seen from the table that the recognition rate of our proposed feature fusion method is better than that of using ResNet and HOG alone, which proves the feasibility of our proposed feature fusion method.

The fusion dataset is consistent with the experiments performed on the single dataset, and the experimental results are shown in Table 4. Table 4 shows the comparison between the proposed feature fusion method and the ResNet method, and it can be seen that the recognition rate of the proposed method is better than that of using ResNet and HOG alone.

Table 4. Recognition rate on the fused dataset after feature fusion.

Methods	Recognition Rate (%)			
	Twenty	Thirty	Forty	Fifty
HOG	67.34	67.36	67.37	67.39
ResNet	83.70	86.83	90.46	91.16
ResNet_HOG (ours)	**85.70**	**89.46**	**92.27**	**93.47**

3.2. Feature Fusion Robustness Experiments

The robustness of the model has always been the focus of cross-database dorsal vein recognition research, and traditional methods are not robust to cross-database dorsal vein images and datasets with Gaussian noise added. Here we conduct experiments on three different datasets using the HOG algorithm. In addition, we use the Partition Local Binary Patterns (PLBP) [24] algorithm for comparison. PLBP is an improvement based on the LBP algorithm. It divides an image into non-overlapping blocks, uses the LBP algorithm for each block, and finally splices the LBP feature statistical histograms of the entire image. The experimental results are shown in Table 5.

Table 5. Comparison of different methods.

Methods	Recognition Rate (%)			
	SDUST	FYO	NCUT	Fusion Dataset
PLBP	60.09	55.50	70.19	61.07
HOG	83.16	82.10	81.35	67.39
ResNet	92.30	93.43	90.06	91.16
ResNet_HOG (ours)	**93.27**	**95.36**	**93.40**	**93.47**

It can be seen from Table 5 that the recognition rate of using the PLBP and HOG algorithms alone not only does not achieve good results but also is much lower than the experimental results of other researchers [10] on the NCUT dataset. The reason for the analysis is that when other researchers conducted experiments on the NCUT dataset, they used the original dataset and did not use Gaussian noise to expand the dataset. Additionally, our experiments are performed on the dataset augmented with Gaussian noise, which also shows that traditional features are not robust to our dataset with Gaussian noise added. Furthermore, it can be seen from Table 5 that in the FYO dataset, the effect of using the PLBP algorithm is particularly low, and we find some categories with the worst recognition results in the FYO dataset, as shown in Figure 10. Most of the categories with poor recognition results are recognized in the category 52. Figure 11 shows a statistical histogram of the texture information extracted from the DHV images using LBP with rotationally invariant consistency pattern. As can be seen from the figure, the texture information for categories 131 and 170 differs significantly from the registered features but differs very little from the registered features for category 52, which leads to DHV images like 131 and 170 being easily identified as the category 52. Analysis of the reasons for this occurrence, by adding Gaussian noise leads to a variation in the intra-class images, where the intra-class distances become larger and are thus misidentified as other classes.

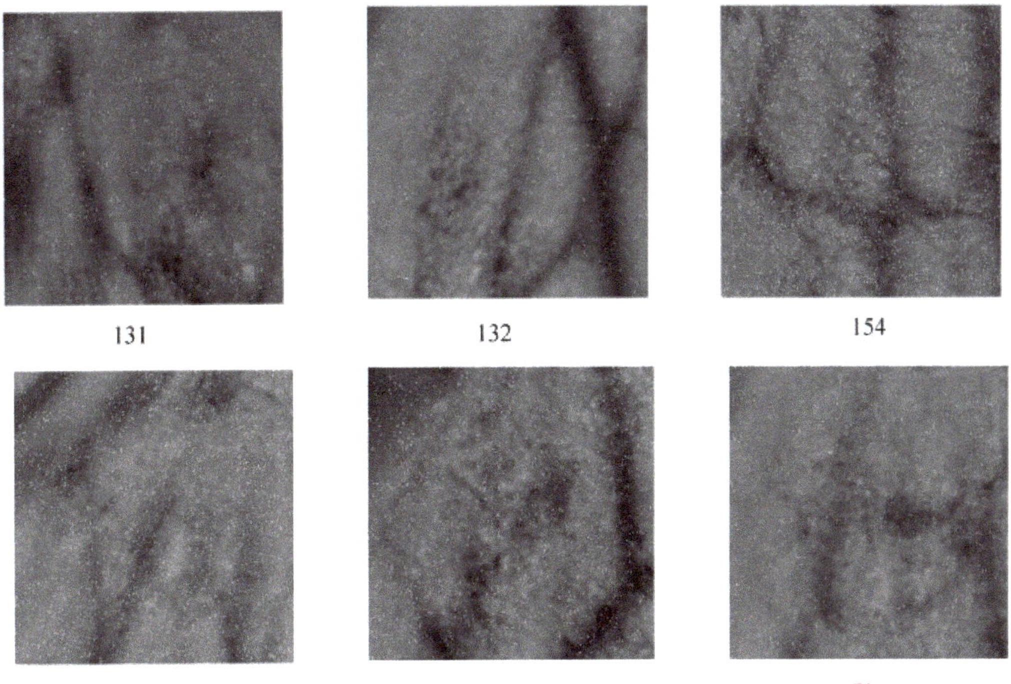

Figure 10. The most misclassified tags in the FYO dataset.

The HOG algorithm can achieve a recognition rate of more than 80% on a single database, but only 67.39% on a fusion dataset. We find some of the worst-recognized classes in the fused dataset, which are mostly data from NCUT and SDUST, as shown in Figure 12. Most of these partially identified worst classes are identified as 316 and 8, which are images in the FYO dataset, as shown in Figure 13.

We found that in the worst-recognized category, the blood vessel information of these images is not obvious, and the Gaussian noise on the images accounts for more. The images of the most misclassified categories have almost no blood vessel information, and most of the information is the hair on the back of the hand. The gap between these two categories cannot be seen visually. We visualize the HOG feature maps of these categories, as shown in Figure 14.

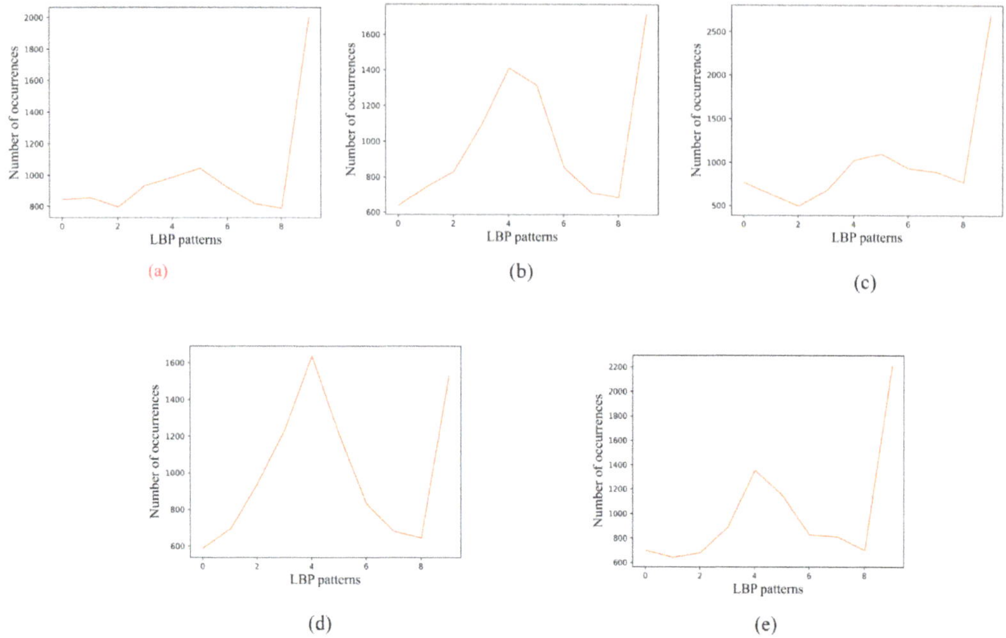

Figure 11. The misclassification of most categories of texture information. (**a**) Registration characteristics of category 52. (**b**) Registration characteristics of category 131. (**c**) Testing characteristics of category 131. (**d**) Registration characteristics of category 170. (**e**) Testing characteristics of category 170.

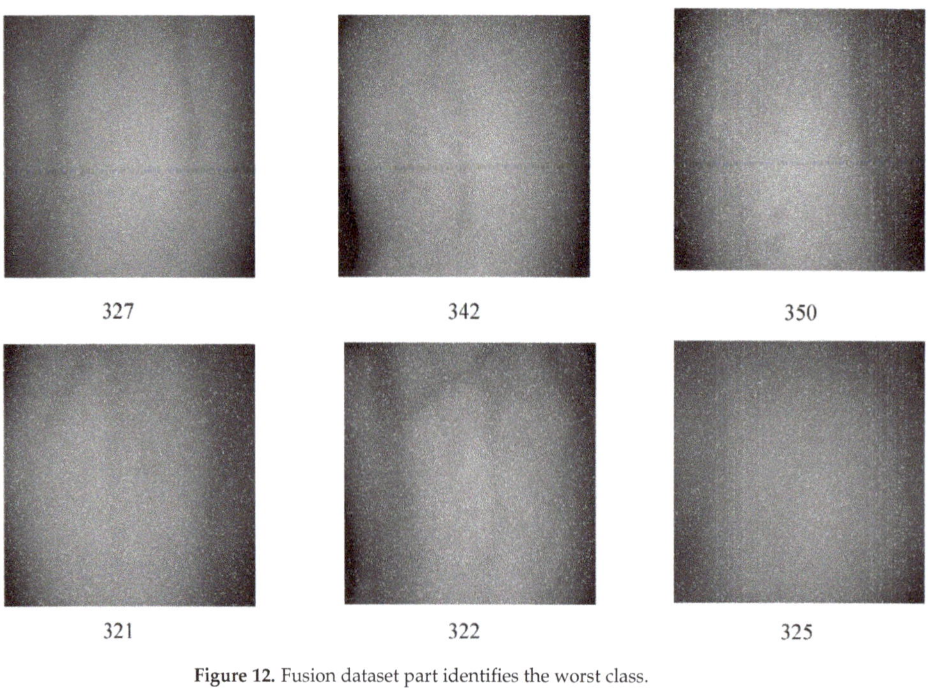

Figure 12. Fusion dataset part identifies the worst class.

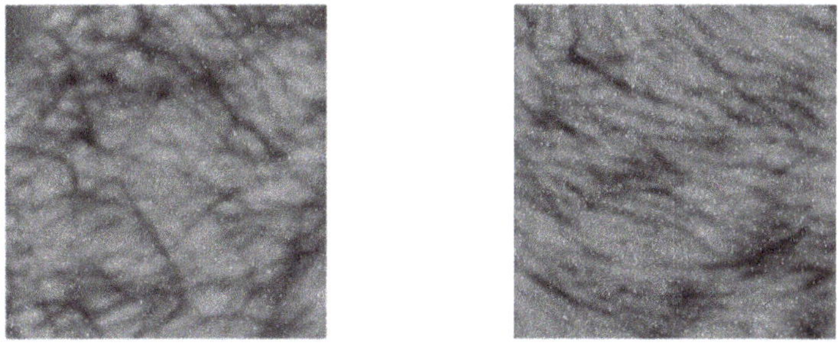

316 8

Figure 13. The most misclassified categories.

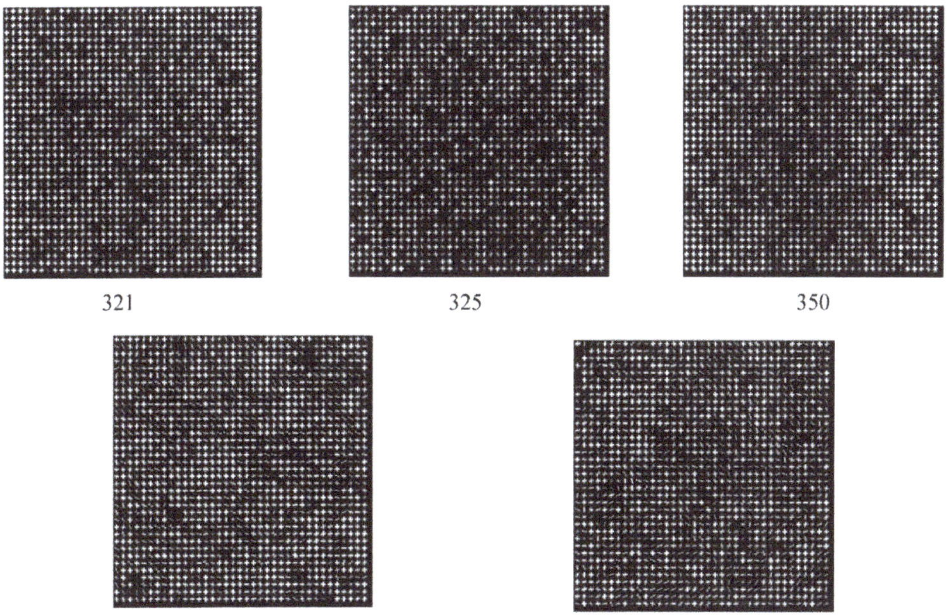

321 325 350

8 316

Figure 14. HOG feature visualization for misclassification.

As can be seen from Figure 14, the images with insignificant vein information in the NCUT and SDUST datasets and the images of these two categories in the FYO dataset have less obvious HOG features. Therefore, errors are prone to occuring when calculating the Euclidean distance between two categories, resulting in a low recognition rate using the HOG algorithm on the fusion dataset.

The recognition rate of the ResNet_HOG method proposed in this paper is significantly higher than that of using ResNet and HOG alone on a single dataset. Moreover, we also achieved good recognition rates on the fusion dataset and have strong robustness.

3.3. Comparison with Other Researchers

Recently, some researchers [25] achieved the current optimal results on the NCUT dataset using CNN and PLBP feature fusion. We adopted this idea and fused PLBP with ResNet for feature fusion, and the experimental results are shown in Table 6. As can be seen from Table 6, our proposed method achieves the current optimal results on the SDUST dataset, but not on the FYO and NCUT. In [18], the authors used a decision-level fusion of palm, dorsal, and wrist biometric features on vein images, which can make full use of hand biometric features, so this method is superior to our proposed method. In [26], the authors used the principal component analysis (PCA) method to expand the DHV dataset to 250 per category, which far exceeded our data volume, so their experimental results were superior to ours. However, when we expand our dataset to 250 per category, the recognition rate is 99.93%, and the recognition results are superior to [26].

Table 6. Recognition rates of different feature fusion methods on a single database.

<table>
<tr><th rowspan="3">Methods</th><th colspan="12">Recognition Rate (%)</th></tr>
<tr><th colspan="4">SDUST</th><th colspan="4">FYO</th><th colspan="4">NCUT</th></tr>
<tr><th>Twenty</th><th>Thirty</th><th>Forty</th><th>Fifty</th><th>Twenty</th><th>Thirty</th><th>Forty</th><th>Fifty</th><th>Twenty</th><th>Thirty</th><th>Forty</th><th>Fifty</th></tr>
<tr><td>ResNet_PLBP</td><td>85.40</td><td>90.87</td><td>91.77</td><td>92.76</td><td>90.27</td><td>91.77</td><td>94.17</td><td>95.20</td><td>85.96</td><td>89.73</td><td>90.17</td><td>92.50</td></tr>
<tr><td>VeinNet [17]</td><td colspan="4">92.28</td><td colspan="4">\</td><td colspan="4">\</td></tr>
<tr><td>Skeleton [27]</td><td colspan="4">\</td><td colspan="4">\</td><td colspan="4">92.75</td></tr>
<tr><td>CNN Model [18]</td><td colspan="4">\</td><td colspan="4">98.90</td><td colspan="4">\</td></tr>
<tr><td>CNN [26]</td><td colspan="4">\</td><td colspan="4">\</td><td colspan="4">99.61</td></tr>
<tr><td>ResNet_HOG (ours)</td><td>86.57</td><td>91.03</td><td>92.70</td><td>93.27</td><td>90.46</td><td>92.40</td><td>94.60</td><td>95.36</td><td>86.53</td><td>90.10</td><td>91.67</td><td>93.40</td></tr>
</table>

Table 7 shows the comparison between our proposed feature fusion method and the current methods of cross-database, from which it can be seen that our method can achieve better results, except for [16]. The reason is that [15,16,28] use datasets collected through different devices and the same subjects, whereas we use datasets with different devices, different subjects, and different ethnicities, and these different factors have a significant impact on DHV identification [29], so our method is slightly below [16].

Table 7. Recognition rates of different feature fusion methods on fusion dataset.

<table>
<tr><th rowspan="2">Methods</th><th colspan="4">Recognition Rate (%)</th></tr>
<tr><th>Twenty</th><th>Thirty</th><th>Forty</th><th>Fifty</th></tr>
<tr><td>ResNet_PLBP</td><td>84.47</td><td>89.26</td><td>92.13</td><td>92.63</td></tr>
<tr><td>Improved SIFT [15]</td><td colspan="4">88.50</td></tr>
<tr><td>SIFT [28]</td><td colspan="4">90.17</td></tr>
<tr><td>Two-stage Coarse-to-fine Matching [16]</td><td colspan="4">96.80</td></tr>
<tr><td>ResNet_HOG (ours)</td><td>85.70</td><td>89.46</td><td>92.27</td><td>93.47</td></tr>
</table>

3.4. Comparison between Feature Fusion and Data Volume

Through the experiments in the previous sections, we can see that increasing the number of samples in the dataset can improve the recognition rate of the dorsal vein of hand, but of the two methods, feature fusion achieved better results in the recognition of dorsal vein in small samples. Figure 15 shows the relationship between feature fusion and the number of samples.

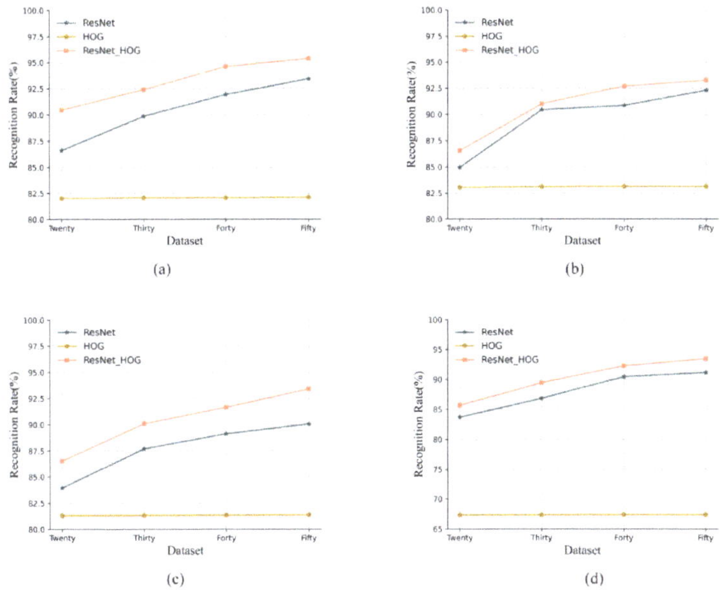

Figure 15. Data volume vs. feature fusion. (**a**) Results of ResNet, HOG, and ResNet_HOG on FYO dataset. (**b**) Results of ResNet, HOG, and ResNet_HOG on SDUST dataset. (**c**) Results of ResNet, HOG, and ResNet_HOG on NCUT dataset. (**d**) Results of ResNet, HOG, and ResNet_HOG on Fusion dataset.

From Figure 15, we can first see that our proposed feature fusion method achieved good results on the small sample DHV dataset. Secondly, we can see that when the amount of data reaches 40 pieces per category, the proposed feature fusion method exceeds the ResNet recognition rate of 50 pieces without feature fusion. This is because the shallow semantic information of the image is extracted using ResNet and then fused with HOG features; after a series of convolution operations, the final features for classification include both the deep semantic information and the gradient information of the image. Through such a feature fusion method, the features of the image can be fully obtained and thus be accurately classified.

4. Discussion

4.1. Gamma Value Influence

When a picture appears too bright or too dark, it leads to poor image contrast, and that is when Gamma correction needs to be performed. Our dataset was grayed out for the experiments, so the color information did not affect the experiments much. The Gamma correction of the images is required to make the black areas of the images appear brighter. When Gamma < 1, in the high gray value area, the dynamic range becomes smaller, the image contrast decreases, the overall gray value of the image becomes larger, and the image becomes brighter. When Gamma > 1, gamma >1 in the low gray value area, the dynamic range becomes smaller, the image contrast decreases, and the overall gray value of the image becomes smaller and darker. To find the appropriate value for Gamma < 1, we conducted four sets of experiments on three different datasets, and the experimental results are shown in the Table 8. From the table, we can see that the value of Gamma does affect different datasets, but overall, the effect of Gamma on the experimental results is minimal, and in most of the literature, Gamma is generally taken as 0.5, so in this paper, we also take 0.5.

Table 8. The influence of different gamma values on the experiment.

Dataset	Gamma Value				
	0.2	0.4	0.5 (ours)	0.6	0.8
NCUT	86.52	86.61	86.53	86.58	86.60
FYO	90.26	90.35	90.46	90.38	90.42
SDUST	86.59	86.51	86.57	86.44	86.49

4.2. *Influence of Gaussian Noise Intensity Model*

How the SNR affects the recognition rate of ResNet, we have carried out three sets of experiments on the Twenty dataset of NCUT, as shown in Table 9. In the table, we can see that the recognition rate of ResNet gets lower and lower as the noise increases. This is mainly because, when adding too much Gaussian noise, the vein information on the image is completely covered by the noise, which makes the model difficult to train and the recognition rate decreases.

Table 9. Influence of Gaussian noise intensity model.

Noise Range	0.1–0.3	0.3–0.5	0.5–0.7
Recognition Rate (%)	77.45	72.94	66.81

4.3. *Influence of Convolution Blocks*

We performed a separate convolution operation before inputting the image and HOG features into the ResNet residual structure, and this operation had an effect on the experimental results, which we performed on three different datasets, and the experimental results are shown in Table 10. From the Table 10, we can see that adding the convolutional block performs significantly on the NCUT and SDUST datasets but not on the FYO dataset, although the recognition rate with the convolutional block is better than that without the convolutional block. This is because without adding the convolutional block, reshaping the HOG feature size will lead to the loss of most of the HOG features, which will affect the recognition results.

Table 10. Experimental comparison with and without convolution blocks.

Methods	Recognition Rate (%)		
	NCUT	FYO	SDUST
With convolution blocks	86.53	90.46	86.57
No convolution blocks	83.60	90.23	85.74

4.4. *The Influence of Increasing the Amount of Data HOG on the Model*

Through the experiments in Section 3, we can see that ResNet_HOG outperforms ResNet alone for small sample dorsal hand vein recognition, but the gain obtained using ResNet_HOG always seems to be around 1–3% regardless of the dataset size, and we analyze the reasons for this as follows.

In this paper, traditional features play an auxiliary role. Traditional features can extract information that cannot be obtained by depth features (such as gradient information, etc.), but this information has a limited impact on depth features, so the recognition results of the model are not significantly improved after performing feature fusion. We also found this problem in [17], where the authors used a fusion of ResNet and LBP features and only achieved a 1.97% higher recognition rate than using ResNet alone. In addition, we conducted four sets of experiments on the NCUT dataset to determine how much HOG affects ResNet. As can be seen from the experiments in the Table 11, the effect of feature fusion seems to become less and less pronounced as the amount of data increases. This is because as the amount of data increases, ResNet has enough samples for training and the trained model becomes more and more robust, so it is possible to obtain a high recognition rate using only ResNet.

Table 11. Experimental comparison of different data volumes.

Methods	Recognition Rate (%)			
	One Hundred Samples	One Hundred and Fifty Samples	Two Hundred Samples	Two Hundred and Fifty Samples
ResNet	96.93	98.07	98.92	99.54
ResNet_HOG	97.89	98.84	99.27	99.93

For the above analysis, we can see that either ResNet and HOG feature fusion or ResNet with LBP for feature fusion can outperform the recognition rate obtained using ResNet alone. However, because the recognition rate obtained using ResNet alone is more than 90%, the model's recognition rate is only improved by 1–3% after feature fusion, which is also effective for a dataset with few samples.

4.5. Other Deep Learning vs. Traditional Methods Discussion

In [17], the authors conducted experiments using the ResNet network. The authors performed two types of data enhancement (increasing the amount of data and changing the image brightness), and the recognition rate obtained by both enhancement methods was better than that obtained by using ResNet alone. In addition, the authors also performed the method of ResNet and LBP feature fusion, and the recognition rate after performing data enhancement using ResNet was 90.31%, while the recognition rate after performing ResNet and LBP feature fusion was 92.28%, which is an improvement but not significant. Our analysis shows that the black background information of the rotated image has an impact on the extraction of LBP features, which leads to the insignificant improvement of recognition results after feature fusion.

We searched many references and found no literature combining DL and HOG, but there are experiments with CNN and PLBP feature fusion. In [25], the authors designed three methods of CNN and PLBP feature fusion, namely serial fusion, decision fusion, and feature fusion. The decision fusion and feature fusion are the best, which are 0.34% higher than the CNN network without fusion.

According to the preceding literature, traditional features and deep learning for feature fusion not only excel in a few sample dorsal hand vein recognition but also improve the recognition rate of large sample data, demonstrating the feasibility of traditional features and deep learning for feature fusion.

5. Conclusions

In this paper, we design a method for the fusion of ResNet and HOG features, which achieves better results on small sample datasets. We adopt the methods of other researchers and conduct experiments on our dataset, and the experimental results show that the recognition rate of the feature fusion method is better than that of using ResNet alone. This proves that the combination of deep learning and traditional features can not only solve the problem that deep learning has a low recognition rate for small samples but also solve the problem that traditional features are not robust to Gaussian noise. It further illustrates the superiority and feasibility of using deep learning and traditional feature fusion in the field of DHV recognition.

At present, our work has achieved good results on the dataset with Gaussian noise. In the future, we will utilize more ways to expand the dataset for verification, such as physical expansion (random rotation, image translation, image exposure, etc.) and deep learning automatically expansion [30,31]. Our fusion dataset now includes three different datasets, and we hope to obtain more datasets in the future to expand the database of DHVs. In addition, to explore the possibility of feature fusion between deep features and traditional features, we will use a variety of traditional features and deep feature fusion methods to verify the DHV dataset.

Author Contributions: Investigation, J.L.; data curation, K.L. (Keming Li), J.W. and Y.Y.; software, K.L. (Kefeng Li) and G.Z.; writing—original draft preparation, J.L. and K.L. (Kefeng Li); writing—review and editing, K.L. (Kefeng Li). All authors have read and agreed to the published version of the manuscript.

Funding: This research received no external funding.

Institutional Review Board Statement: The study was conducted in accordance with the Declaration of Helsinki, and approved by the The Ethics Committee of Shandong Jiaotong University(protocol code 3701063670893 and date of approval 27 July 2022).

Informed Consent Statement: Informed consent was obtained from all subjects involved in the study. Written informed consent has been obtained from the participants to publish this paper.

Data Availability Statement: The datasets used in this paper are all available on request from the authors of the following three articles. (1) **SDUST:** [17]. (2) **FYO:** [18]. (3) **NCUT:** [11].

Conflicts of Interest: The authors declare no conflict of interest.

References

1. Putte, T.V.D.; Keuning, J. Biometrical Fingerprint Recognition: Don't Get Your Fingers Burned. In *Smart Card Research and Advanced Applications*; Springer: Boston, MA, USA, 2000; pp. 289–303.
2. Lei, J.; Pei, Q.; Liu, X.; Sun, W. A Practical Privacy-Preserving Face Authentication Scheme with Revocability and Reusability. In Proceedings of the International Conference on Algorithms and Architectures for Parallel Processing, Guangzhou, China, 15–17 November 2018; Springer: Cham, Switzerland, 2018; pp. 193–203.
3. Chao, H.; He, Y.; Zhang, J.; Feng, J. Gaitset: Regarding Gait as a Set for Cross-View Gait Recognition. In Proceedings of the AAAI Conference on Artificial Intelligence, Honolulu, HI, USA, 27 January–1 February 2019; pp. 8126–8133.
4. Kiss, P.J.; Klimkó, G. Authentication of Electronic Legal Statements by a Trust Service Provider Using Two-Factor Dynamic Handwritten Signature Verification. In Proceedings of the International Conference on Electronic Government and the Information Systems Perspective, Bratislava, Slovakia, 14–17 September 2020; Springer: Cham, Switzerland, 2020; pp. 147–158.
5. Wang, J.; Wang, G. Quality-specific hand vein recognition system. *IEEE Trans. Inf. Forensics Secur.* **2017**, *12*, 2599–2610. [CrossRef]
6. Huang, D.; Tang, Y.; Wang, Y.; Chen, L.; Wang, Y. Hand-dorsa vein recognition by matching local features of multisource keypoints. *IEEE Trans. Cybern.* **2014**, *45*, 1823–1837. [CrossRef] [PubMed]
7. Kosmala, J.; Saeed, K. Human Identification by Vascular Patterns. In *Biometrics and Kansei Engineering*; Springer: New York, NY, USA, 2012; pp. 67–87.
8. Wang, L.; Leedham, G.; Cho, S.-Y. Infrared imaging of hand vein patterns for biometric purposes. *IET Comput. Vis.* **2007**, *1*, 113–122. [CrossRef]
9. Driscoll, P. Gray's Anatomy. *Emerg. Med. J. EMJ* **2006**, *23*, 492. [CrossRef]
10. Vairavel, K.; Ikram, N.; Mekala, S. Performance analysis on feature extraction using dorsal hand vein image. *Soft Comput.* **2019**, *23*, 8349–8358. [CrossRef]
11. Wang, Y.; Li, K.; Cui, J. Hand-Dorsa Vein Recognition Based on Partition Local Binary Pattern. In Proceedings of the IEEE 10th International Conference on Signal Processing Proceedings, Beijing, China, 24–28 October 2010; pp. 1671–1674.
12. Liu, F.; Jiang, S.; Kang, B.; Hou, T. A recognition system for partially occluded dorsal hand vein using improved biometric graph matching. *IEEE Access* **2020**, *8*, 74525–74534. [CrossRef]
13. Wang, J.; Pan, Z.; Wang, G.; Li, M.; Li, Y. Spatial pyramid pooling of selective convolutional features for vein recognition. *IEEE Access* **2018**, *6*, 28563–28572. [CrossRef]
14. Zhong, D.; Shao, H.; Du, X. A hand-based multi-biometrics via deep hashing network and biometric graph matching. *IEEE Trans. Inf. Forensics Secur.* **2019**, *14*, 3140–3150. [CrossRef]
15. Wang, Y.; Zheng, X. Cross-device hand vein recognition based on improved SIFT. *Int. J. Wavelets Multiresolution Inf. Process.* **2018**, *16*, 1840010. [CrossRef]
16. Wang, Y.; Cao, X.; Miao, X. Cross-device recognition of dorsal hand vein images by two-stage coarse-to-fine matching. *Vis. Comput.* **2021**, 1–16. [CrossRef]
17. Gu, G.; Bai, P.; Li, H.; Liu, Q.; Han, C.; Min, X.; Ren, Y. Dorsal Hand Vein Recognition Based on Transfer Learning with Fusion of LBP Feature. In Proceedings of the Chinese Conference on Biometric Recognition, Shanghai, China, 10–12 September 2021; Springer: Cham, Switzerland, 2021; pp. 221–230.

18. Toygar, Ö.; Babalola, F.O.; Bitirim, Y. FYO: A novel multimodal vein database with palmar, dorsal and wrist biometrics. *IEEE Access* **2020**, *8*, 82461–82470. [CrossRef]
19. Wang, L.; Leedham, G.; Cho, D.S.-Y. Minutiae feature analysis for infrared hand vein pattern biometrics. *Pattern Recognit.* **2008**, *41*, 920–929. [CrossRef]
20. Kumar, A.; Prathyusha, K.V. Personal authentication using hand vein triangulation and knuckle shape. *IEEE Trans. Image Process.* **2009**, *18*, 2127–2136. [CrossRef] [PubMed]
21. Li, K. Biometric Person Identification Using Near-infrared Hand-dorsa Vein Images. Ph.D. Thesis, University of Central Lancashire, Preston, UK, 2013.
22. Dalal, N.; Triggs, B. Histograms of Oriented Gradients for Human Detection. In Proceedings of the 2005 IEEE Computer Society Conference on Computer Vision and Pattern Recognition (CVPR'05), San Diego, CA, USA, 20–25 June 2005; pp. 886–893.
23. He, K.; Zhang, X.; Ren, S.; Sun, J. Deep Residual Learning for Image Recognition. In Proceedings of the IEEE Conference on Computer Vision and Pattern Recognition, Las Vegas, NV, USA, 27–30 June 2016; pp. 770–778.
24. Wang, Y.; Li, K.; Cui, J.; Shark, L.-K.; Varley, M. Study of Hand-Dorsa Vein Recognition. In Proceedings of the International Conference on Intelligent Computing, Changsha, China, 18–21 August 2010; pp. 490–498.
25. Li, K.; Liu, Q.; Zhang, G. Fusion of Partition Local Binary Patterns and Convolutional Neural Networks for Dorsal Hand Vein Recognition. In Proceedings of the Chinese Conference on Biometric Recognition, Shanghai, China, 10–12 September 2021; Springer: Cham, Switzerland, 2021; pp. 177–184.
26. Li, K.; Zhang, G.; Wang, P. Hand-Dorsa Vein Recognition Based on Deep Learning. In Proceedings of the 2018 International Conference on Security, Pattern Analysis, and Cybernetics (SPAC), Jinan, China, 14–17 December 2018; pp. 203–207.
27. Huang, D.; Zhu, X.; Wang, Y.; Zhang, D. Dorsal hand vein recognition via hierarchical combination of texture and shape clues. *Neurocomputing* **2016**, *214*, 815–828. [CrossRef]
28. Wang, Y.; Zheng, X.; Wang, C. Dorsal Hand Vein Recognition across Different Devices. In Proceedings of the Chinese Conference on Biometric Recognition, Chengdu, China, 14–16 October 2016; Springer: Cham, Switzerland, 2016; pp. 307–316.
29. Jia, W.; Xia, W.; Zhang, B.; Zhao, Y.; Fei, L.; Kang, W.; Huang, D.; Guo, G. A survey on dorsal hand vein biometrics. *Pattern Recognit.* **2021**, *120*, 108122. [CrossRef]
30. Zhao, A.; Balakrishnan, G.; Durand, F.; Guttag, J.V.; Dalca, A.V. Data Augmentation Using Learned Transformations for One-shot Medical Image Segmentation. In Proceedings of the IEEE/CVF Conference on Computer Vision and Pattern Recognition, Long Beach, CA, USA, 15–20 June 2019; pp. 8543–8553.
31. Suzuki, T. TeachAugment: Data Augmentation Optimization Using Teacher Knowledge. In Proceedings of the IEEE/CVF Conference on Computer Vision and Pattern Recognition, New Orleans, LA, USA, 21 June 2022; pp. 10904–10914.

electronics

MDPI

Article

Research on Path-Planning Algorithm Integrating Optimization A-Star Algorithm and Artificial Potential Field Method

Lisang Liu [1,2], Bin Wang [1,2,*] and Hui Xu [1,2]

[1] School of Electronic, Electrical Engineering and Physics, Fujian University of Technology, Fuzhou 350118, China

[2] National Demonstration Center for Experimental Electronic Information and Electrical Technology Education, Fujian University of Technology, Fuzhou 350118, China

* Correspondence: 2201905138@smail.fjut.edu.cn

Abstract: A fusion pathfinding algorithm based on the optimized A-star algorithm, the artificial potential field method and the least squares method is proposed to meet the performance requirements of path smoothing, response speed and computation time for the path planning of home cleaning robots. The fusion algorithm improves the operation rules of the traditional A-star algorithm, enabling global path planning to be completed quickly. At the same time, the operating rules of the artificial potential field method are changed according to the path points found by the optimal A-star algorithm, thus greatly avoiding the dilemma of being trapped in local optima. Finally, the least squares method is applied to fit the complete path to obtain a smooth path trajectory. Experiments show that the fusion algorithm significantly improves pathfinding efficiency and produces smoother and more continuous paths. Through simulation comparison experiments, the optimized A-star algorithm reduced path-planning time by 60% compared to the traditional A-star algorithm and 65.2% compared to the bidirectional A-star algorithm path-planning time. The fusion algorithm reduced the path-planning time by 65.2% compared to the ant colony algorithm and 83.64% compared to the RRT algorithm path-planning time.

Keywords: A-star algorithm; artificial potential field method; least squares method; path planning

Citation: Liu, L.; Wang, B.; Xu, H. Research on Path-Planning Algorithm Integrating Optimization A-Star Algorithm and Artificial Potential Field Method. *Electronics* **2022**, *11*, 3660. https://doi.org/10.3390/electronics11223660

Academic Editor: Akshya Swain

Received: 10 October 2022
Accepted: 4 November 2022
Published: 9 November 2022

1. Introduction

Thanks to the rapid development of artificial intelligence technology, cleaning robots are now being introduced into ordinary households. One of the key parts of a cleaning robot is the path planning of the cleaning area [1,2]. Path-planning technology involves following certain pathfinding rules in an environment with obstacles to obtain a collision-free path from the starting point to the target point that satisfies the evaluation metrics [3,4]. As the complexity of the working environment of mobile robots continues to increase, it also places higher demands on path-planning techniques. Depending on the environment in which the robot works and the work requirements, it can be divided into global path planning and local path planning [5,6]. Global path planning can be divided into graph-based search algorithms, sampling-based planning algorithms, biomimetic-based algorithms, neural network algorithms, etc. Local path planning is divided into the dynamic window method, time-elastic band method, artificial potential field method, etc.

In 1959, Dijkstra, a Dutch scientist, was the first to propose an algorithm to solve the single-source shortest path problem, which was one of the earliest global path-planning algorithms [7]. The algorithm is centered on the starting point using a breadth-first search strategy to continuously expand outwards and then continuously search for the shortest path between the starting point and each expanded node in the map until it finds the target node, completing the path planning. In 1968, P. Hart, N. Nilsson and B. Raphael first proposed the heuristic A-star algorithm to solve the global path optimal problem [8]. The traditional A-star algorithm starts from the starting point and calculates the cost of moving the current node to the starting point and the ending point under the constraint of the evaluation function and extends radially to the target point. When an obstacle is encountered, it returns to the vicinity of the starting point to resume pathfinding and repeats until the target point is reached. Therefore, the algorithm generates a large number of useless nodes to be computed in the process of application, which leads to problems such as too much computation, too much memory occupation and too long pathfinding time. X. Zhang [9] optimized the algorithm by introducing a time factor to the A-star algorithm and combining it with a time window and priority strategy. Although this method reduces the number of turns and improves the efficiency of the system, it greatly increases the computational effort, and the chosen obstacle avoidance strategy tends to cause the algorithm to fall into a dead loop.

The artificial potential field method was proposed by Khatib in 1986. The idea is to simulate the environment in which the mobile robot is located as the "gravitational force" in physics, called a virtual potential field [10]. A virtual artificial potential field is formed by the repulsive field of an obstacle and the gravitational field of the target location, in which the mobile robot is influenced by the potential field to automatically search for a suitable collision-free path. As the robot moves, the potential field it is subjected to varies continuously, along a gradient from the repulsive field of a particular obstacle or the gravitational field of a target point alone. Due to its real-time nature, the artificial potential field method has been applied to the field of dynamic obstacles by many scholars. However, for scenarios where multiple obstacles exist at the same time, the problem of becoming caught in local minima that cannot be dislodged and the phenomenon of oscillation near the target point easily occur. Y. Wang et al. [11] addressed the problem that the potential field method tends to fall into local optima by improving the gravitational formulation of the traditional potential field method by adding new variables and also expanding the obstacle to a circular obstacle. The optimized algorithm solves the problem of the manual potential field method not being able to avoid large obstacles, greatly reduces the scanning time and reduces the working cost of the mobile robot. However, in an environment with irregular obstacles, it easily divides the passable paths between obstacles into obstacle regions, thus failing to obtain the optimal path. When the map is updated quickly, its real-time obstacle avoidance capability will be greatly reduced. H. Liu et al. [12] introduced the idea of fuzzy control into the path planning of mobile robots, dividing the environment in which the robot moves into two parts: a global safety region and a local danger, according to the location of obstacles and their influence range. In safe areas, the artificial potential field method acts on the robot to guide it towards the target; in dangerous areas, the artificial potential field method is combined with fuzzy control to guide the robot to avoid obstacles and move towards the target. The precise control of the deflection angle of the mobile robot effectively reduces the problem of unreachable targets and local minima. However, the fusion algorithm is limited in its application, and in dynamic environments, it relies mainly on the artificial potential field method, which does not practically solve the shortcomings of the artificial potential field method in dynamic environments.

This paper first optimizes the structure of the traditional A-star algorithm, then optimizes the potential field method by adding an intermittent point search strategy and constructing an intermittent point judgment function, and then combines the artificial potential field method and the least squares method to propose a fused path-planning algorithm. On the one hand, the structure of the traditional A-star algorithm is optimized to improve its speed in global path planning, and on the other hand, the artificial potential field method is optimized to solve the problem of the potential field method tending to fall into local optimality. The fusion algorithm overcomes the drawbacks of the original algorithm well and improves the efficiency and success rate of path planning. The paper concludes with a comparative simulation analysis of the optimized A-star algorithm and the traditional and bidirectional A-star algorithms for different starting points and different map environments. The fusion algorithm is analyzed and compared with the ant colony algorithm and the RRT algorithm. The simulation comparison and data analysis confirm the fast, robust and advanced nature of the optimized fusion algorithm.

2. Global Path Planning

2.1. Traditional A-Star Algorithm

The A-star algorithm is a heuristic search algorithm for finding optimal paths in static obstacle environments [13]. It combines the advantages of Dijkstra's algorithm to find the shortest path well and the heuristic search algorithm breadth-first search (BFS) to search upwards at the most probable places first [14], to which a cost evaluation function is added to find the optimal path point. The principle is shown in Figure 1.

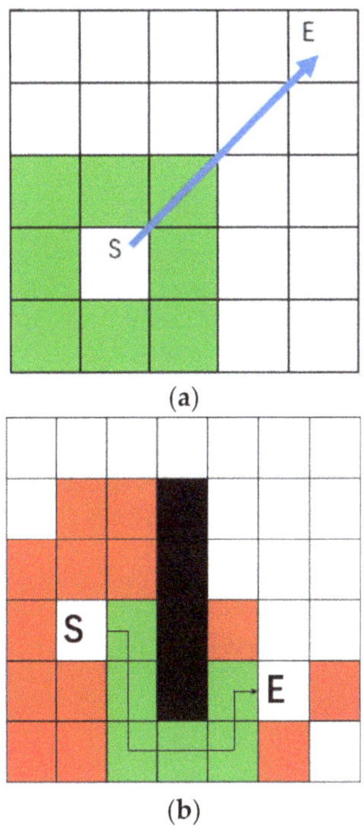

Figure 1. Schematic diagram of the traditional A-star algorithm. (**a**) Conventional A-star algorithm pathfinding in the absence of obstacles; (**b**) conventional A-star algorithm pathfinding in the presence of obstacles.

As shown in Figure 1a, S is the starting point and E is the end point. If the path encounters an obstacle, the algorithm returns to the starting point and searches again until the search width exceeds the width of the obstacle, then continues to search and so on until it reaches the target point. In practice, only a small number of nodes are relevant to the path, but many nodes need to be computed. For this reason, a large number of useless nodes are searched, creating problems such as too many calculations, more useless memory usage and longer pathfinding times. In addition, there are too many turning points in the planned path. The traditional A-star algorithm does not smooth the path, so the planned path turns rigidly, and the robot needs to accelerate and decelerate frequently to perform the turning action when walking, which is not conducive to the robot's path tracking.

The set of optimal path points then forms the optimal path, where the cost evaluation function is as follows:

$$f(n) = h(n) + g(n) \tag{1}$$

where $f(n)$ is the cost function of the current position, $g(n)$ is the actual cost of the mobile robot from the starting point to the current position and $h(n)$ is the estimated cost of the mobile robot from the current position to the position of the target point [15].

For the cost function, the more commonly used metric is the Euclidean or Manhattan distance. The absolute value of the difference between the x-coordinates of two points and the sum of the absolute values of the differences between the y-coordinates of two points is called the Manhattan distance [16]. In this paper, the Euclidean distance is used, i.e.,

$$g(n) = \sqrt{(X_n - X_s)^2 + (Y_n - Y_s)^2} \tag{2}$$

$$h(n) = \sqrt{(X_t - X_n)^2 + (Y_t - Y_n)^2} \tag{3}$$

where (X_n, Y_n) is the position of the current point, (X_s, Y_s) is the position of the starting point and (X_t, Y_t) is the position of the target point. The closer the value of the function $h(n)$ is to the actual value, the more efficient and accurate the search will be.

2.2. Optimization of the A-Star Algorithm

Based on the shortcomings of the traditional A-star algorithm, the algorithm structure of the A-star algorithm is improved so that when an obstacle is encountered during the pathfinding process, instead of returning to the vicinity of the starting point for a new pathfinding instance, the algorithm defines the node that the mobile robot is currently on as the parent node. The open list is used to store the data of the parent node and the neighboring points with the parent node as the core and to filter the next walkable path point of the mobile robot based on the open list. The close list is defined to store the entire set of walkable path points for the mobile robot. As this paper uses Euclidean distances, the F-value is the distance from the current position of the mobile robot to the target point. The G-value is the distance from the current position of the mobile robot to the starting point. These are the values of $f(n)$ and $g(n)$ at the current point as stated above.

The optimized A-star algorithm sets the evaluation function of the obstacle node in situ to infinity, indicating unreachability, and then finds the best node among the nodes around the parent node as the parent node for the next cycle. This continues until the path found leaves the obstacle node, and then the path search continues forward. The principle is shown in Figure 2.

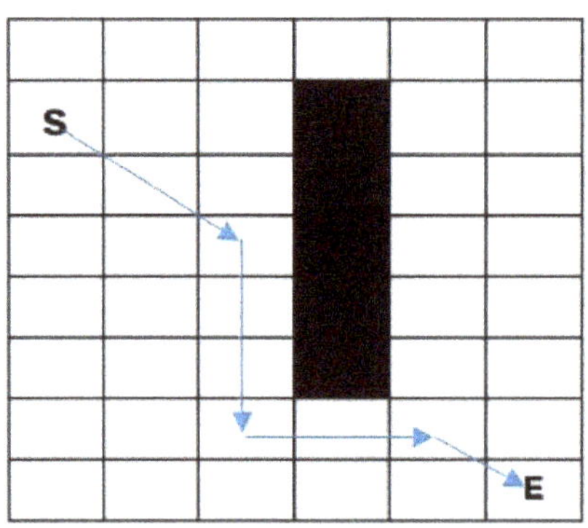

Figure 2. Optimization A-star algorithm schematic.

The algorithm steps are as shown in Figure 3:

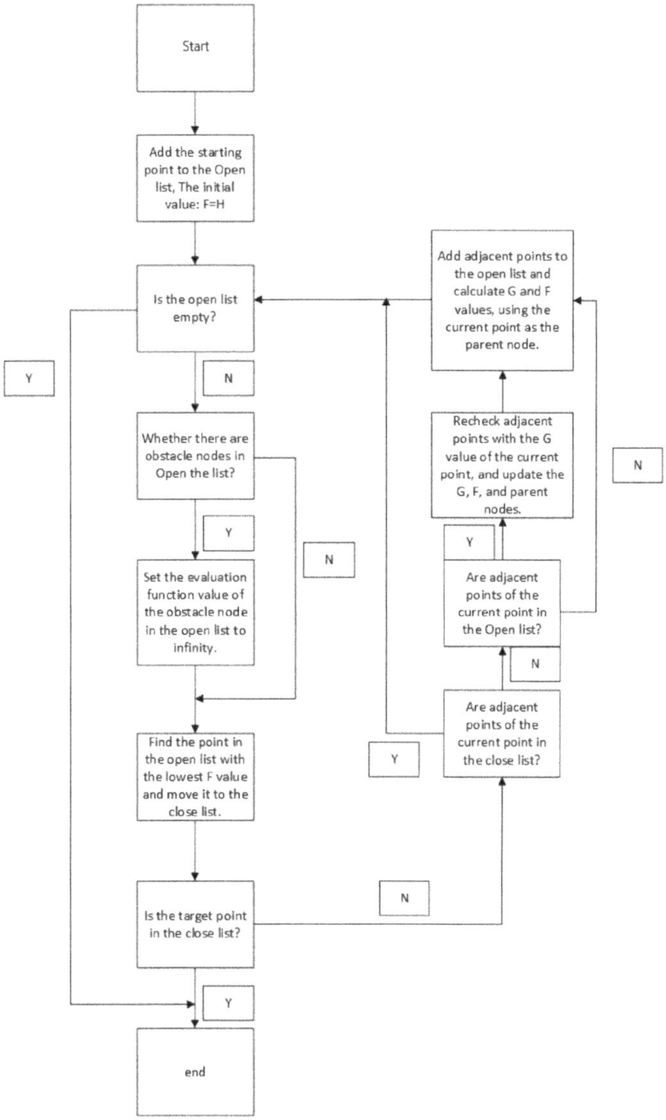

Figure 3. Optimization of the A-star calculation hair path-planning process.

The optimized A-star algorithm modifies the path-planning rules of the traditional A-star algorithm. When an obstacle is encountered, the optimized A-star algorithm no longer returns to the vicinity of the starting point to re-run the path planning. When an obstacle is encountered, the optimized A-star algorithm stays put and sets the value of the cost function of the obstacle node from the open list to infinity. The best of the child nodes is then selected as the parent node for the next cycle. Compared to traditional A-star algorithms, the optimized A-star algorithm will reduce the amount of computation and data redundancy, thus reducing path-planning time. As the complexity of the environment increases, the benefits of the optimized A-star algorithm will become more apparent. Detailed simulation comparisons and data analysis will be shown in Section 4.1 of this paper.

3. Local Route Planning

3.1. Artificial Potential Field Method

The artificial potential field method relies on the repulsive force $F_{rep}(x)$ (shown in Equation (7)), which is directed from the obstacle to the mobile robot, and the gravitational force $F_{aat}(x)$ (shown in Equation (5)), which is directed from the mobile robot to the target point, to construct the gravitational field [17–19]. The gravitational field varies with the distance between the vehicle and the target point. The gravitational field is proportional to the linear distance between the moving vehicle and the target point, as shown below.

$$U_{att}(x) = \frac{1}{2}K\rho^2(P_S, P_E) \tag{4}$$

where $U_{att}(x)$ is the gravitational potential field generated by the target on the mobile robot, K is the coefficient of action of the gravitational field and $\rho(P_S, P_E)$ is the Euclidean distance from the starting point to the endpoint.

The gravitational force is the negative gradient of the gravitational potential field, as follows:

$$F_{aat}(x) = -\nabla U_{att}(x) = -K\rho(P_S, P_E) \tag{5}$$

The magnitude of the repulsive field is inversely proportional to the distance between the mobile robot and the target point, as follows:

$$U_{rep}(x) = \begin{cases} \frac{1}{2}K_{rep}\left[\frac{1}{\rho(P,P_{obs})} - \frac{1}{P_0}\right]^2 & ,\rho(P, P_{obs}) \leq P_0 \\ 0 & ,\rho(P, P_{obs}) \geq P_0 \end{cases} \tag{6}$$

where $U_{rep}(x)$ is the repulsive field of the obstacle, K_{rep} is the coefficient of action of the repulsive field, $\rho(P, P_{obs})$ is the Euclidean distance between the mobile robot and the obstacle and P_0 is the critical distance of the repulsive force on the obstacle. When the distance P_0 between the trolley and the obstacle is greater, the repulsive force on the trolley is zero [20]. Meanwhile, the repulsive force is the negative gradient of the repulsive field, as follows:

$$F_{rep}(x) = -\nabla U_{rep}(x) = \begin{cases} K_{rep}\left[\frac{1}{\rho(P,P_{obs})} - \frac{1}{P_0}\right]\frac{1}{\rho^2(P,P_{obs})} & ,\rho(P, P_{obs}) \leq P_0 \\ 0 & ,\rho(P, P_{obs}) \geq P_0 \end{cases} \tag{7}$$

When there are N obstacles on the map, the combined force on them is as follows:

$$F_{sum}(x) = F_{aat}(x) + \sum_{i=1}^{N} F_{rep}(x) \tag{8}$$

where $F_{sum}(x)$ is the repulsive force, $\sum_{i=1}^{N} F_{rep}(x)$ is the combined gravitational force and $F_{sum}(x)$ is the set of repulsive forces.

As shown in Figure 4, the artificial potential field approach to path planning involves the mobile robot following the direction of the combined force F_{sum}. The combined force is generated by the combination of multiple repulsive forces F_{rep} exerted by the obstacle on the mobile robot and gravitational forces F_{aat} exerted by the target point on the mobile robot. The repulsive, gravitational and combined forces all follow the rule of vector addition and subtraction. As shown by Equations (5) and (7), the repulsive and gravitational forces change as the position of the mobile robot in the map environment changes. This, therefore, causes the combined forces to change as well. It is the real-time nature of the artificial potential field method that allows the artificial potential field method to be used for local path planning.

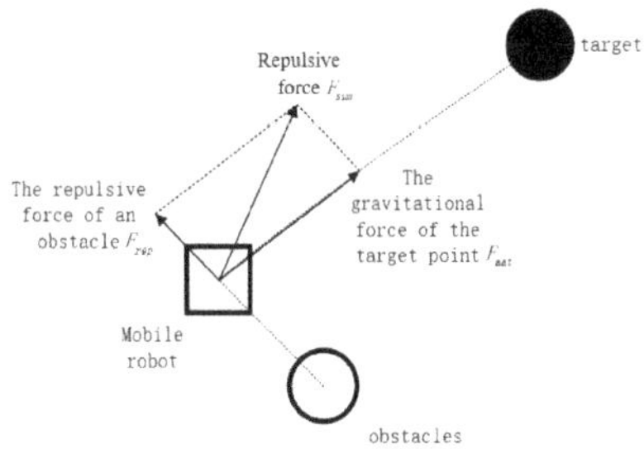

Figure 4. Force analysis of robot in artificial potential field.

Due to the pathfinding rules of the potential field method, there is an inherent problem with the traditional artificial potential field approach to path planning; when the robot is moving near the target point, if there are no obstacles near the target point, the gravitational force will be close to zero and the repulsive force should also be close to zero at this point. If there is an obstacle near the target, the robot will be subjected to a large repulsive force, when the gravitational force on the robot at the target is less, which will cause the robot to move away from the target, preventing it from ever reaching it. Secondly, when the robot moves to certain locations on the map, it may be that the combined gravitational and repulsive forces at that location are zero and the robot will fall into a local optimum.

3.2. Optimization of the Artificial Potential Field Method

3.2.1. Interruption Point Selection

In this paper, an interruption search strategy is proposed to improve the speed of path planning by the artificial potential field method. The optimized artificial potential field method optimizes the A-star algorithm on the basis of the global path-planning data obtained. The optimized potential field method uses the turning points of the global paths as intermittent points. The turning point of the global path is the place where the path obtained from the global path planning takes a turn. The judgment function is shown in Equations (6) and (7).

$$K_1 = \frac{X_c - X_{c-1}}{Y_c - Y_{c-1}} \tag{9}$$

$$K_2 = \frac{X_{c+1} - X_c}{Y_{c+1} - Y_c} \tag{10}$$

where (X_c, Y_c) are the coordinates of the current point. As the global path-planning data are stored on a stack, (X_{c-1}, Y_{c-1}) are specified as the coordinates of the point after the current point. (X_{c+1}, Y_{c+1}) are the coordinates of the point before the current point. K_1, K_2 are the slope of the line connecting the current point to the two adjacent points before and after it. When K_1, K_2 are not equal, the current point is the turning point.

As shown in Figure 5. Starting from point 1, point 3 is the temporary endpoint of point 1, and point 4 is the temporary endpoint of point 2. In the artificial potential field method of smoothing, point 2 is the starting point when the distance from the robot's position to 1 is greater than the spacing from 1 to 2. When it reaches the endpoint, the endpoint is used as the temporary endpoint and the penultimate path point is used as the temporary start point. For this reason, in the manual potential field method for local pathfinding, this paper uses the Manhattan distance for determining whether an intermediate point has been passed. To improve the efficiency of the optimal potential field method and reduce data redundancy, this paper proposes an adaptive number of iterations, which is formulated as follows:

$$I = L * \sqrt{(R_X - T_{EX})^2 + (R_Y - T_{EY})^2} \tag{11}$$

where L = 10, and each reference value corresponds to a Euclidean distance of 1. I is the iteration parameter. (R_X, R_Y) are the coordinates of the robot's position. (T_{EX}, T_{EY})

are the coordinates of the current temporary target point. Iteration parameters are used in the optimized potential field method when performing path planning. The iteration parameters determine whether the optimized potential field method path planning will reach the target point or not. The Euclidean distance between the current position of the robot and the temporary target point is rounded upwards. When the mobile robot changes its temporary start points and temporary endpoints, the number of iterations is adjusted.

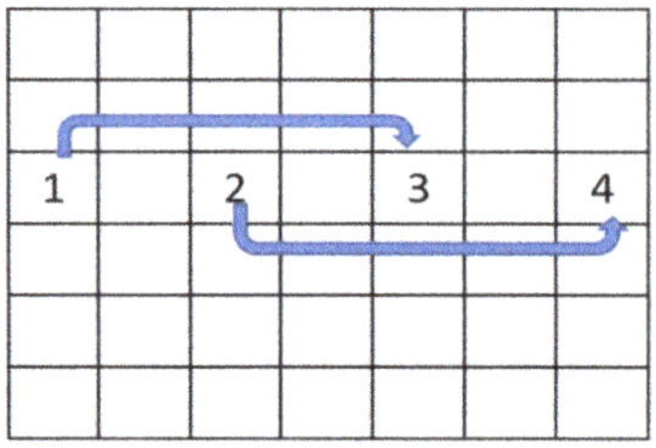

Figure 5. Potential Field Method Wayfinding.

3.2.2. Least Squares Method

The path obtained by the potential field method will look like a discontinuous path because the interval between the temporary start point and the temporary endpoint is a turning point in the path-planning process of the optimal potential field method. The principle is shown in Figure 6, points 1, 2, 3 and 4 are turning points. Where points 2 and 3 are also turning points. The path discontinuity appears when 2 is the temporary starting point. Therefore, whenever there are turning points in the global path, the most dominant field method will produce path discontinuities after smoothing.

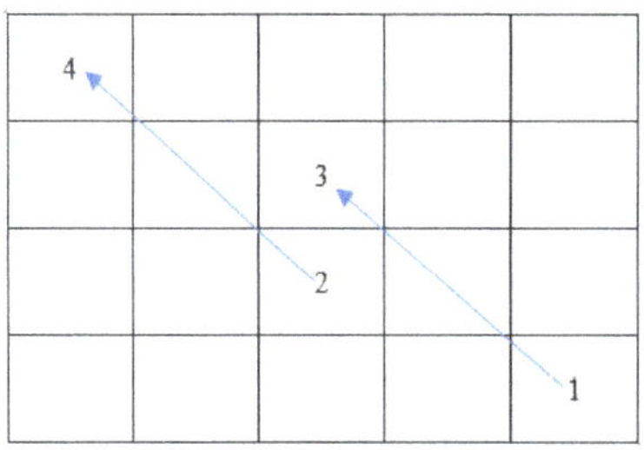

Figure 6. Potential field method misalignment principle.

Path fitting is currently more commonly used in the interpolation and least squares methods. The interpolation method is used in environments where the accuracy and reliability of the observed data are high. The interpolation method seeks high accuracy, which results in large data redundancy and can lead to longer path-planning times, resulting in failure to avoid dynamic obstacles in a timely manner. The least squares method is suitable for situations where the observed data already contain unavoidable errors and it is only necessary to reach as close to them as possible. The least squares method is fast and can also fulfill path-planning requirements, so this paper uses least squares for path fitting [21,22].

To solve the path discontinuity problem of the optimal field method, this paper proposes the combination of the least squares method for path fitting, so that the mobile robot can obtain a smooth and continuous path trajectory. The variance between the hypothetical regression results and the actual values is expressed as follows:

$$\phi(x) = a_0 + a_1 x + a_2 x^2 + \cdots + a_k x^k \tag{12}$$

where a is the polynomial's indeterminate coefficient and x is the path point's abscissa. The total distance between each point and this curve is as follows:

$$R^2 = \sum_{i=1}^{n}\left[y_i - \left(a_0 + a_1x_i + a_2{x_i}^2 + \cdots + a_k{x_i}^k\right)\right]^2 \tag{13}$$

where n is the polynomial's highest order and k is the highest order of the system. We can obtain the following results by deriving the indeterminate coefficients in the regression equation:

$$\begin{bmatrix} 1 & x_1 & \cdots & x_1^k \\ 1 & x_2 & \cdots & x_2^k \\ \vdots & \vdots & \ddots & \vdots \\ 1 & x_n & \cdots & x_n^k \end{bmatrix}\begin{bmatrix} a_0 \\ a_1 \\ \vdots \\ a_k \end{bmatrix} = \begin{bmatrix} y_1 \\ y_2 \\ \vdots \\ y_n \end{bmatrix} \tag{14}$$

The coefficient matrix A and the fitting curve can be generated simultaneously using matrix operation, as shown below.

$$X^*A = Y \Rightarrow A = (X^*X)^{-1}X^*Y \tag{15}$$

Therefore, the algorithmic flow for optimizing the artificial potential field method is as shown in Figure 7:

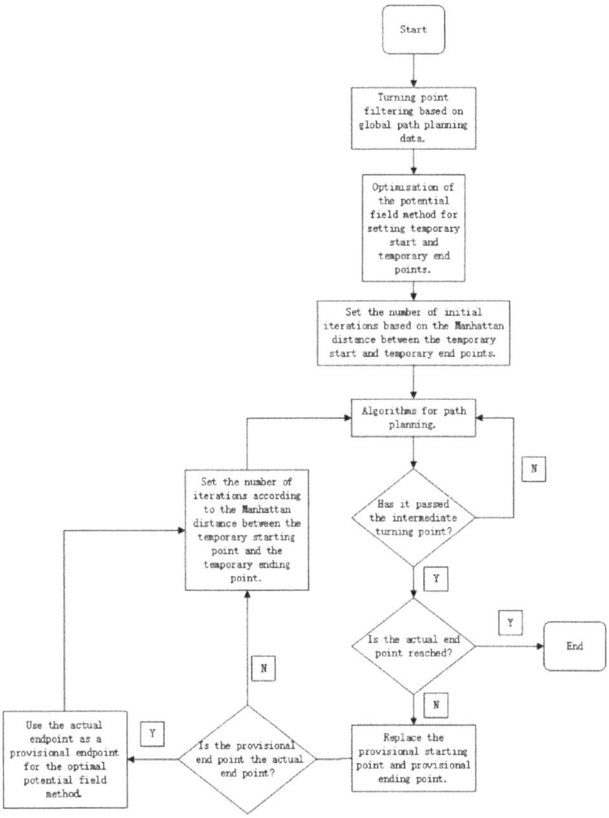

Figure 7. Optimization of the potential field method path-planning process.

The optimized potential field method changes the start and end points in pathfinding and solves the problem of falling into local optima and unreachable target points by setting parameters such as step size reasonably. At the same time, the optimized potential field method takes the path obtained from global planning as the general direction, so it also limits the pathfinding range of local paths and avoids crossing between multiple obstacles. At the same time, since the potential field method gives the mobile robot a repulsive force during path planning, pushing the robot away from the obstacle appropriately, the fusion algorithm page solves the problem of narrower paths for the mobile robot to pass through. In addition, as the optimized A-star algorithm can only perform global path planning, moving obstacles may appear on the map when the robot moves, so given the nature of the potential field method updating the map in real time, the potential field method can perform path smoothing while also performing dynamic obstacle avoidance.

4. Simulation and Analysis

4.1. Comparative Analysis of Optimization Algorithms and Traditional Algorithms

To verify the effectiveness and generalization of the fusion algorithm based on the optimized A-star algorithm and the artificial potential field method proposed in this paper, MATLAB simulations of the traditional A-star algorithm and the optimized A-star algorithm, the traditional potential field method and the optimized potential field method were carried out in simple and complex environments to verify the performance of the optimization algorithm as proposed in this paper.

A simple environment mapping is shown in Figure 8. A raster map of three different environments was constructed in MATLAB, with a map size of 20 × 20 and black x's indicating obstacles. The red boxed points are the calculated path points, the green boxed points are the points to be included in the open list to be checked and the connecting lines are the optimal paths found. From Figure 8, it can be seen that the optimized A-star algorithm can obtain the same path as the traditional A-star algorithm under the same map environment. A comparison of the path-planning times for the 10 groups based on Figure 8a,b is shown in Table 1, where the optimized algorithm reduces the path-planning time by 60% compared to the traditional algorithm. In addition, based on Figure 8b,d,e,f, the optimized A-star algorithm can obtain a feasible path quickly and accurately in the same map environment with different start and end point settings.

Table 1. Comparison of path-planning times based on Figure 8a,b (unit: s).

Time	1	2	3	4	5	6	7	8	9	10
Traditional A-star algorithm	0.623	0.583	0.635	0.592	0.606	0.581	0.604	0.600	0.596	0.603
Optimization of the A-star algorithm	0.368	0.225	0.208	0.190	0.218	0.175	0.179	0.172	0.170	0.174

As shown in Figure 9, after the map environment is changed, the optimized algorithm can still meet the path-planning requirements. Compared with the traditional A-star algorithm, the optimized A-star algorithm searches a much smaller range of path points than the traditional algorithm while obtaining the same path in the same map environment. As shown in Figure 9b,d, the optimized A-star algorithm can complete the path-planning requirements in the new map environment with different start and end points replaced. As shown in Figure 9a,b and the path-planning time comparison in Table 2, the path-planning time of the optimized A-star algorithm is reduced by more than 50% compared to the traditional A-star algorithm.

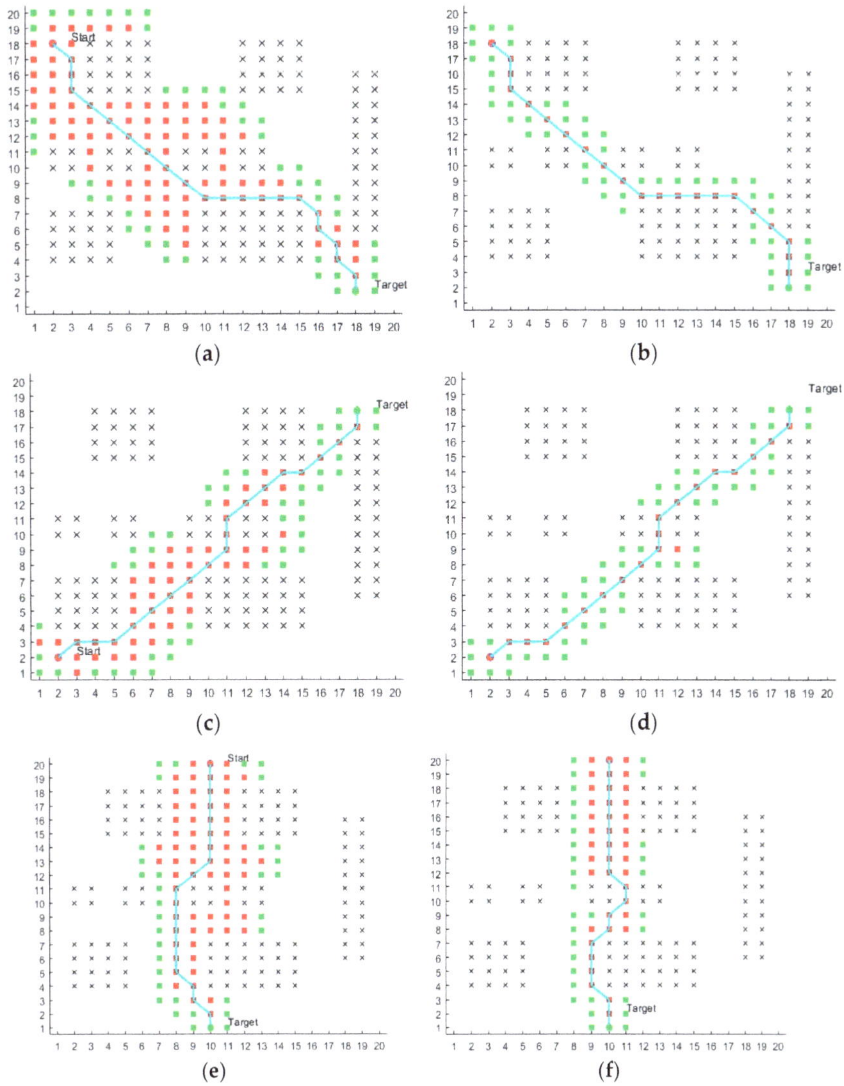

Figure 8. Global path comparison. (**a**,**c**,**e**) Traditional A-star algorithm; (**b**,**d**,**f**) optimization of the A-star algorithm.

Table 2. Comparison of path-planning times before and after algorithm optimization (unit: s).

Time	1	2	3	4	5	6	7	8	9	10
Traditional A-star algorithm	0.309	0.307	0.308	0.305	0.303	0.299	0.304	0.300	0.296	0.299
Optimization of the A-star algorithm	0.140	0.139	0.139	0.139	0.140	0.138	0.136	0.134	0.133	0.133

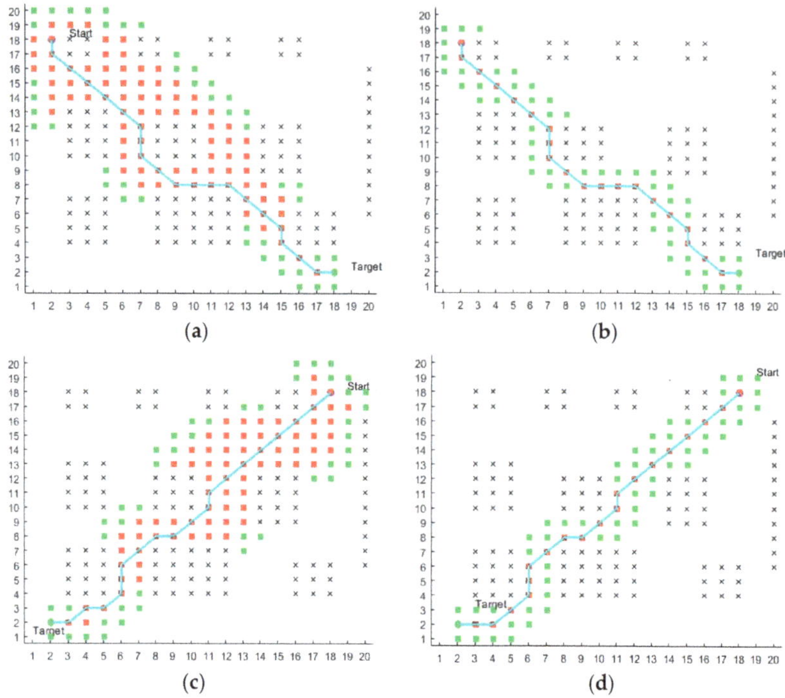

Figure 9. Global path comparison for map replacement. (**a**,**c**) Traditional A-star algorithm; (**b**,**d**) optimization of the A-star algorithm.

The complex map is shown in Figure 10, and a 40 × 40 grid map was created in MATLAB. The optimization algorithm is still able to obtain a feasible path when performing path planning in a complex map environment.

Compared with the traditional A-star algorithm, the advantage of less computation of the optimized A-star algorithm is more obvious. A comparison of path-planning times based on Figure 10a,b, as shown in Table 3, shows that the optimized A-star algorithm reduces the pathfinding time by nearly 70% compared to the traditional algorithm.

Table 3. Comparison of path-planning times for complex maps (unit: s).

Time	1	2	3	4	5	6	7	8	9	10
Traditional A-star algorithm	2.427	2.436	2.419	2.371	2.402	2.552	2.349	2.359	2.371	2.368
Optimization of the A-star algorithm	0.683	0.732	0.703	0.689	0.691	0.782	0.704	0.699	0.685	0.692

Based on the analysis of the path-planning time and the path accessibility, the optimized A-star algorithm achieves a significant improvement in path accessibility and speed over the traditional algorithm, while ensuring that a complete global path can be obtained. As the complexity of the pathfinding environment increases, the efficiency of the optimized algorithm becomes more pronounced than that of the traditional algorithm. Nevertheless, the optimized A-star algorithm does not solve the problem of insufficient smoothness at path transitions.

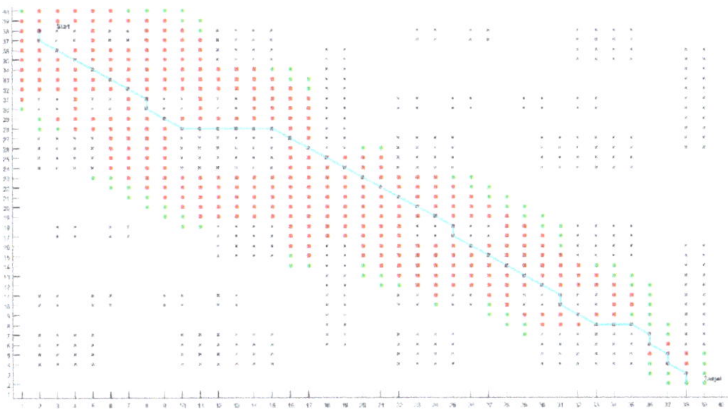

(a)

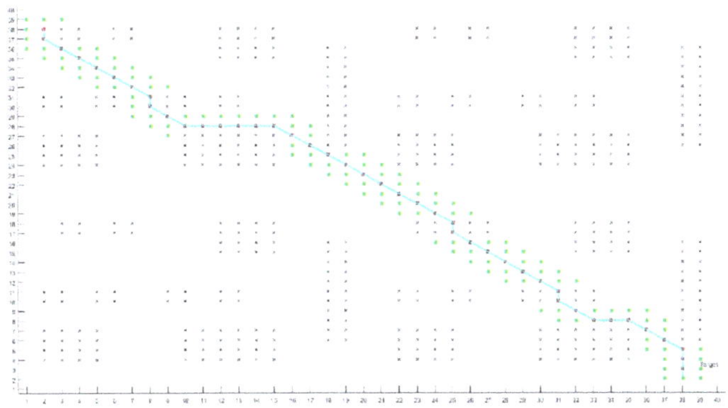

(b)

Figure 10. Comparison of global path planning for complex map environments. (**a**) Traditional A-star algorithm; (**b**) optimization of the A-star algorithm.

To address the problem of the global path-planning transitions not being smooth enough to facilitate smooth robot tracking, an optimized potential field method is proposed for smoothing, as described in the previous section. The simulation diagram is shown in Figure 11.

Figure 11 also shows that the optimization algorithm is still able to meet the pathfinding requirements when different starting points and different endpoints are set in the same environment. Moreover, from Figure 11b,d, it can be seen that the optimization algorithm can effectively complete the library path planning when the same start and end points are set in different map environments. It can also be seen from Figure 11 that the algorithm has good robustness and generalizability. The comparison between Figures 11 and 12 shows that the smoothing process reduces a large number of inflection points compared to global path planning, thus improving the path-tracking capability of the robot and increasing the movement speed of the mobile robot. In addition, as the repulsive force of the obstacle in the artificial potential field method acts on the cart, it will cause the cart to move away from the obstacle appropriately, making the path of the cart more reasonable and solving the global path-planning problem of walking along the edge of the obstacle.

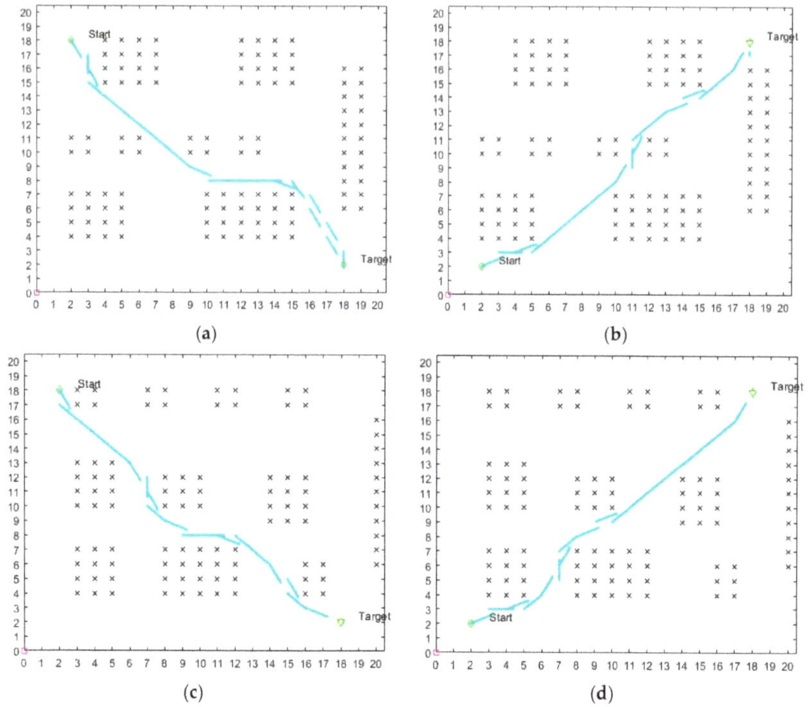

Figure 11. Use of the optimized potential field method in different map environments to set different start and end points of the path planning. (**a**,**b**) are route plans for different starting points and different end points in the same environment; (**c**,**d**) are route plans for different starting points and different end points after changing maps.

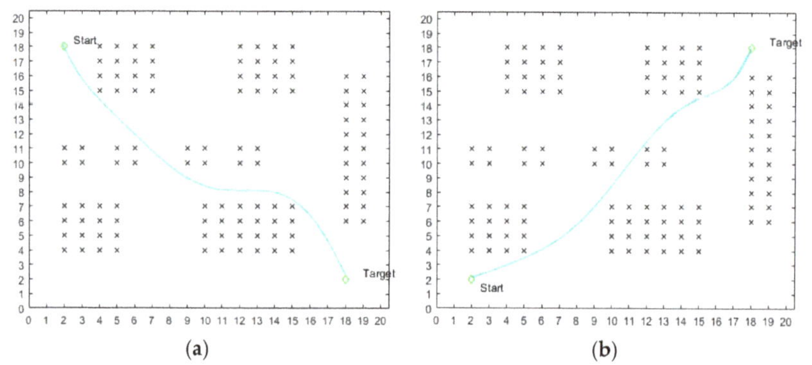

Figure 12. Path fitting results. (**a**) Figure 11a path fit; (**b**) Figure 11b path fit.

As shown in Figure 11, the path obtained by the artificial potential field method is discontinuous in the global path steering (see Section 3.2.2 above for the rationale). We, therefore, used the least squares method of path fitting to obtain Figure 12. Figure 12 gives the fitted paths planned by the algorithm based on different starting and ending points in the same environment.

In the case of local path planning, the repulsive force from the obstacles is only applied to the moving car within a certain range with the moving car as the center of the circle, and the repulsive force from the obstacles outside the range is 0. In addition, considering that dynamic obstacles are inevitable in the real environment, 20 random dynamic obstacles were included, indicated by the circles in Figure 13.

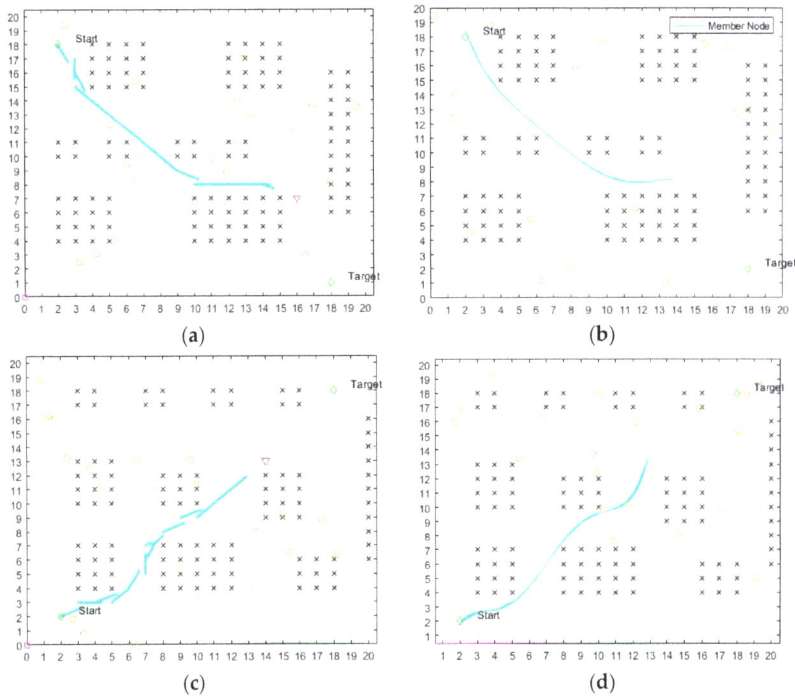

Figure 13. Addition of dynamic obstacles.

From the simulation results in Figure 13a, it can be seen that during the pathfinding process, when dynamic obstacles appear within the force range of the vehicle, the mobile robot will react quickly to avoid the obstacles. It is guaranteed to move towards the target point while moving towards the safe area as far as possible, as shown in Figure 13a at points (8, 9) and (12, 8). The vehicle based on global path planning can achieve real-time dynamic obstacle avoidance and reach the target point smoothly and safely. Meanwhile, as shown in Figure 13c,d, after the map environment is changed, the mobile robot path-planning process can still maintain the ability to quickly avoid dynamic obstacles while traveling towards the target point. As shown in Figure 13b, the vehicle can achieve real-time dynamic obstacle avoidance based on global path planning and reach the target point smoothly and safely.

4.2. Comparative Analysis of Optimized A-Star Algorithm and Bidirectional A-Star Algorithm

Currently, many scholars have proposed improving the bidirectional A-star algorithm for path planning [23–25]. The principle of the bidirectional A-star algorithm is to select a virtual endpoint in the middle of the straight line distance between the starting point and the ending point [26]. If the virtual endpoint lies in an obstacle area, the nearest obstacle edge is chosen as the virtual endpoint, while the endpoint at the other end is used as the starting point, and then path planning is performed towards the virtual endpoint.

As shown in Figure 14a,c, (10.9) is defined as the midpoint of the bidirectional A-star, indicated by the red "☆" in the diagram. A comparison of the path planning of the optimized A-star algorithm and the bidirectional A-star algorithm proposed in this paper shows that the optimized A-star algorithm has better throughput, that the path-planning efficiency of the optimized A-star algorithm is higher (as shown in Table 4), and that the path-planning time of the optimized A-star algorithm is 65.2% less than the path-planning time of the bidirectional A-star algorithm. The number of computing nodes is 103 in Figure 14a and 70 in Figure 14b. The optimized A-star algorithm reduces the number of computing nodes by approximately 32% compared to the bidirectional A-star algorithm. Meanwhile, as shown in Figure 14a,c, the bidirectional A-star algorithm is affected by various factors such as obstacle size and map environment complexity when selecting virtual endpoints, which indirectly affects the pathfinding efficiency of the bidirectional A-star algorithm. The pathfinding efficiency of the optimized A-star algorithm is only affected by the complexity of the map environment, and therefore the performance of the optimized A-star algorithm is more stable than that of the bidirectional A-star algorithm.

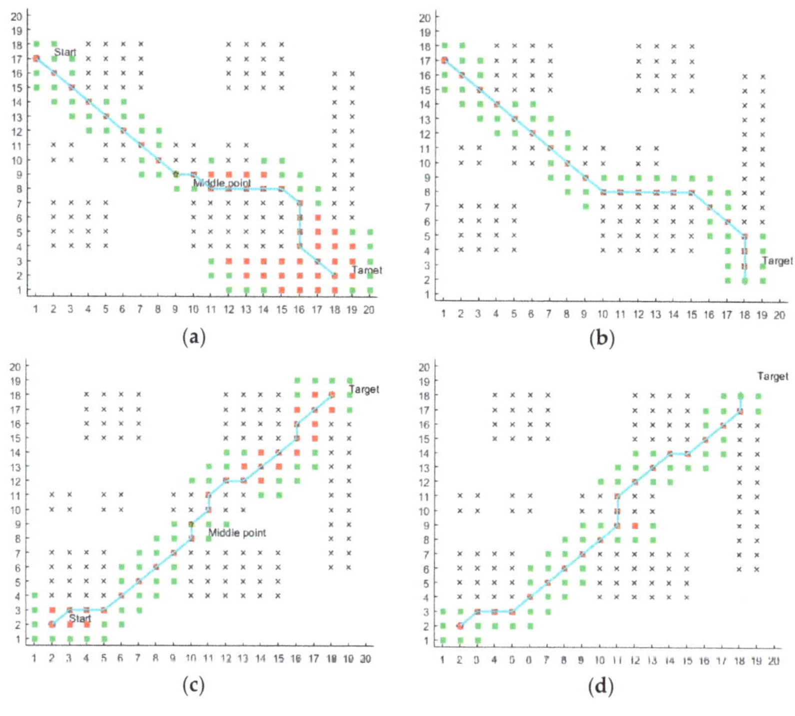

Figure 14. Optimized A-star algorithm vs. the bidirectional A-star algorithm. (**a**,**c**) The bidirectional A-star algorithm; (**b**,**d**) optimization of the A-star algorithm.

Table 4. Comparison of path-planning times based on Figure 14a,b (unit: s).

Time	1	2	3	4	5	6	7	8	9	10
Two-way exploration A-star algorithm	0.486	0.493	0.497	0.452	0.462	0.489	0.473	0.472	0.458	0.454
Optimization of the A-star algorithm	0.163	0.157	0.161	0.158	0.154	0.162	0.169	0.181	0.183	0.158

4.3. Simulation Analysis of the Effect of Different L Values on the Potential Field Method

In this paper, the algorithm rules for path planning in the optimized potential field method are changed, replacing the fixed starting point and fixed endpoint of the traditional potential field method with temporary starting points and temporary endpoints that change as the position of the mobile robot changes. As a result, the number of iterations of the traditional potential field method is no longer suitable for the optimized potential field method. For this reason, an adaptive iteration number setting is proposed in this paper, which has been theoretically derived in Section 3.2.1. The experimental part was set up with L values of 1, 10 and 100 for comparison, and the results of the comparison are shown below.

From Figure 15a,b, it can be seen that when the value of L is too small, it leads to too few iterations, and therefore it is difficult for the optimized potential field method to reach the interim endpoint. In addition, as shown in Figure 16, the path-planning time of the optimized potential field method is only reduced by 30% when the value of L is 1 compared to when the value of L is 10. From Figure 15b,c, the path-planning results are almost the same for L values of 10 and 100. However, as can be seen in Figure 16, the optimized potential field method time increases by 288% for the L value of 100 compared to the L value of 10. Therefore, it can be concluded that when the L value is too large, it increases the path-planning time of the optimized potential field method significantly, but there is no significant improvement in the final path-planning result.

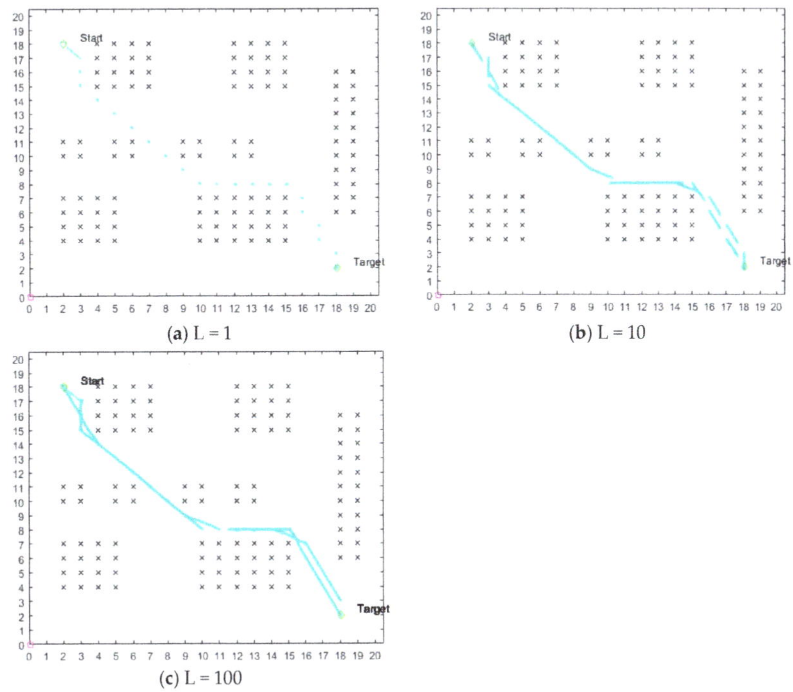

Figure 15. Effect on the optimized potential field method for different L values.

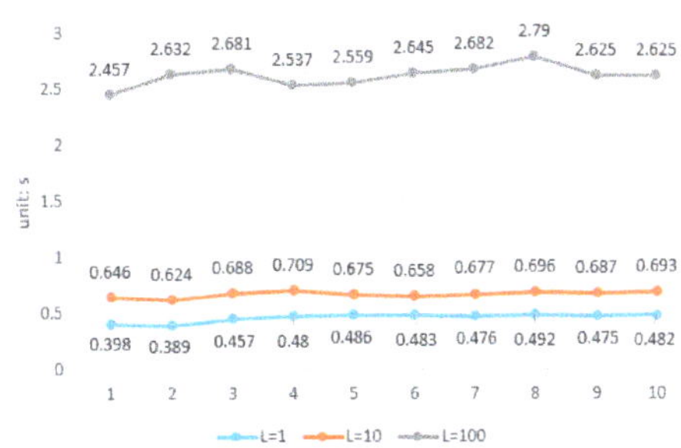

Figure 16. Optimal potential field method path-planning time for three L values.

4.4. Comparative Analysis of Fusion and Ant Colony Algorithms

The principle of the ant colony algorithm represents the feasible solution to the problem to be optimized in terms of the paths taken by ants, with all the paths of the entire ant colony forming the solution space of the problem to be optimized. The ants with shorter paths release more pheromones, and as time passes, the concentration of pheromones accumulated on the shorter paths gradually increases, and the number of ants choosing that path increases. Eventually, the entire ant population will concentrate on the best path under the effect of positive feedback, which then corresponds to the optimal solution of the problem to be optimized [27,28].

A comparison of the path-planning times from Figures 11b and 17b is shown in Table 5. The average planning time of the fusion algorithm is 0.722 s, and the average path-planning time of the ant colony algorithm is 2.073 s. The path-planning time of the fusion algorithm is reduced by 65.2% relative to the ant colony algorithm, which indicates that the fusion algorithm and the ant colony algorithm have the same path-planning requirements while completing the same path. This indicates that the fusion algorithm and the ant colony algorithm have the same path-planning time advantage while meeting the same path-planning requirements.

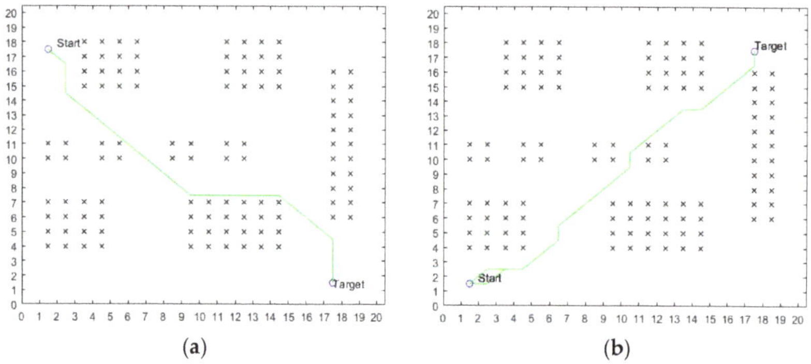

Figure 17. Ant colony algorithm path planning.

Table 5. Comparison of path-planning time based on Figures 11b and 17b (unit: s).

Time	1	2	3	4	5	6	7	8	9	10
Ant colony algorithm	2.185	2.092	2.029	2.047	2.053	2.064	2.066	2.049	2.055	2.092
Fusion algorithm	0.693	0.712	0.706	0.718	0.700	0.750	0.740	0.736	0.738	0.726

4.5. Comparative Analysis of Fusion Algorithms and the RRT Algorithm

The RRT algorithm takes the starting point as the root node and adds leaf nodes by random sampling to generate a randomly expanded tree that is able to find a path from the starting point to the target point when the target point lies on the randomly expanded tree [29,30].

A comparison of the path-planning times for Figures 11a and 18a is shown in Table 6, with the same guaranteed obstacle environment and the same start and end points. The red circled area indicates the target point area. When the path is planned to this area, the path planning is completed. The average planning time of the fusion algorithm is 0.655 s. The average path-planning time of the RRT algorithm is 4.003 s. The path-planning time of the fusion algorithm is reduced by 83.64% relative to the path-planning time of the RRT algorithm. This indicates that the fusion algorithm and the RRT algorithm have a greater path-planning time advantage while meeting the same path-planning requirements.

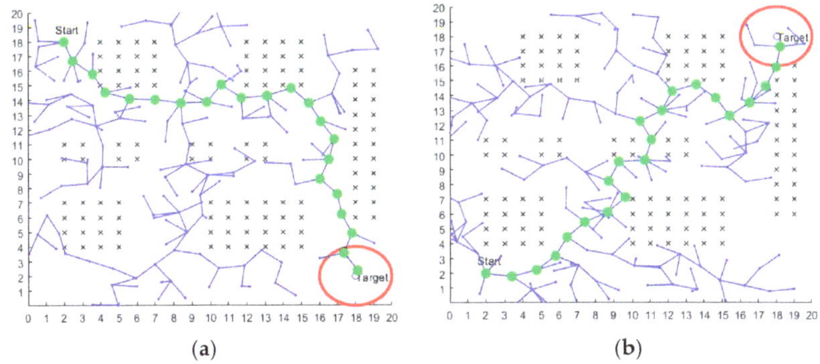

Figure 18. The RRT algorithm path planning.

Table 6. Comparison of path-planning time based on Figures 11a and 18a (unit: s).

Time	1	2	3	4	5	6	7	8	9	10
The RRT algorithm	3.779	3.871	3.996	4.063	3.921	4.012	4.039	4.182	4.094	4.074
Fusion algorithm	0.593	0.633	0.677	0.715	0.640	0.647	0.658	0.648	0.680	0.660

5. Results and Discussion

Due to the increasingly complex obstacle environments faced by mobile robots, it has been difficult for traditional path-planning algorithms to meet the path-planning needs of mobile robots, so this paper proposes a path-planning algorithm that incorporates the optimized A-star algorithm and the artificial potential field method. For the traditional A-star algorithm, as of now, many scholars are choosing to incorporate function constraints or implement algorithmic parallel pathfinding methods such as the bidirectional A-star algorithm. Such optimization provides very limited performance improvement to the traditional A-star algorithm, and a comparative analysis is also presented in the simulation section of this paper. Therefore, future optimization directions for the A-star algorithm should be able to significantly reduce the amount of algorithmic computation and increase path optimality.

The artificial potential field method has the problem of not being able to perform optimization and easily falling into local optimality, so the artificial potential field method can complete path planning but not necessarily find the optimal path. This leads to the fact that the artificial potential field method is capable of path planning, but may not necessarily find the optimal path, or may fall into a local optimum at some location in the map. This is an important issue in the current study of artificial potential field methods for path planning. This paper, therefore, uses the data obtained from the global path planning of the optimized A-star algorithm to optimize the artificial potential field method, thus limiting the path-planning space of the artificial potential field method. The problem of the artificial potential field method tending to fall into local optimality and path non-optimality is thus solved.

Each algorithm has its own strengths and weaknesses and limitations in the use of scenarios. In order to achieve complementary advantages, researchers at home and abroad in recent years have preferred to fuse multiple algorithms, which will also be the development direction for path planning for a long time.

6. Conclusions

In this paper, the traditional A-star algorithm is optimized, and new pathfinding rules and algorithm structures are designed to obtain global path-planning information. The data obtained by the optimized A-star algorithm are applied to local path planning, a new pathfinding rule for the potential field method is proposed, an intermittent point pathfinding strategy is added and an intermittent point judgment function is constructed. The results show that the fusion algorithm can reduce the pathfinding time by about 40% while guaranteeing the same path as the traditional algorithm. The fusion algorithm also reduces the probability of the potential field method falling into local optima, improves the smoothness of the planned path at the turn, and enables the robot to find a safe path quickly even in complex environments. In this paper, the fusion-optimized A-star algorithm is compared with the more advanced bidirectional A-star algorithm, the ant colony algorithm and the RRT algorithm for path-planning time to demonstrate the advanced nature of the optimization algorithm. Through a series of experiments and data analyses, it can be concluded that the fusion algorithm in this paper can effectively improve the path-planning capability of mobile robots in complex scenarios, but its operation speed still needs to be improved, which is the focus of the next step.

Author Contributions: All of the authors contributed extensively to the work. B.W. proposed the key ideas, analyzed the key contents using a simulation and wrote the manuscript; L.L. obtained the financial support for the project leading to this publication; H.X. modified the manuscript. All authors have read and agreed to the published version of the manuscript.

Funding: This work was supported in part by the Initial Scientific Research Fund of FJUT under Grant GY-Z12079, Grant GY-Z21036 and Grant GY-Z20067 and in part by Fujian Provincial Science and Technology Department (Grant 2022H6005 and Grant 2022J01952).

Acknowledgments: The authors would like to thank the anonymous reviewers for their valuable comments.

Conflicts of Interest: The authors declare no conflict of interest.

References

1. Liu, J.; Zhang, L.; Li, H. The study on path planning of stair cleaning robot rest platform. *J. Jiangxi Norm. Univ. (Nat. Sci.)* **2022**, *46*, 67–74.
2. Yang, M.; Xie, M.; Zhang, X. The Design of Obstacle Avoidance System and Path Planning of Cleaning Robot. *Chang. Inf. Commun.* **2021**, *34*, 14–17.
3. Li, X.; Ma, X.; Wang, X. A Survey of Path Planning Algorithms for Mobile Robots. *Comput. Meas. Control* **2022**, *30*, 9–19.
4. Zhang, K. Overview of Path Planning Algorithms for Unmanned Vehicles. *Equip. Manuf. Technol.* **2021**, *6*, 111–113. [CrossRef]
5. Yang, Y. Overview of Global Path Planning Algorithms for Mobile Robots. *Inf. Rec. Mater.* **2022**, *23*, 29–32.
6. Wang, Z.; Hu, X.; Li, X.; Du, Z. Overview of Global Path Planning Algorithms for Mobile Robots. *Comput. Sci.* **2021**, *48*, 19–29.
7. Dijkstra, E. A Note on Two Problems in Connexion With Graphs. In *Numerische Mathematik*; Springer: New York, NY, USA, 1959; pp. 269–271.
8. Hart, P.E.; Nilsson, N.J.; Raphael, B. A formal basis for the heuristic determination of minimum cost paths in graphs. *IEEE Trans. Syst. Sci. Cybern.* **1968**, *4*, 100–107. [CrossRef]
9. Zhang, X.; Zou, Y. Collision-free path planning for automated guided vehicles based on improved A-star algorithm. *Syst. Eng. Theory Pract.* **2021**, *41*, 240–246.
10. Khatib, O. *The Potential Field Approach and Operational Space Formulation in Robot Control*; Springer: New York, NY, USA, 1986; pp. 367–377.
11. Wang, Y. Improvement of Artificial Potential Field Algorithm for Robots in Different Environments. *Nanjing Univ. Inf. Sci. Technol.* **2020**, *2*, 45–53.
12. Liu, H.; Wang, D.; Wang, Y.; Lu, X. Research of Path Planning for Mobile Robots Based on Fuzzy Artificial Potential Field Method. *Control Eng. China* **2022**, *29*, 33–38.
13. Sheng, L.; Bao, L.; Wu, P. Application of Heuristic Approaches in the Robot Path Planning and Optimization: A Review. *Electron. Opt. Control* **2018**, *25*, 58–64.
14. Hen, J.; Wen, J.; Xie, G. Mobile robot path planning based on improved A-star algorithm. *J. Guangxi Univ. Sci. Technol.* **2022**, *33*, 78–84.
15. Lin, M.; Yuan, K.; Shi, C.; Wang, Y. Path planning of Mobile robot based on improved A-star algorithm. *Mech. Sci. Technol. Aerosp. Eng.* **2022**, *41*, 795–800.
16. Zhou, J.; Yang, L.; Zhang, C. Indoor robot path planning based on improved A-star algorithm. *Mod. Electron. Tech.* **2022**, *32*, 202–206.
17. Shi, Z.; Mei, S.; Shao, Y.; Wan, R.; Song, Z.; Xie, M.; Li, Y. Research status and prospect of path planning for mobile robots based on artificial potential field method. *J. Chin. Agric. Mech.* **2022**, *42*, 182–188.
18. Liu, X. Review on UAV obstacle avoidance methods. *J. Ordnance Equip. Eng.* **2022**, *43*, 40–47.

19. Wu, Q.; Zeng, Q.; Luo, J.; Kuang, X.; Huang, H. Application research on improved artificial potential field method in UAV path planning. *J. Chongqing Univ. Technol. (Nat. Sci.)* **2022**, *36*, 144–151.
20. Sun, L. Obstacle Avoidance Algorithm of Autonomous Vehicle Based on an Improved Artificial Potential Field. *J. Henan Univ. Sci. Technol. (Nat. Sci.)* **2022**, *43*, 5–6, 28–34, 41.
21. Gao, Q. Research on Least Square Curve Fitting and Optimization Algorithm. *Ind. Control Comput.* **2021**, *34*, 100–101.
22. Wang, R. Research of Least Square Curve Fitting and Simplified Algorithm. *Sens. World* **2021**, *27*, 8–10, 25.
23. Zhao, J.; Wang, J.; Lu, Z.; Sun, H. Research on path planning of medical inspection robot based on improved bidirectional exploration A-star algorithm. *J. Jilin Norm. Univ. (Nat. Sci. Ed.)* **2022**, *43*, 121–127.
24. Wang, Z.; Zeng, G.; Huang, B. Mobile robot path planning algorithm based on improved bidirectional A star. *Transducer Microsyst. Technol.* **2020**, *39*, 141–143, 147.
25. Yue, G.; Zhang, M.; Shen, C.; Guan, X. Bi-directional smooth A-star algorithm for navigation planning of mobile robots. *Sci. Sin. Technol.* **2021**, *51*, 459–468. [CrossRef]
26. Chen, D.; Liu, X.; Liu, S. Improved A-star algorithm based on two-way search for path planning of automated guided vehicle. *J. Comput. Appl.* **2021**, *41*, 309–313.
27. Wang, Z.; Xia, X. Application of adaptive ant colony algorithm in robot path planning. *J. Minnan Norm. Univ. (Nat. Sci.)* **2022**, *35*, 38–45.
28. Yue, C.; Huang, J.; Deng, L. Research on improved ant colony algorithm in AGV path planning. *Comput. Eng. Des.* **2022**, *43*, 2533–2541.
29. Wang, H.; Cui, Y.; Li, M.; Li, G. Mobile Robot Path Planning Algorithm Based on Improved RRT∗FN. *J. Northeast Univ. (Nat. Sci.)* **2022**, *43*, 1217–1224, 1249.
30. Chen, H.; Wang, L. A Path Planning Algorithm Based on Two-Way Simultaneous No-Collision Goal RRT. *J. Airf. Eng. Univ. (Nat. Sci. Ed.)* **2022**, *23*, 60–67.

electronics

MDPI

Article

Improvement of Image Stitching Using Binocular Camera Calibration Model

Mengfan Tang [1], Qian Zhou [1], Ming Yang [1,*], Yifan Jiang [1] and Boyan Zhao [2]

[1] College of Computer & Information Science, Southwest University, Chongqing 400715, China
[2] Scivatar Technologits Co., Ltd., Chongqing 401220, China
* Correspondence: yangming@swu.edu.cn

Abstract: Image stitching is the process of stitching several images that overlap each other into a single, larger image. The traditional image stitching algorithm searches the feature points of the image, performs alignments, and constructs the projection transformation relationship. The traditional algorithm has a strong dependence on feature points; as such, if feature points are sparse or unevenly distributed in the scene, the stitching will be misaligned or even fail completely. In scenes with obvious parallaxes, the global homography projection transformation relationship cannot be used for image alignment. To address these problems, this paper proposes a method of image stitching based on fixed camera positions and a hierarchical projection method based on depth information. The method does not depend on the number and distribution of feature points, so it avoids the complexity of feature point detection. Additionally, the effect of parallax on stitching is eliminated to a certain extent. Our experiments showed that the proposed method based on the camera calibration model can achieve more robust stitching results when a scene has few feature points, uneven feature point distribution, or significant parallax.

Keywords: image stitching; camera calibration; layered projection; binocular ranging; stereo correction

Citation: Tang, M.; Zhou, Q.; Yang, M.; Jiang, Y.; Zhao, B. Improvement of Image Stitching Using Binocular Camera Calibration Model. *Electronics* **2022**, *11*, 2691. https://doi.org/10.3390/electronics11172691

Academic Editor: Oscar Deniz Suarez

Received: 13 July 2022
Accepted: 24 August 2022
Published: 27 August 2022

1. Introduction

Image stitching technology is widely used in medical, aerial photography [1], assisted driving, surveillance, virtual reality (VR), and other fields [2], but there are still some problems to be solved. Traditional image stitching algorithms have the following shortcomings: firstly, the dependence on the scene feature points is strong, which can easily lead to stitching misalignments or even complete failure, and the robustness is relatively low. Secondly, the stitching effect is different in different scenes, and lighting and parallax have obvious effects. Additionally, it is impossible to use the global homography projection change relationship for image alignment [3–6].

To address the shortcomings of traditional methods, the following methods are proposed in this paper: (1) An image stitching method based on a special plane. This method takes advantage of the fact that the relative positions of the cameras are invariant and places a checker pattern on the special plane for camera calibration. The obtained internal parameters and the external parameters relative to the pattern can then construct an accurate projection relationship between the two cameras about this plane. (2) We further introduce a hierarchical projection method based on depth information. This method uses the internal and external parameters obtained from camera calibration for stereo correction and to project images taken by the binocular cameras into a form with parallel optical axes, a co-planar imaging plane, and identical internal parameters. It is then possible to obtain the horizontal parallax of the corresponding pixel point in the overlapping area of the image by stereo matching and to calculate the depth information of the point according to the focal length and baseline length of the binocular lens. The depth information is used to layer the original image, and each layer is mapped using different relationships.

Finally, the image stitching results can be obtained by superimposing all projections. The experimental results showed that our method is more robust than other algorithms based on feature points when a scene has few feature points, uneven distribution of feature points, or significant parallax.

2. Related Work

2.1. Image Stitching

Image stitching refers to the process of seamlessly stitching several overlapping pictures into a new picture with higher resolution and a wider viewing angle through pixel alignment. In 1996, Richard Szeliski [7] proposed the Levenberg Marquardt (LM) algorithm to improve the quality of stitched panoramic images. In recent years, to solve the most critical parallax problem in image stitching, scholars in the industry have proposed algorithms such as Global Similarity Priority (GSP) [8] and Seam-guided Local Alignment (SEAGULL) [9]. Most of these algorithms are based on the meshing concept of As-Projective-As-Possible Image Stitching (APAP) [10]; on this basis, mechanisms such as line alignment constraint and contour detection were added to improve the stitching performance. However, such algorithms usually have higher requirements for stitching images. In addition, some scholars have combined image stitching with deep learning, giving rise to Learned Invariant Feature Transform (LIFT) [11]. This algorithm is based on a convolutional neural network (CNN) [12] and uses backward propagation for end-to-end training. Its training data adopts the feature points detected by Structure-from-Motion (SFM) [11]. The feature point detection performance of this model comprehensively exceeds that of Scale Invariant Feature Transform (SIFT) [13,14]. However, due to the cumbersome training process, it is still unable to be put into practical application.

2.2. Camera Calibration

To determine the relationship between the coordinates of objects in the real world and their pixel coordinates on a camera imaging plane, a geometric camera imaging model must be established, and in real-world cases, the parameters of the camera must be obtained through experiments and calculations [3], i.e., camera calibration. In 1971, Abdel-Aziz [15] first proposed a camera calibration method based on Direct Linear Transform (DLT) [16] transformation and developed a linear equation as a mathematical model of camera imaging through the corresponding relationship between three-dimensional space points and two-dimensional pixels. However, because the linear equation can only calculate linear relationships and cannot consider the distortion effect of the camera, the parameters obtained by this method are only applicable to some scenes. In 1992, Faugeras and Luong [17] proposed a camera self-calibration method which does not need a fixed reference object; instead, it is only necessary to change the camera viewpoint to shoot multiple images and establish a connection according to the same points within the images. Although this method is not limited by a reference object, the calibration process is complex and its use has not been extensive. In 1999, Zhang Zhengyou [18] proposed a camera calibration method based on a planar pattern which uses a nonlinear model for the calculation to solve the optimal results regarding the camera parameters. This method not only has high precision and low manufacturing cost of the selected reference, but also is suitable for various calibration scenes in daily life.

3. Method

3.1. Establishment of Camera Calibration Model

In this section, we first introduce the image stitching method based on the camera parameters, as image stitching methods based on the camera calibration model depend on the internal and external parameters of the camera, which may be obtained by offline calibration. Using the checker pattern for monocular calibration, we fixed the camera position, took twenty pictures of the pattern at different positions and angles, and then measured and recorded the horizontal distance between adjacent corner points on the pattern. To construct the world coordinate system, we took the plane of the pattern as the $Z = 0$ plane, the corner point at the top left of the pattern as the origin, and the vertical outward direction of the pattern as the Z-axis. At the same time, the world coordinates of all corner points in the figure were constructed according to the measured real distance; there were then stored in a list. Then, using the SIFI algorithm, we detected all of the corner points of the figure and recorded their pixel coordinates in a separate list. The calibration process is shown in Figure 1.

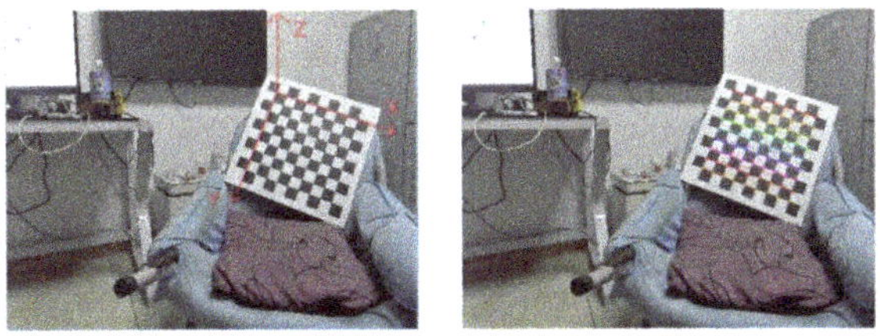

Figure 1. World coordinate system construction and corner detection.

We took 20 pictures of the pattern shown in Figure 1 at different positions and angles, recorded the world coordinates and pixel coordinates of their corner points, and solved the internal and external parameters using the perspective projection model using Equation (1):

$$Z_c \begin{bmatrix} u \\ v \\ 1 \end{bmatrix} = \begin{bmatrix} f_x & 0 & u_0 & 0 \\ 0 & f_y & v_0 & 0 \\ 0 & 0 & 1 & 0 \end{bmatrix} \begin{bmatrix} R & t \\ 0_3^T & 1 \end{bmatrix} \begin{bmatrix} X_w \\ Y_w \\ Z_w \\ 1 \end{bmatrix} \tag{1}$$

where Z_c is the scale factor, which represents the distance from the corner in the figure to the camera imaging plane, (u, v) is the pixel coordinate of the corner point, (X_w, Y_w, Z_w) is the world coordinate of the corner point, and the right side of (1) is the internal parameter matrix of the camera and the external parameter matrix of the relative pattern, respectively. Since we specified $Z = 0$ as the plane of the pattern, the Z_w value of all corner points was 0; as such, Equation (1) could be simplified follows:

$$Z_c \begin{bmatrix} u \\ v \\ 1 \end{bmatrix} = \begin{bmatrix} f_x & 0 & u_0 \\ 0 & f_y & v_0 \\ 0 & 0 & 1 \end{bmatrix} \begin{bmatrix} r_1 & r_2 & t \end{bmatrix} \begin{bmatrix} X_w \\ Y_w \\ 1 \end{bmatrix} \tag{2}$$

where r_1, r_2 are the first and second column components of the rotation matrix R. Letting homography matrix H be the product of internal parameter matrix and external parameter matrix:

$$H = \begin{bmatrix} h_{11} & h_{12} & h_{13} \\ h_{21} & h_{22} & h_{23} \\ h_{31} & h_{32} & h_{33} \end{bmatrix} = \begin{bmatrix} f_x & 0 & u_0 \\ 0 & f_y & v_0 \\ 0 & 0 & 1 \end{bmatrix} \begin{bmatrix} r_1 & r_2 & t \end{bmatrix} \tag{3}$$

Since the degree of freedom of homogeneous matrix H was 8, we set $h_{33} = 1$ and then substituted it into Equation (2):

$$\begin{cases} h_{31}uX_w + h_{32}uY_w + u - h_{11}X_w - h_{12}Y_w - h_{13} = 0 \\ h_{31}vX_w + h_{32}vY_w + v - h_{21}X_w - h_{22}Y_w - h_{23} = 0 \end{cases} \tag{4}$$

There are eight unknown parameters in (4); at the very least, the pixel coordinates and world coordinates of the four diagonal points are needed to construct the linear equations and solve them. Using the constraints of the orthogonality of r_1 and r_2 units, the internal and external parameter matrices in each picture were obtained. After obtaining the above parameter matrix, the transformation relationship between world coordinates and pixel coordinates could be constructed. However, the essence of image stitching is to project a floating image onto a plane where the target image is located. Therefore, the transformation relationship of the pixel coordinates between two images is required. Taking a binocular camera as an example, we let the internal parameter matrices of the left and right cameras be K_l and K_r, the external parameter matrices be E_l and E_r, the world coordinates of point P_w on the pattern be $(X_w, Y_w, Z_w, 1)$, and its projection points on the imaging surfaces of the left and right cameras be $p_l(u_l, v_l, 1)$ and $p_r\ (u_r, v_r, 1)$, which can be listed as (5):

$$\begin{cases} p_l = K_l E_l P_w \\ p_r = K_r E_r P_w \end{cases} \tag{5}$$

where $E_l = \begin{bmatrix} r_{l1} & r_{l2} & t_l \end{bmatrix}$, $E_r = \begin{bmatrix} r_{r1} & r_{r2} & t_r \end{bmatrix}$. After transforming Equation (5), the result was as shown in Equation (6):

$$p_l = K_l E_l E_r^{-1} K_r^{-1} p_r \tag{6}$$

In Equation (6), E_l and E_r are the external parameters of the camera imaging surface relative to the pattern, and there was a constraint condition of $Z_w = 0$. Therefore, the coordinate transformation relationship could only produce a good splicing effect on the plane where the pattern was located. When the model was used for image registration directly, obvious ghosting, dislocation, and even deformation occurred, as shown in Figures 2 and 3.

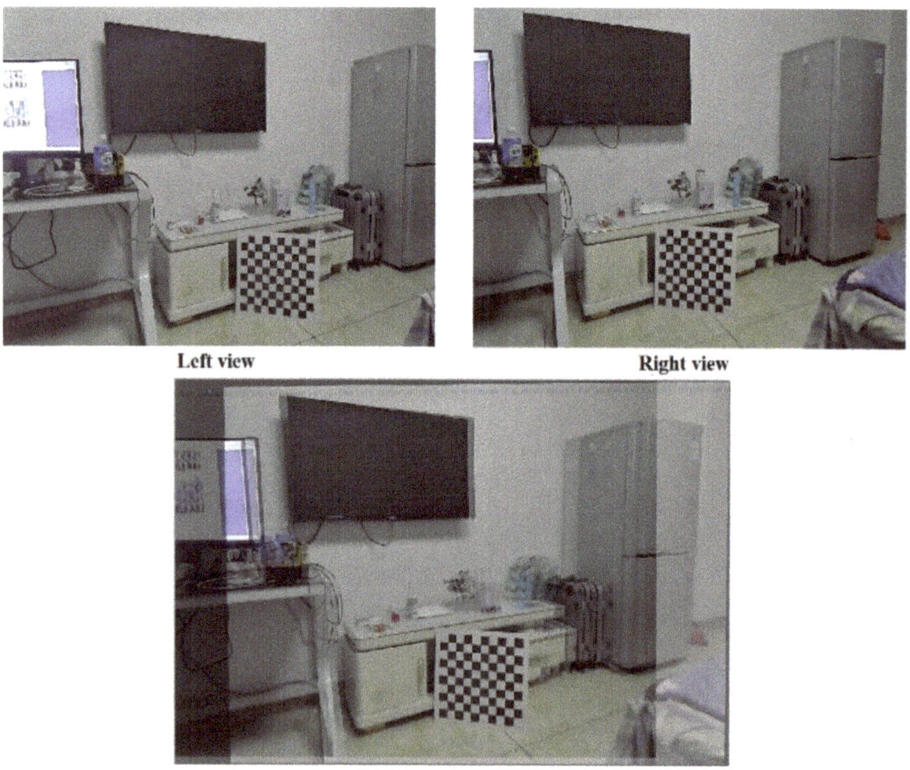

Figure 2. Image stitching results.

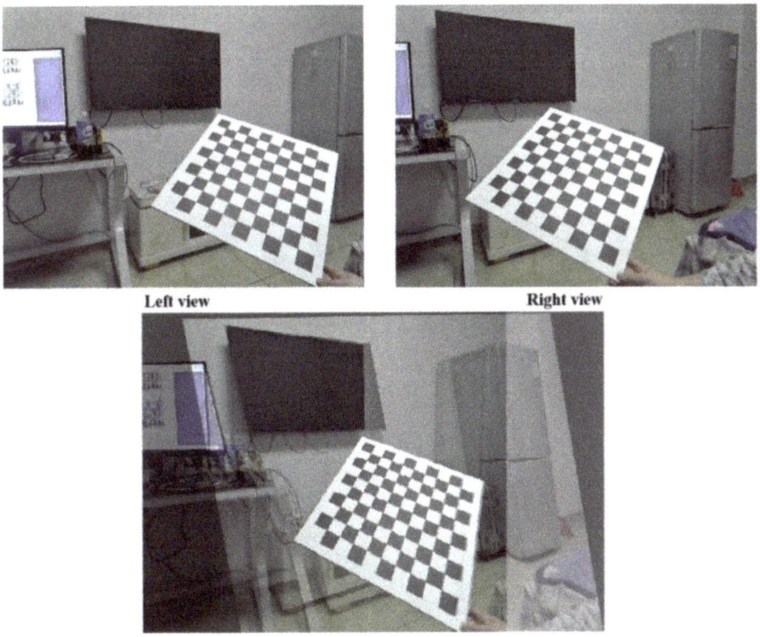

Figure 3. Image stitching results.

3.2. Design of Camera Calibration Method for Special Plane

In the camera calibration scene mentioned in the previous section, because the background object and the pattern were in different planes, the same coordinate transformation relationship could not be applied for alignment, so a more stable projection transformation model was required. The design was as follows: we fixed the binocular camera, made the line of the left and right camera optical center parallel to the wall, and fixed the pattern on the wall or vertical plane. We then took a set of pictures of the pattern in which the left and right camera imaging surface were approximately parallel to each other, as shown in Figure 4.

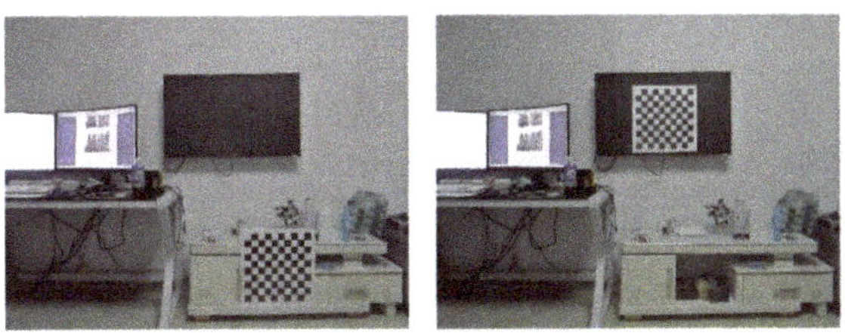

Figure 4. Camera calibration method for the special plane.

As can be seen in Figure 4, the checker pattern was fixed on the vertical plane of the cabinet and TV cabinet respectively. The binocular camera was placed at a position whereby the baseline was parallel to the wall. Therefore, it was considered that a stable projection model had been established. The external parameters were calibrated when the reference object was in this attitude and were used to construct the coordinate transformation relationship for image registration; the effect of this is shown in Figures 5 and 6. Although there were still a number of ghost dislocation phenomena in the stitching results, the plane behind the pattern achieved a relatively good stitching effect. This was because although the pattern and the object behind it were not on the same planes, the two planes were approximately parallel. It could be determined that the TV plane and the pattern plane had the same rotation matrices relative to the camera imaging plane, and that the translation vector was only slightly different in the t_z component. Therefore, using the external parameters calibrated by the checker pattern to construct the projection transformation relationship can also result in a better image stitching effect for the plane behind it.

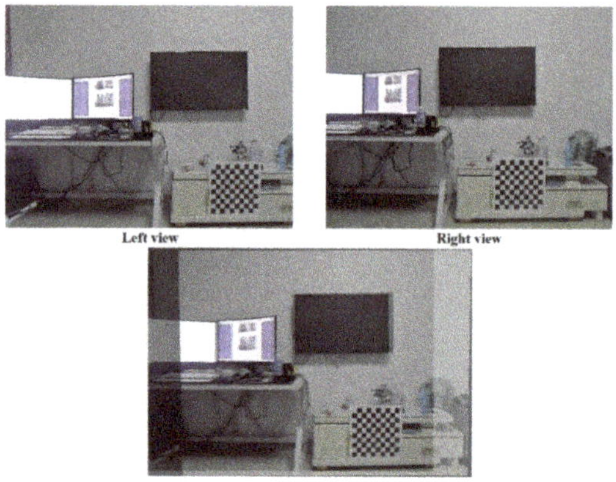

Figure 5. Image stitching results.

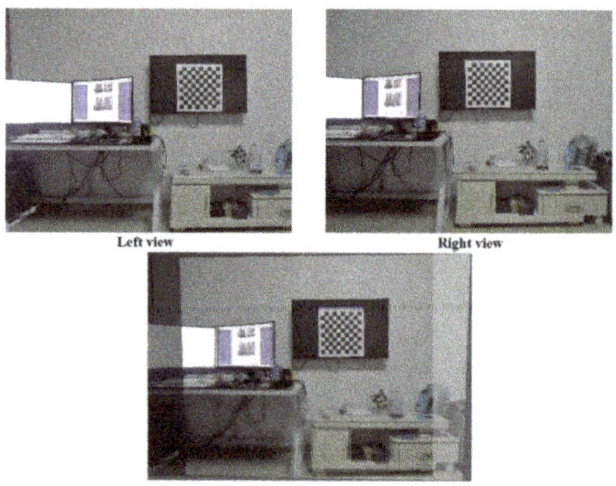

Figure 6. Image stitching results.

3.3. Design of Hierarchical Projection Method for Depth Information

A camera calibration method based on a special plane and a spatially layered image registration model was constructed with the external parameters obtained from the pattern. Although the model achieved a good stitching effect at each distance from the scene, the split image registration model could not be directly put into practical application. Therefore, it was necessary to segment the image according to the distance; to this end, an image layering method based on depth information was proposed. To obtain the depth information in the scene, we first had to calibrate the camera and obtain the relative external parameters between the left and right cameras, including the rotation matrix and translation vector. We then used this parameter to stereo correct the image and obtain the horizontal parallax of pixels in the overlapping area through stereo matching. Finally, we calculated the depth information of pixels in the scene using the corrected focal length and baseline length. The process of binocular calibration was similar to that of monocular calibration. Based on monocular calibration, it was only necessary to take additional pictures of multiple groups of the pattern in the overlapping area for use as input data, and then to substitute the data into the perspective projection model in order to obtain the external parameters of left and right cameras relative to the pattern. The first component t_x of translation vector t is the distance between the optical centers of the two cameras and the length of the baseline of the binocular camera. After obtaining the rotation matrix between the two cameras, according to the stereo correction principle, it was decomposed into the rotation matrix of half the rotation of the left and right views, and the overall rotation matrix was constructed through the translation matrix. The image could be corrected to the attitude whereby the imaging surfaces of the two cameras were coplanar and the optical axis was parallel by using the matrix for coordinate transformation. The result is shown in Figure 7.

Figure 7. Stereo correction effect.

After stereo correction, each pixel in the left and right viewing angles was almost on the same horizontal line. The stereo matching algorithm then searched for the matching pixel on the corresponding horizontal line in the right-hand side image in Figure 7. The search method involved setting an odd-size sliding window, using the minimum and maximum parallax in the two images to determine the starting point and endpoint of the search, and calculating the sum of the absolute value of the gray value difference of the corresponding pixel points in the two image windows as the matching basis and selecting the point with the minimum value in the process from the start to the endpoint as the best matching point. Subsequently, the pixel coordinates of the pixel points corresponding to the left and right views in the overlapping area could be obtained. Parallax d of the point could be obtained by subtracting the abscissa of the two points. The calculated disparity map is shown in Figure 8 with the parameters and the stereo-corrected image as input.

Figure 8. Overlapping area disparity map.

After obtaining the parallax map of the overlapping area, the depth information was calculated by remapping the stereo-corrected pose to the original pose of the right perspective using Equation (7).

$$Z = \frac{f_x \times Baseline}{d} \tag{7}$$

where f_x is the number of pixels in the horizontal direction occupied by the focal length of the two cameras after stereo correction, *Baseline* is the distance between the optical centers of the two cameras, and d is the parallax of the pixel points. After converting the disparity map into a depth map, the image was layered; the effect is shown in Figure 9. The pixels in the scene were divided into several layers according to the depth information, and the average value of the depth information in each layer was recorded. At the same time, the original image of the camera angle on the right was also layered in this way, and the parts outside the overlapping area were incorporated into the layer of adjacent pixels in behavioral units. The effect is shown in Figure 10.

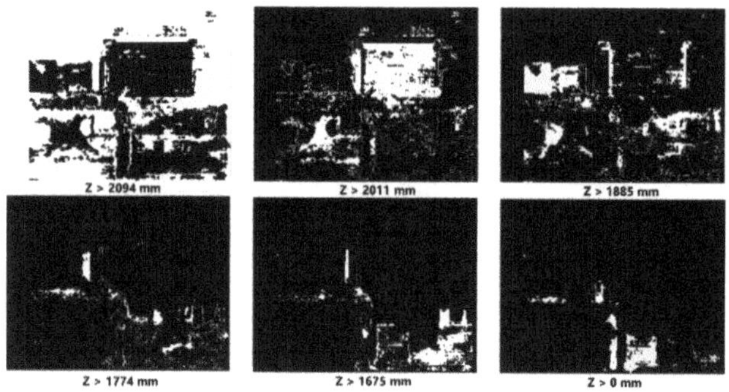

Figure 9. Image layering method based on depth information.

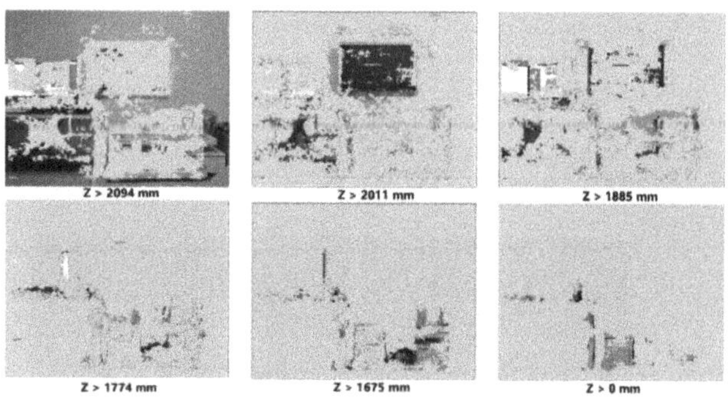

Figure 10. The original image layering effect.

After layering the original image, each layer used the pre-built projection transformation model based on the special plane and substituted the depth information of the layer into t_z in the model for calculation. Each layer used the coordinate transformation relationship calculated independently for projection, and finally, superimposed all the projection results onto the plane where the target image was located. The effect is shown in Figure 11.

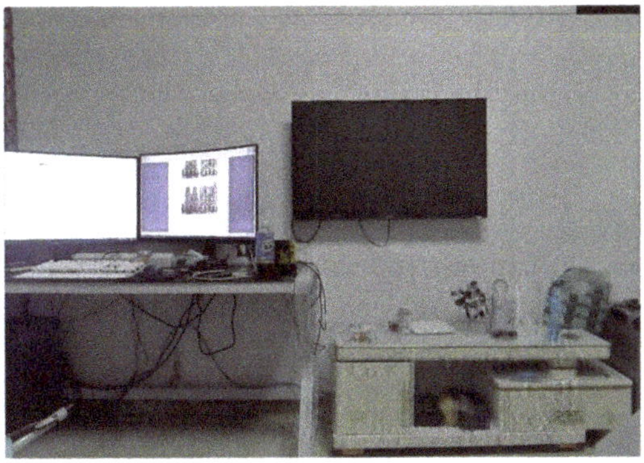

Figure 11. Stitching results based on image layering.

3.4. Process of Image Stitching Method Based on Camera Calibration

To begin, the internal parameters and the external parameters were solved by calibrating the camera. Then, using the camera calibration method based on the special plane, an additional group of pictures parallel to the camera imaging plane were taken and used as input data to solve the external parameter matrix representing the pose relationship between the pattern and the camera imaging plane. Using the external parameter matrix and the internal parameters, the coordinate transformation relationship of the binocular camera about the point on the distance plane could be constructed. In addition, the stitched image had to be layered through the layered projection method based on depth information. We then used the external parameters of the calibrated binocular camera to stereo correct the left and right viewing angles so that the corresponding pixels in the image would fall on the same horizontal line. Next, we searched the corresponding pixels in one-dimensional space using the stereo matching algorithm and obtained their horizontal parallax. After obtaining the disparity map in the above way, we calculated the depth map according to the focal length of the camera and the length of the baseline and divided the depth map according to the specific situation. Finally, the original floating map was layered according to the layered model of the depth map, and the corresponding depth information was substituted into the coordinate transformation relationship so that each layer of the image was projected according to its registration model. All projection results could then be superimposed to obtain the final image stitching result. Since the parallax of objects at the same distance imaged on the camera plane was the same, the layered projection method based on depth information could maximally eliminate the impact of parallax. The process is shown in Figure 12.

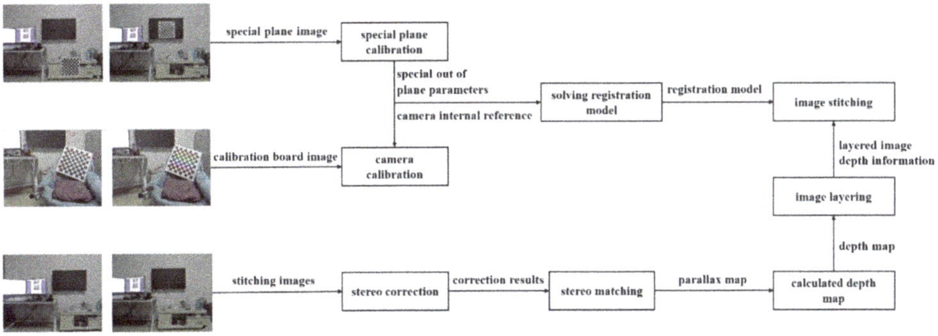

Figure 12. Image stitching process based on camera calibration model.

4. Experiment

4.1. Realization of Camera Calibration Based on Special Plane

The camera calibration algorithm flow is shown in Figure 13.

A pattern was composed of 10 × 10 black-and-white squares. The center of area 4 × 4 in the top left corner was the first corner point. There were 81 corner points on the surface of the reference object, and the horizontal distance between the points was found to be 40 mm. Before the experiment, the camera had to be monocularly calibrated many times to obtain accurate internal parameters. The internal parameter matrix of the binocular camera calibrated in the above way is shown in Table 1.

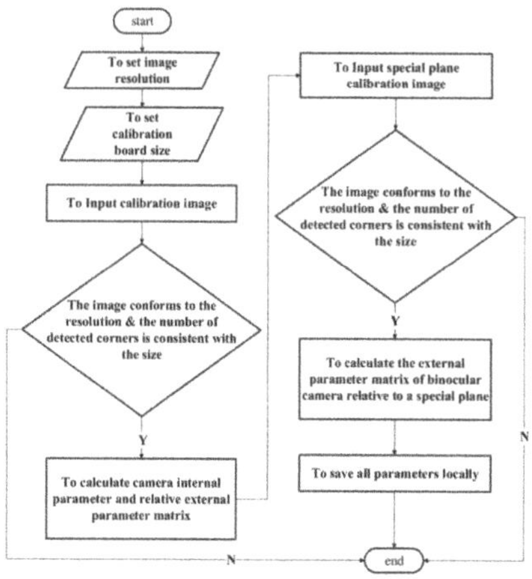

Figure 13. Camera calibration algorithm flow based on a special plane.

Table 1. Internal parameters of binocular camera.

Equipment Name	Internal Parameter Matrix	Distortion Parameter Matrix
Left camera	$\begin{bmatrix} 903.0102 & 0 & 750.2198 \\ 0 & 901.3962 & 451.7124 \\ 0 & 0 & 1 \end{bmatrix}$	$\begin{bmatrix} -0.0159 & 0.0353 & -0.00127 & -0.0008 & -0.045 \end{bmatrix}$
Right camera	$\begin{bmatrix} 898.9699 & 0 & 688.2513 \\ 0 & 900.6719 & 432.5414 \\ 0 & 0 & 1 \end{bmatrix}$	$\begin{bmatrix} 0.0022 & -0.027 & -0.0017 & 0.0011 & 0.0090 \end{bmatrix}$

After monocular calibration, the relative external parameters of the binocular camera could be obtained by taking the obtained camera internal parameters and the picture group with complete reference objects in the overlapping area of the left and right viewing angles as input, as shown in Table 2.

After binocular calibration, camera calibration based on the special plane was carried out. The binocular camera was placed in a position whereby the imaging surface was parallel to the wall, and the pattern was fixed on the vertical plane in the overlapping area for photographing. The method is shown in Figure 14.

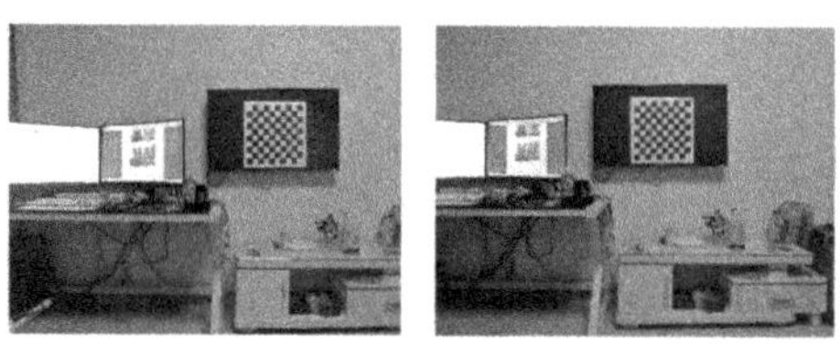

Figure 14. Camera calibration based on a special plane.

Table 2. Relative external reference of binocular camera.

Transformation Mode	Rotation Matrix	Translation Vector
From left to right	$\begin{bmatrix} 0.9999 & -0.0019 & -0.0026 \\ 0.0019 & 0.9999 & -0.0024 \\ 0.0026 & 0.0024 & 0.9999 \end{bmatrix}$	$\begin{bmatrix} -16.7401 & -0.0788 & -0.0569 \end{bmatrix}$
From right to left	$\begin{bmatrix} 0.9999 & -0.0016 & 0.0079 \\ -0.0017 & 0.9999 & 0.0042 \\ -0.0079 & -0.0042 & 0.9999 \end{bmatrix}$	$\begin{bmatrix} 16.7664 & 0.0526 & 0.0939 \end{bmatrix}$

The plane of the pattern was approximately parallel to the imaging plane of the camera. The external parameters of the camera, relative to the plane of the pattern obtained using the image and known internal parameters, are shown in Table 3.

Table 3. External parameters of the binocular camera relative to the special plane.

Equipment Name	Rotation Matrix	Translation Vector
Left camera	$\begin{bmatrix} 0.0109 & 0.9998 & 0.0093 \\ 0.9993 & -0.0113 & 0.0369 \\ 0.0370 & 0.0089 & -0.9993 \end{bmatrix}$	$\begin{bmatrix} 9.8148 & -40.2235 & 198.7077 \end{bmatrix}$
Right camera	$\begin{bmatrix} 0.0098 & 0.9997 & 0.0227 \\ 0.9999 & -0.0099 & 0.0072 \\ 0.0074 & 0.0026 & -0.9997 \end{bmatrix}$	$\begin{bmatrix} -7.5794 & -40.2643 & 198.7469 \end{bmatrix}$

4.2. Implementation of Hierarchical Projection of Depth Information

The implementation flow of hierarchical projection of depth information is shown in Figure 15.

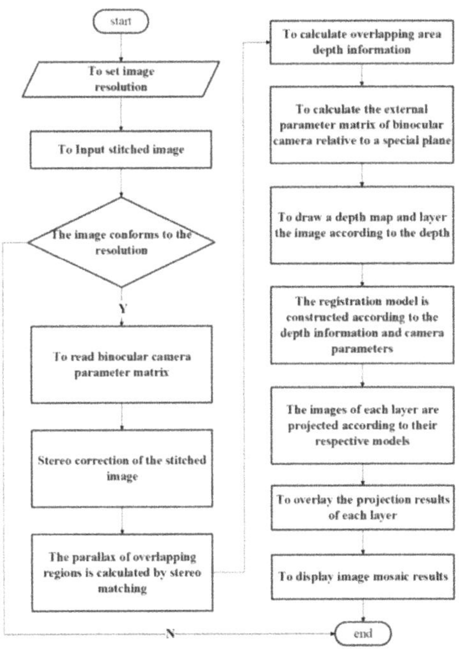

Figure 15. Hierarchical projection algorithm flow based on depth information.

To verify the hierarchical projection based on depth information, an experiment was carried out in which the variables were controlled. Firstly, the left and right camera internal parameters and special out-of-plane parameters obtained in Tables 1 and 2 were substituted into the model in Equation (6) to calculate the coordinate transformation relationship. In the experiment, different values were input for image registration. Based on the internal and external parameters of the camera, the correction matrix and projection matrix of the left and right views could be deduced, as shown in Table 4.

Table 4. Stereo correction and projection matrix.

Pespective Name	Correction Matrix	Projection Matrix
Left camera view	$\begin{bmatrix} 0.9999 & 0.0028 & 0.0008 \\ -0.0028 & 0.9999 & -0.0012 \\ -0.0008 & 0.0012 & 0.9999 \end{bmatrix}$	$\begin{bmatrix} 901.0340 & 0 & 721.6889 & 0 \\ 0 & 901.0340 & 438.3938 & 0 \\ 0 & 0 & 1 & 0 \end{bmatrix}$
Right camera view	$\begin{bmatrix} 0.9999 & 0.0047 & 0.0034 \\ -0.0047 & 0.9999 & -0.0012 \\ -0.0034 & 0.0012 & 0.9999 \end{bmatrix}$	$\begin{bmatrix} 901.0340 & 0 & 721.6889 & -15083.6593 \\ 0 & 901.0340 & 438.3938 & 0 \\ 0 & 0 & 1 & 0 \end{bmatrix}$

The left and right viewing angles of the image to be stitched could then be transformed into a horizontally aligned attitude through the correction matrix, and the stereo correction results could be obtained by mapping the projection matrix to the new coordinate system. Next, we searched the corresponding points of the left and right viewing angles in the one-dimensional space, calculated the parallax and depth information, and obtained a depth map which we used to layer the image. The effect is shown in Figure 16.

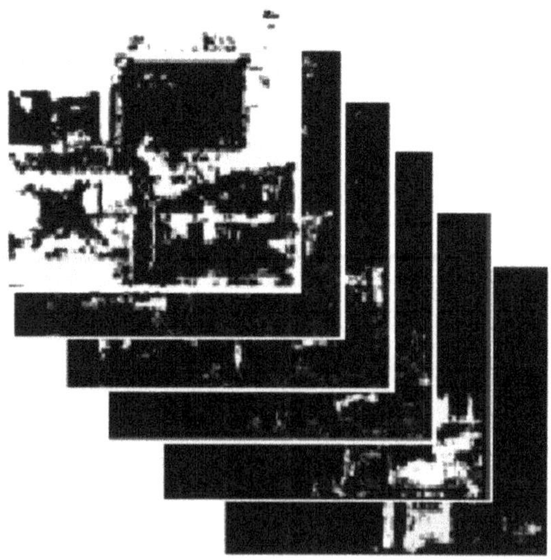

Figure 16. Image layering effect based on depth information.

The image was layered in such a way that the outermost layer was greater than 2 m and each layer was separated by 100 mm. We took the average depth information of the current layer as the input to solve the respective image registration models. All layers were aligned with the corresponding models and superimposed upon one another to obtain the final stitching results.

4.3. Effect Analysis Experiment after Image Stitching

The experiments were conducted with binocular cameras. We captured close scenes several times; these were then stitched together using the method proposed in this paper, with the following results.

As shown in Figure 17, good stitching results were achieved for various planes at different distances, such as TVs, monitors, and chairs; however, objects such as table legs, which are long and thin and have different overall depth information still showed a certain degree of overlap and misalignment, and there was still blurring due to the layering in the stitched image.

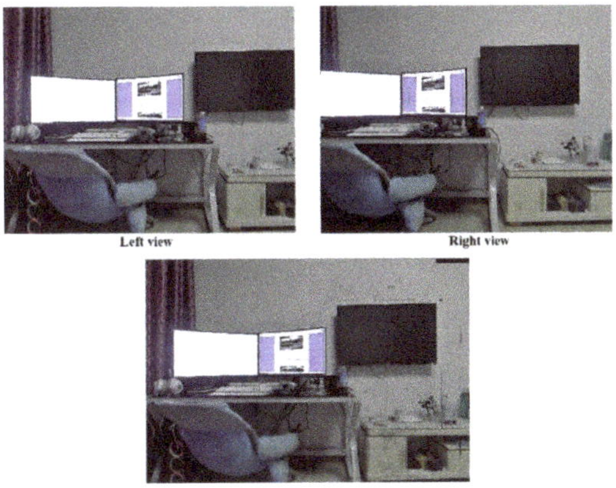

Figure 17. Image stitching results.

Figure 18 illustrates a depth span of a scene. Except for the door seam of the closet, a good stitching effect was achieved for all objects despite some blurring phenomena.

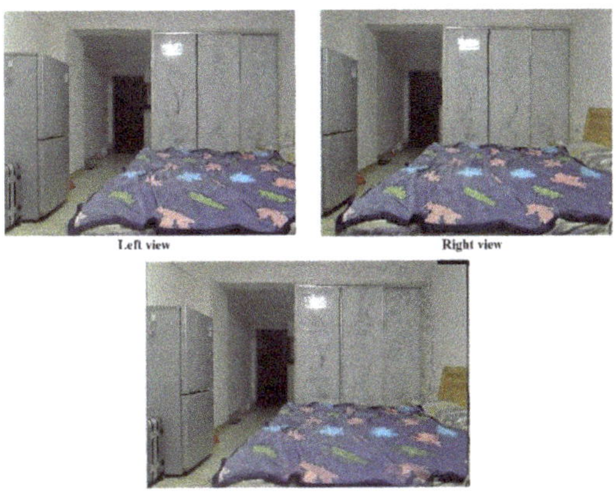

Figure 18. Image stitching results.

Similarly, Figure 19 presents to a scene with a large depth span. Once again, except for the door seam of the closet, a good stitching effect was achieved for all objects despite some blurring phenomena.

Comparative experiments were conducted to stitch the images using the method proposed in this paper and the feature point-based stitching method, respectively. The latter applies the SURF [19] and ORB (Oriented FAST and Rotated BRIEF) algorithms [20] to detect and match the feature points, purifies the matching results via the random sampling consistency method, obtains the best matching results, constructs a global homography model to align the images, and then uses the fading-in and fading-out method to obtain the final image. The experimental results are shown in Figures 20–22.

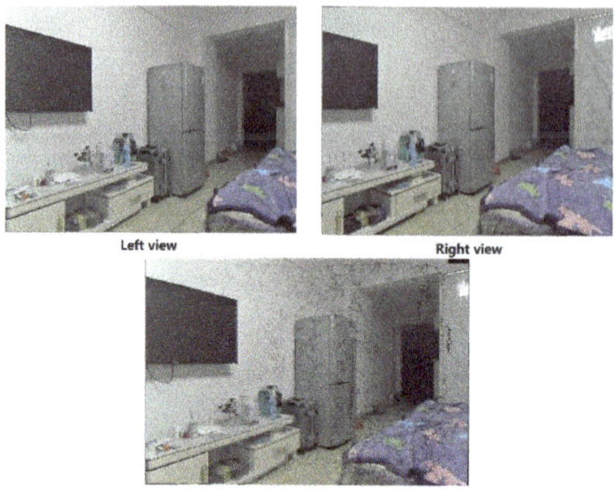

Figure 19. Image stitching results.

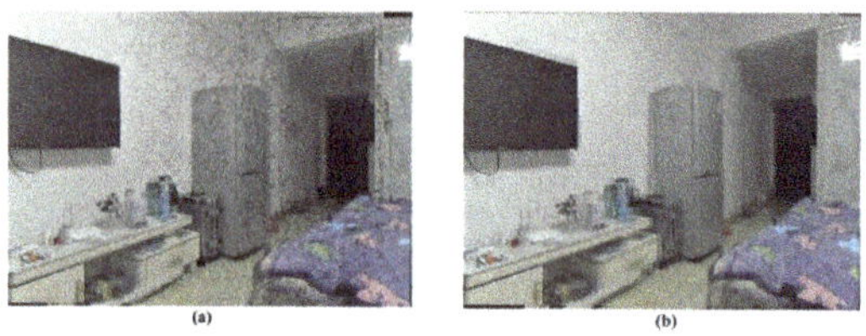

Figure 20. Image stitching experiment. (**a**) Image stitching effect based on depth information layering; (**b**) Image stitching effect based on SURF feature points.

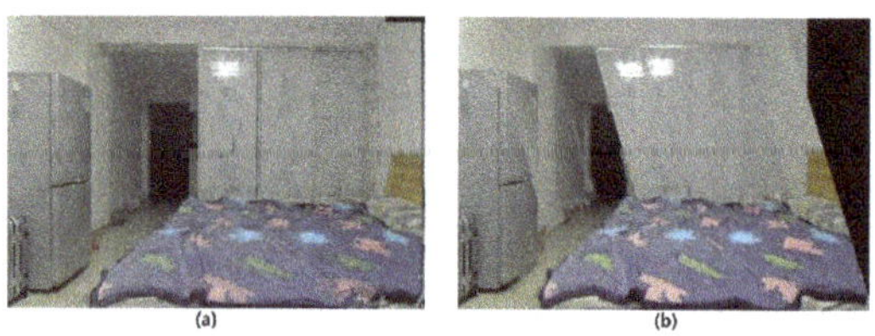

Figure 21. Image stitching experiment. (**a**) Image stitching effect based on depth information layering; (**b**) Image stitching effect based on ORB feature points.

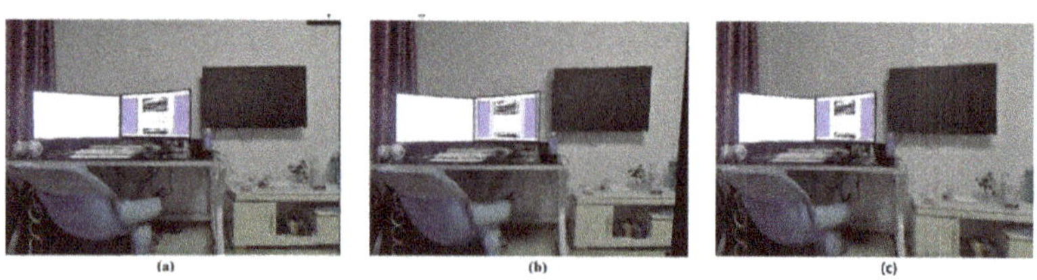

Figure 22. Image stitching experiment. (**a**) Image stitching effect based on depth information layering; (**b**) Image stitching effect based on SURF feature points; (**c**) Image stitching effect based on ORB feature points.

Since Figure 20 is an oblique scene, the depth information of the object surface varies linearly; as such, the objects with different depths achieved better stitching results using the method proposed in this paper (Figure 20a). In the stitching results obtained using the SURF feature points (Figure 20b), the TV, the refrigerator and the shelf all showed different degrees of misalignment.

As shown in Figure 21, when the proposed method was used for stitching (Figure 21a), good results were obtained except for the closet gap. However, in the stitching results based on ORB feature points (Figure 21b), there were mismatches, because the feature points detected for the edges of the three closet doors were too similar. As such, the result was not satisfactory.

As shown in Figure 22, using the method based on depth information layering (Figure 22a), most of the objects in the scene achieved good results, with only the seat closest to the camera showing a small amount of ghosting. Meanwhile, in the results obtained using the two image stitching method based on feature points (Figure 22b,c), most objects showed ghosting, and the upper right corner of the TV set had significant deformation.

Although the results of the method proposed in this paper demonstrated less ghosting and fewer errors, some fuzzy edge noise appeared. This may have been because the coordinate transformation relationship between the layers was not accurate enough when the image was layered. Assuming that the present registration model is not accurate enough, eliminating such edge noise will be a focus in subsequent research.

5. Conclusions

To maintain high robustness in cases of sparse feature points, uneven distribution, or obvious parallax, an image stitching method based on the camera calibration model is proposed in this paper. Based on the general camera calibration, an additional set of pictures with a vertical pattern were taken. Using the external parameters obtained from the camera calibration and the internal parameters of the camera, a spatially layered image registration model could be constructed. By adjusting the depth information in the model, the coordinate transformation relationship between the viewing angles of the two cameras concerning the vertical plane at any distance could be obtained. To apply the spatially layered image registration model, this paper also proposed an image layered projection method based on depth information. The depth information of the overlapping area in the scene was obtained through stereo correction and matching. According to this information, the original image was layered, and each layer was registered according to the coordinate transformation relationship based on the current depth information. By superimposing all the projection results, image stitching results that were resistant to parallax disturbances could be obtained.

Author Contributions: Conceptualization, M.T., Q.Z., M.Y. and B.Z.; methodology, M.T.; validation, M.T., Q.Z. and Y.J.; formal analysis, M.T.; investigation, Q.Z., M.Y., Y.J. and B.Z.; resources, M.T. and Q.Z.; data curation, M.T.; writing—original draft preparation, M.T.; writing—review and editing, M.T., Q.Z., M.Y. and Y.J. All authors have read and agreed to the published version of the manuscript.

Funding: This work was supported by science and technology to boost the economy 2020 key project (No. SQ2020YFFO4107-66), Chongqing technology innovation and application development special general project (No. cstc2020jscx-msxmX0147), and research fund supported projects of Southwest University (No. SWU2008045).

Institutional Review Board Statement: Not applicable.

Informed Consent Statement: Not applicable.

Data Availability Statement: Not applicable.

Conflicts of Interest: The authors declare no conflict of interest.

References

1. Yuan, Y.; Fang, F.; Zhang, G. Superpixel-Based Seamless Image Stitching for UAV Images. *IEEE Trans. Geosci. Remote Sens.* **2021**, *59*, 1565–1576. [CrossRef]
2. Wang, Z.; Yang, Z. Review on image-stitching techniques. *Multimed. Syst.* **2020**, *26*, 413–430. [CrossRef]
3. Yang, Z.; Dan, T.; Yang, Y. Multi-Temporal Remote Sensing Image Registration Using Deep Convolutional Features. *IEEE Access* **2018**, *6*, 38544–38555. [CrossRef]
4. Knops, Z.F.; Maintz, J.A.; Viergever, M.A.; Pluim, J.P. Normalized mutual information based registration using k-means clustering and shading correction. *Med. Image Anal.* **2006**, *10*, 432–439. [CrossRef] [PubMed]
5. Ojansivu, V.; Heikkila, J. Image Registration Using Blur-Invariant Phase Correlation. *IEEE Signal Processing Lett.* **2007**, *14*, 449–452. [CrossRef]
6. Lucchese, L.; Leorin, S.; Cortelazzo, G.M. Estimation of Two-Dimensional Affine Transformations Through Polar Curve Matching and Its Application to Image Mosaicking and Remote-Sensing Data Registration. *IEEE Trans. Image Processing* **2006**, *15*, 3008–3019. [CrossRef] [PubMed]
7. Szeliski, R. Video mosaics for virtual environments. *IEEE Comput. Graph. Appl.* **1996**, *16*, 22–30. [CrossRef]
8. Chen, Y.; Chuang, Y. Natural image stitching with the global similarity prior. In *Computer Vision—ECCV 2016*; Springer International Publishing: Cham, Switzerland, 2016; pp. 186–201.
9. Lin, K.; Jiang, N.; Cheong, L.F.; Do, M.; Lu, J. SEAGULL: Seam-guided local alignment for parallax-tolerant image stitching. In *Computer Vision—ECCV 2016*; Springer International Publishing: Cham, Switzerland, 2016; pp. 370–385.
10. Zaragoza, J.; Chin, T.J.; Brown, M.S.; Suter, D. As-Projective-As-Possible Image Stitching with Moving DLT. *IEEE Trans. Pattern Anal. Mach. Intell.* **2014**, *36*, 1285–1298. [PubMed]
11. Yi, K.M.; Trulls, E.; Lepetit, V.; Fua, P. LIFT: Learned invariant feature transform. In *Computer Vision—ECCV 2016*; Springer International Publishing: Cham, Switzerland, 2016; pp. 467–483.
12. Lecun, Y.; Bottou, L.; Bengio, Y.; Haffner, P. Gradient-based learning applied to document recognition. *Proc. IEEE* **1998**, *86*, 2278–2324. [CrossRef]
13. Lowe, D.G. Object recognition from local scale-invariant features. In Proceedings of the Seventh IEEE International Conference on Computer Vision, Kerkyra, Greece, 20–27 September 1999. [CrossRef]
14. Lowe, D.G. Distinctive Image Features from Scale-Invariant Keypoints. *Int. J. Comput. Vis.* **2004**, *60*, 91–110. [CrossRef]
15. Abdel-Aziz, Y.I.; Karara, H.M.; Hauck, M. Direct Linear Transformation from Comparator Coordinates into Object Space Coordinates in Close-Range Photogrammetry. *Photogramm. Eng. Remote Sens.* **2015**, *81*, 103–107. [CrossRef]
16. Chien, H.; Geng, H.; Klette, R. Bundle adjustment with implicit structure modeling using a direct linear transform. In *Computer Analysis of Images and Patterns*; Springer International Publishing: Cham, Switzerland, 2015; pp. 411–422.
17. Faugeras, O.D.; Luong, Q.T.; Maybank, S.J. Camera Self-Calibration-Theory And Experiments. *Lect. Notes Comput. Sci.* **1992**, *588*, 321–334.
18. Zhang, Z. Flexible camera calibration by viewing a plane from unknown orientations. In Proceedings of the Seventh IEEE International Conference on Computer Vision, Kerkyra, Greece, 20–27 September 1999. [CrossRef]
19. Dewanti, F.; Sumiharto, R. Purwarupa Sistem Penggabungan Foto Udara Pada UAV Menggunakan Algoritma Surf (Speeded-Up Robust Features). *IJEIS (Indones. J. Electron. Instrum. Syst.) (Online)* **2015**, *5*, 165–176. [CrossRef]
20. Rublee, E.; Rabaud, V.; Konolige, K.; Bradski, G. ORB: An efficient alternative to SIFT or SURF. In Proceedings of the 2011 International Conference on Computer Vision, Barcelona, Spain, 6–13 November 2011. [CrossRef]

electronics

MDPI

Article

A Study on Particle Swarm Algorithm Based on Restart Strategy and Adaptive Dynamic Mechanism

Lisang Liu [1,2], Hui Xu [1,2,*], Bin Wang [1,2], Rongsheng Zhang [1,2] and Jionghui Chen [1]

[1] School of Electronic, Electrical Engineering and Physics, Fujian University of Technology, Fuzhou 350118, China; liulisang@fjut.edu.cn (L.L.); 2201905138@smail.fjut.edu.cn (B.W.); rongsheng@smail.fjut.edu.cn (R.Z.); 2221905020@smail.fjut.edu.cn (J.C.)

[2] National Demonstration Center for Experimental Electronic Information and Electrical Technology Education, Fujian University of Technology, Fuzhou 350118, China

* Correspondence: xuhui@smail.fjut.edu.cn

Abstract: Aiming at the problems of low path success rate, easy precocious maturity, and easily falling into local extremums in the complex environment of path planning of mobile robots, this paper proposes a new particle swarm algorithm (RDS-PSO) based on restart strategy and adaptive dynamic adjustment mechanism. When the population falls into local optimal or premature convergence, the restart strategy is activated to expand the search range by re-randomly initializing the group particles. An inverted S-type decreasing inertia weight and adaptive dynamic adjustment learning factor are proposed to balance the ability of local search and global search. Finally, the new RDS-PSO algorithm is combined with cubic spline interpolation to apply to the path planning and smoothing processing of mobile robots, and the coding mode based on the path node as a particle individual is constructed, and the penalty function is selected as the fitness function to solve the shortest collision-free path. The comparative results of simulation experiments show that the RDS-PSO algorithm proposed in this paper solves the problem of falling into local extremums and precocious puberty, significantly improves the optimization, speed, and effectiveness of the path, and the simulation experiments in different environments also show that the algorithm has good robustness and generalization.

Keywords: restart strategy; adaptive adjustment; particle swarm optimization; spline interpolation

Citation: Liu, L.; Xu, H.; Wang, B.; Zhang, R.; Chen, J. A Study on Particle Swarm Algorithm Based on Restart Strategy and Adaptive Dynamic Mechanism. *Electronics* **2022**, *11*, 2339. https://doi.org/10.3390/electronics11152339

Academic Editor: Savvas A. Chatzichristofis

Received: 20 June 2022
Accepted: 21 July 2022
Published: 27 July 2022

1. Introduction

With the development of robot technology, the environment is becoming more and more complex, and people's performance requirements for robots are also getting higher and higher. In a complex environment to complete the task autonomously, navigation technology is more important, and path planning is an important part of navigation technology; a good planning algorithm not only can plan the shortest path, but the cost of time, robot mechanical loss costs, maintenance costs, etc. also need to be reduced to a minimum [1]. The formatter will need to create these components, incorporating the applicable criteria that follow.

Researchers have been studying the path planning problem for many years, and have been constantly exploring and improving, with some good results. For example, the A* algorithm [2], Dijkstra [3] algorithm, RRT [4], etc. can achieve some good results in simple environments, but with the increase in environmental complexity and requirements, there will be problems such as larger computation and more memory occupation. With the emergence of intelligent optimization algorithms, more and more researchers apply intelligent optimization algorithms and their improved algorithms to path planning problems. Liu Jingsen et al. [5] proposed a bat algorithm with reverse learning and tangent random exploration mechanism, combined with cubic spline interpolation to define a smooth path based on node coding. Sun Huihui et al. [6] started from the three types of reinforcement learning motion planning methods based on value, strategy, and actor-critic, and deeply analyzed the characteristics and practical application scenarios of deep reinforcement learning planning methods, and experimentally proved that although intelligent optimization algorithms such as gray wolf algorithm [7], ant colony algorithm [8], particle swarm algorithm, and genetic algorithm [9] can initially solve the path planning problem, these algorithms have their own shortcomings. The accuracy of the search cannot be guaranteed, and it is easy to fall into the problem of local optimization.

The particle swarm algorithm, proposed by Kennedy and Eberhart in 1995 [10], is widely used to solve various engineering problems because of its fast convergence speed, ease of implementation, and few parameters for simple modeling [11–14]. However, it also has defects such as precocious puberty, low precision, and easily falling into local optimization. Thus, many improved algorithms have been proposed in recent years. In Kang Yuxiang et al. [15], in view of the problems of precocious particle swarm algorithm and low optimization accuracy, the speed update model was improved, the adaptive particle position update coefficient was increased, and a greedy strategy was added to the algorithm process. In Panda et al. [16], in view of the rapid loss of particle swarm diversity and the problem of premature convergence, they proposed that the hybrid crossover algorithm be combined with the particle swarm algorithm to enhance the ability to explore particles and surrounding space. Ouyang Haibin et al. [17] proposed a hierarchical path planning method based on the mixed genetic particle swarm optimization algorithm, which first used the genetic algorithm improved by the artificial potential field method for primary path planning, and then used the particle swarm algorithm to optimize the path for secondary optimization. However, the method does not do a good job of fusing the two algorithms. Song et al. [18] proposed a new path smoothing method. An adaptive fractional-order velocity is introduced to enforce some disturbances on the particle. A new strategy is developed to plan the smooth path for mobile robots through an improved PSO algorithm in combination with the continuous high-degree Bezier curve. Miao et al. [19] proposed a new particle swarm optimization method. The algorithm merges two strategies, the static exploitation (SE, a velocity updating strategy considering inertia-free velocity) and the direction search (DS) of Rosenbrock method, into the original PSO.

In this paper, a particle swarm optimization algorithm (PSO) based on parameter and restart strategy improvement is proposed, and it is applied to the path planning problem. We named the proposed algorithm RDS-PSO, where R represents restart strategy, D represents dynamic adjustment, and S is for spline interpolation. The uniform distribution, inverted S-type inertia weight coefficient, cubic spline interpolation function, and enhanced control learning factor are introduced in the PSO algorithm, and a restart strategy is added to enhance the global optimization performance of the algorithm. Finally, its effectiveness was verified in an experimental environment with obstacles. Experimental results show that, compared with other path planning algorithms, the proposed RDS-PSO can achieve better results in both complex and simple environments.

2. RDS-PSO Algorithm

2.1. Standard Particle Swarm Algorithm

The PSO algorithm is a population-based optimization problem heuristic strategy proposed by Kennedy and Eberhard in 1995. The core of the PSO algorithm is to share information through individuals in the group, so that the motion of the entire group is transformed from disorder to order in the solution space problem, so as to obtain the optimal solution of the problem. The result of each optimization problem is performed by Equations (1) and (2). The first term of the velocity update Formula (1) is the inertia part, which indicates that the next move of the particle is influenced by the size and direction of the velocity of the last flight, and the inertia weight w determines how much information is inherited from the previous generation, thus balancing the global and local search; the second term indicates that the subsequent move of the particle is influenced by the particle's own historical experience, and the closer the particle is to its own historical best position, the smaller the difference between the second term and the smaller the velocity. From

Formula (2), it can be seen that the next step position distance is also smaller, which at this time is conducive to local search; the third term indicates that the next action of the particle is influenced by the best particle in the group, the same as the second part, the farther the particle is from the best position in the group, the larger the difference; at this time the speed is larger, the step length in Formula (2) is also larger, which is conducive to global search. Therefore, the next step of the particle is determined by three parts: the inertial part, its own historical experience, and the group historical experience.

Particle velocity update formula:

$$V_{id}^{t+1} = wV_{id}^{t} + c_1 r_1 (Pbest_{id}^{t} - x_{id}^{t}) + c_2 r_2 (Gbest_{id}^{t} - x_{id}^{t}) \tag{1}$$

Position update formula:

$$x_{id}^{t+1} = x_{id}^{t} + V_{id}^{t+1} \tag{2}$$

where V_{id}^{t} is the speed at which the ith particle flies; t is the number of iterations; d denotes dimensionality; c_1 and c_2 are the learning factor; r_1 and r_2 are random numbers within [0, 1] to enhance randomness; $Pbest_{id}^{t}$ indicates the best position of particle i in the t iteration; $Gbest_{id}^{t}$ represents the best position of the particle population in the t iteration; and w is the inertia weight coefficient that adjusts the search space searchability.

2.2. Improved Particle Swarm Algorithm

Inertia Weights

Adaptive tuning parameters have always been the focus of research on PSO algorithms. The change of inertia weight w affects the position of particles, the larger the value of w, the stronger the global search ability, the weaker the local search ability. Several studies show that the dynamic adjustment of w can improve the convergence and search accuracy of PSO. The value of w can vary linearly during a PSO search [20] or dynamically as an adaptability function based on PSO performance [21]. Since the fixed and simple linear decrement strategy is not conducive to the global search of particles, this paper proposes an adaptive and dynamic weight adjustment method, that is, the inertial weight based on the sin function is introduced in the linear decrement strategy, which makes w take a larger value in the early iteration period, which strengthens the algorithm's global search capability; at the same time, it takes a smaller value in the later stage, and strengthens the algorithm's local search capability.

The improved inertia weight formula is:

$$w = w_{\max} - (w_{\max} - w_{\min})\, sin(\frac{\pi * t}{2Itmax})^2 \tag{3}$$

where $w_{\max}$ is the maximum inertia weight, $w_{\min}$ is the minimum inertia weight, $Itmax$ is the maximum number of iterations, and t is the current number of iterations.

As can be seen from the above Figure 1, this improved strategy makes the inertia weights show an inverted S-shaped decreasing trend throughout the iterative process, keeping larger values in the early part of the process for a longer time, decreasing faster in the middle, and keeping smaller values in the later part of the process for a longer time. This can balance the global search and local search well.

2.3. Learning Factors

As important parameters in PSO, learning factors c_1 and c_2 have the effect of regulating the performance of the algorithm, which determines the influence of the particle's own historical experience and group experience on the particle motion trajectory, reflecting the information exchange between particles. c_1 and c_2 are too large or too small to facilitate particle search [22]. This paper adopts the power function to perform symmetric treatment of c_1 and c_2. The specific formula is as follows:

$$c_1 = \alpha e^{w} \tag{4}$$

$$c_1 = \beta e^{-w} \quad (5)$$

In order to achieve the symmetry effect, after several experiments, the two coefficients in the equation are taken as α = 0.83 and β = 2. In the improved learning factor Formulas (4) and (5), it can be found that c_1 is decreasing while c_2 is increasing. The early focus on individual information exploration is a feasible solution. The later stage focuses on the rapid convergence of global information, which not only makes PSO have good learning ability in the optimization process, but also turns the inertia weight and learning factor into a variable, which is convenient for practical application and also strengthens the uniformity in the process of algorithm optimization.

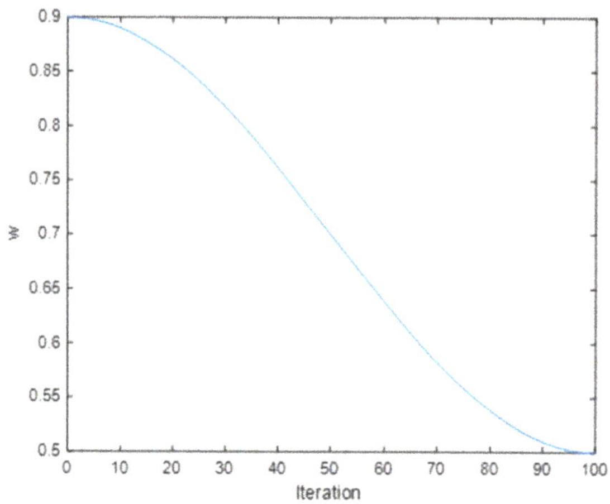

Figure 1. Inertia weight curve graph.

2.4. Cubic Spline Interpolation

In the simulation experiment, it was found that the path of the classical PSO program has many turning points, the path is not smooth enough, and the dynamic characteristics are poor during sharp turns. Thus, it is necessary to further improve the algorithm to make the algorithm more in line with the dynamic adaptability requirements of the robot.

Cubic spline interpolation is a piecewise interpolation method that can be fitted by multiple interpolation intervals based on cubic polynomials to form a smooth curve, and the robot movement path fitted with the cubic spline interpolation method is smoother.

The definition and algorithm of cubic spline interpolation are as follows:

In the interval [a, b], there are n + 1 data nodes $(x_1, y_1), (x_2, y_2), \ldots (x_n, y_n)$ that are called cubic spline functions if the following conditions are met.

Each interval (x_i, x_{i+1}), where i = 0, 1, . . . , n, satisfies the second cubic polynomial:

$$f_i(x) = a_i + b_i(x - x_i) + c_i\ (x - x_i)^2 + d_i(x - x_i)^3 \quad (6)$$

The function and its first and second derivatives are continuous at the interpolation point.

$f(x)$ commonly uses endpoint conditions that can satisfy the following three requirements:

- Free boundary: the second derivative at the endpoint is zero.
- Fixed limitation: the range value of the differential function from the beginning to the end is specified.
- Non-node boundary: the third derivative at the 2nd to the last node is continuous.

The Algorithm 1 process is:

Algorithm 1 Triple spline interpolation

```
For each of these intervals it is necessary to satisfy:
S_i(x) = a_i + b_i(x − x_i) + c_i (x − x_i)^2 + d_i(x − x_i)^3
S'_i(x) = b_i + 2c_i (x − x_i) + 3d_i(x − x_i)^2
S''_i(x) = 2c_i + 6d_i(x − x_i)
Input parameters x, y Interpolation point n.
Calculate step size: h_i = x_{i+1} − x_i
for i = 1: n − 1
  Substituting the parameters into the above matrix equation
  A system of linear equations with m as the unknown is obtained
   Solve the matrix equation to find the quadratic differential value m_i
   Find a, b, c, d.
   In the interval (x_i, x_{i+1}), the Equation (6) is obtained.
   end
```

2.5. Particle Coding

The junction of each segment is termed a path node, and the spline curve of each segment is distinct. Cubic spline interpolation is a segmental interpolation method. The cubic spline curve is first-order continuous in nature and second-order continuous at the node; the number of path nodes denotes the maximum number of turns in the entire path; in the most challenging instance, obstacles can be avoided after 3 to 4 turns. As a result, the particle encoding in this paper is based on path nodes.

Assuming that there are path nodes (x_{m1}, x_{m1}), (x_{m2}, x_{m2}), ... , (x_{mm}, x_{mm}), the coordinates of the start point and end point are (x_s, x_s), (x_t, x_t), and n interpolation points are obtained on the interval $(x_s, x_{m1}, x_{m2}, \ldots, x_t)$ and $(y_s, y_{m1}, y_{m2}, \ldots, y_t)$ by cubic spline interpolation, and the coordinates of the interpolation points are (x_1, x_1), (x_2, x_2), ... , (x_m, x_m). Finally, the line consisting of the path nodes, interpolation points, and the start and end points are the robot motion path we require.

2.6. Evaluation Function

In the path planning problem, two conditions are generally satisfied to determine whether a path is optimal or not: (i) it cannot collide with an obstacle; (ii) the path is required as short as possible.

The fitness function F constructed in this article is shown in Equation (7), where L represents the planned path length, and its mathematical expression is Equation (8), where (x_i, x_i) is the coordinate of the i interpolation point, and a is a weight coefficient set to 100, which is used to exclude illegal paths. P is a barrier avoidance constraint function that is used to determine the safety distance; the calculation formula is shown in (9), where R_m is the radius of the m-th obstacle, m is the number of obstacles, and c, d is the obstacle's center coordinate; the smaller the value of P, the higher the final path's safety factor.

$$F = L \times (1 + a \times P) \tag{7}$$

$$L = \sum_{i+1}^{n} \sqrt{(x_{(i+1)} - x_i)^2 + (y_{(i+1)} - y_i)^2} \tag{8}$$

$$P = \sum_{m=1}^{m} (MAX(1 - \frac{\sqrt{((x_i - c)^2 - (y_i - d)^2)}}{R_m}, 0)) \tag{9}$$

2.7. Restart Strategy

A restart strategy is introduced under the above improvement circumstances in order to increase the algorithm's optimization abilities and overcome the problems of local optimization and precocious puberty. Huberman et al. were the first to use the restart technique to a stochastic optimization algorithm in 1997 [23]. It has become a standard strategy in stochastic optimization algorithms, and it is frequently used to boost algorithm performance [24]. By reinitializing the generation of fresh potential particles, you can avoid getting into a local ideal scenario.

In this paper, an iteration threshold is set in the process of the algorithm. If the optimal solution is not improved in the process of successive H-generation iterations, the optimal solution will be retained at this time and reinitialized into the next iteration. The improvement strategy enables the algorithm to effectively jump out of the local optimum, enhance the global search capability of the algorithm, and avoid premature maturity of the algorithm.

2.8. RDS-PSO Algorithm

Through the above comprehensive improvements, the inverted S-type inertia weights better balance the global and local search ability of the algorithm, and the dynamic learning factor not only strengthens the learning ability of the algorithm in the optimization process, but also combines the inertia weights and the learning factor into one variable, which is convenient for practical applications. On this basis, the cubic spline interpolation method is introduced to smooth the path, which improves the defect of the unsmooth path and enables the robot to better adapt to the real environment. For the problem that PSO is prone to falling into local optimum and premature maturity, a restart strategy is introduced by combining the above improved parameters, and the improved strategy enhances the algorithm's optimization-seeking ability and improves the problems of premature maturity and falling into local optimum. We call the proposed algorithm RDS-PSO.

The basic steps of the RDS-PSO algorithm are as follows.

Step 1: The number of path nodes and the number of interpolation points are determined according to the specific environment, and the starting and ending points are determined.

Step 2: Set the parameters, initialize the population and particle velocity, and initialize the population distribution.

Step 3: The coordinates of the interpolation points in the x and y directions are calculated for each particle using the cubic spline interpolation method.

Step 4: Calculate the adaptation value using Equation (7)

Step 5: The parameters are updated according to Equations (1)–(5), respectively, and update the local optimal value $Pbest_{id}^{t}$ and the global optimal value $Gbest_{id}^{t}$ and save it.

Step 6: According to Equation (9), we confirm whether the updated particle intersects with the obstacle, and apply algorithm 1 to obtain a path consisting of path nodes, interpolation points, and start-end connections after the update.

Step 7: In the iteration process, determine whether the restart condition is met. If the restart condition is met, the optimal path is kept at this time, reinitialized, and steps 1 to 6 are executed again; if not, the number of iterations is increased by 1 until the maximum number of restarts is reached, the algorithm ends, and the path is output.

The specific flowchart is shown in Figure 2.

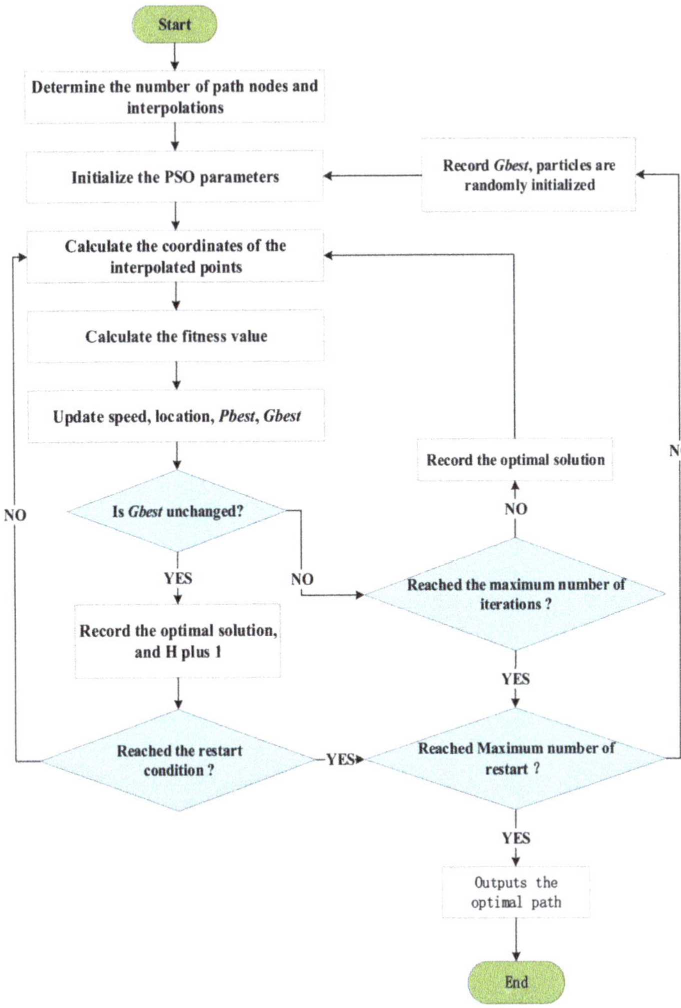

Figure 2. Flowchart of the RDS-PSO algorithm.

3. Experiments and Analysis of Results

3.1. Experimental Environment and Parameter Settings

The RDS-PSO algorithm and the standard particle swarm algorithm (PSO), the Improved PSO (RandWPSO-SP) based on random inertia weights and cubic spline interpolation [25], and the improved particle swarm optimization algorithm (IPSO) proposed in the literature [26] were experimentally compared and analyzed to verify the effectiveness and advancedness of the proposed algorithm in solving the robot path programming problem. This evaluates the algorithm's performance in terms of path planning for robots.

In order to ensure the objectivity and fairness of the experiment, all algorithms use the same software and hardware platform for experimentation, the simulation environment is Windows 10, Core i5, CPU (2.4 GHz), memory 12 GB, programming environment MATLAB R2019b. In order to ensure the authenticity of experimental data, 30 independent experiments on each algorithm, the experimental data were averaged.

In the simulation experiment, the parameters of the four algorithms, such as population size and maximum number of iterations, were consistent with $Itmax = 100$,

$Npop = 150$, In the standard PSO, the inertia weights and learning factors, $w = 0.9$, $c_1 = 1.5$, $c_2 = 1.5$, RandPSO-SP and the same parameter settings in this algorithm are consistent, $w_{\max} = 0.9$, $w_{\min} = 0.4$, the number of cubic spline interpolation points is set to 100, and the boundary is non-node boundary. Among them, the learning factor regulation parameters in the algorithm of this paper are $\alpha = 2$, $\beta = 0.83$.

In order to verify the universality of the algorithm in the path planning problem, the simulation experiment is carried out on MATLAB.

3.2. Experiments in Map 1

There are many obstacles in map 1, where obstacles are represented by blue circles. As can be seen from Figure 3, compared with the other three algorithms, the RDS-PSO of this algorithm has a shorter path, the least inflection point, and because the obstacles are more scattered, the best path is almost straight, and the other paths are smoother, which is due to the use of cubic spline interpolation, so the path is smoother.

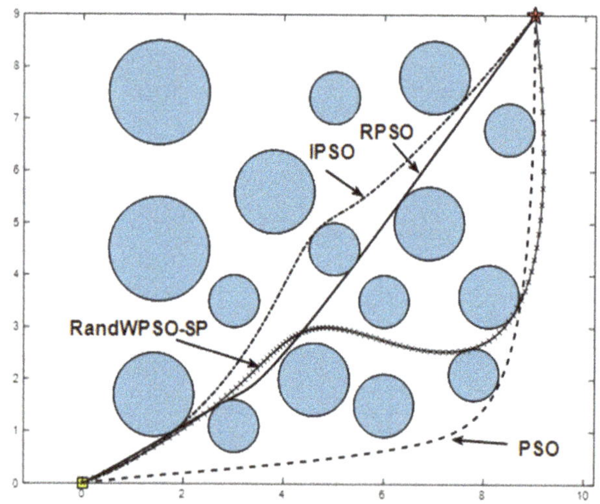

Figure 3. Comparison of path planning.

The iterative process of RDS-PSO is shown in Figure 4, and it can be seen that the algorithm has performed two restarts and finally found the optimal path because the algorithm has added a restart strategy. When the algorithm stagnates, it can be considered that the algorithm falls into local optimization; at this time, a new randomly distributed particle is added, combined with the inverted S-type inertia weight and the learning factor improvement method to improve the algorithm search ability, and also uses the characteristics of PSO convergence speed to shorten the iteration time; and restart multiple times to find the optimal path to achieve the purpose of jumping out of the local optimal.

The fastest convergence of IPSO in iterative Figure 5 is due to the addition of enhanced learning factors, but it can be seen in Table 1 that the algorithm is less robust and difficult to jump out when it falls into local optimality. RandWPSO-SP and PSO converge at the same rate, converging around 10 generations, but the optimal path was not found. RandWPSO-SP is too random; although the perturbation is obvious, it is easy to miss the optimal solution, and when the particles converge, it is not easy to jump out of the local optimal.

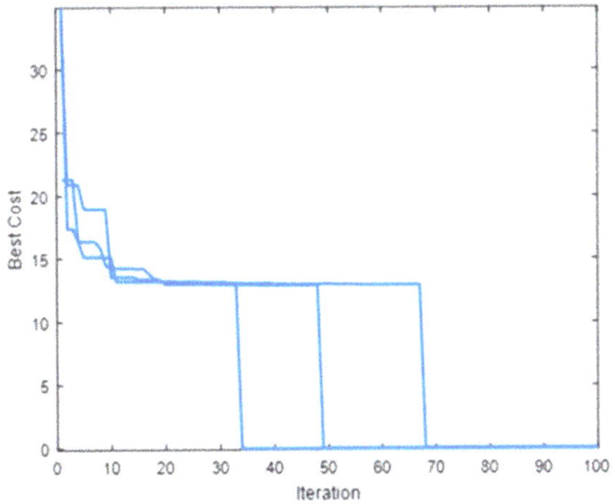

Figure 4. RDS-PSO iteration.

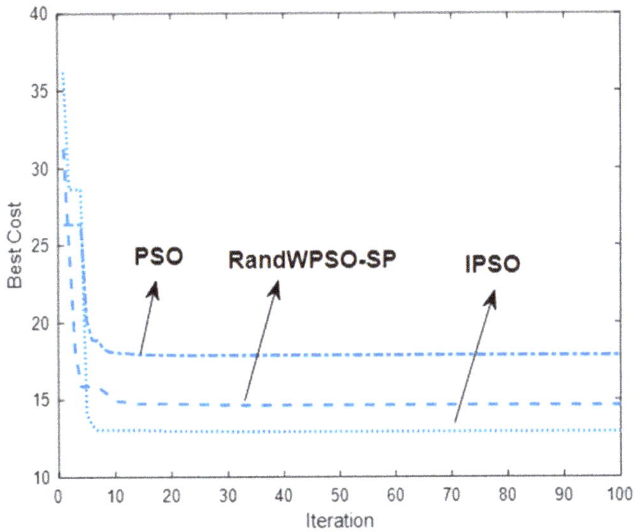

Figure 5. Iteration of three algorithms.

The data in Table 1 are the path results obtained by running the algorithm independently for 30 times in map 1, and an accuracy rate is introduced in the table as an evaluation index to judge the stability of the algorithm, that is, to find the optimal solution or the suboptimal solution is to find the correct path. It can be seen from the table that the average path length, the worst path, and the average simulation run time of the RDS-PSO are better than the other three algorithms, and the four algorithms have found the optimal path, but the RDS-PSO has the highest accuracy rate, only once did not reach the optimal value, which is due to the introduction of the restart strategy. When it falls into the local optimal, you can find a new solution in time, combined with the improved inverted S-type inertia weight and symmetric learning factor to enhance the search ability while improving the convergence speed. In this way, many optimizations are sought in a short period of time, which greatly enhances the optimization ability of the algorithm, and the optimization results are more stable.

Table 1. Comparison of algorithm performance.

Algorithm	Longest Path	Shortest Path	Average Path	Average Time (s)	Accuracy
PSO	12.89	15.34	13.6	29.86	47%
IPSO	12.9	15.85	13.96	30.7	57%
RandWPSO-SP	12.89	15.45	13.47	29.46	60%
RDS-PSO	12.89	13.16	12.94	29.10	94%

3.3. Experiments in Map 2

In the experimental map 2, the environment is more complex. With continuous obstacles, there is less room at the beginning, fewer paths to choose from, and it is easier to fall into local extremums; therefore, the ideal path must span a tighter area.

As can be seen in Figures 6 and 7, the path prepared after two RDS-PSO restarts is the shortest and smoothest. As can be seen in Figure 8, RandWPSO-SP has multiple jumps out of the native extremum, which is due to the addition of random inertia weights, which strengthens particle randomness. While the ultimate designing path is also shorter, the shortest path is not found, indicating that the algorithmic rule is ineffective in improving performance. Around the twentieth generation, IPSO and PSO merged. IPSO discovered a more robust path, owing to the employment of linear decreasing inertia weights and unified learning factors to improve algorithmic rule search performance. However, the convergence speed is swift, and the algorithmic rule search performance is improved.

Table 2 shows the path results of the four algorithms running independently 30 times in map 2. It can be seen from the table that the optimal solutions of the four algorithms are the same, but the worst solutions are very different, reflecting the difference in the optimization ability of the algorithms. Compared with experimental map 1, experimental map 2 is more complex, so the accuracy of the four algorithms is reduced. The average time of RDS-PSO is slightly longer, which is caused by the restart mechanism, but the average path length and accuracy rate are the best of the four algorithms, indicating that the optimization performance and robustness of the algorithm have been greatly improved.

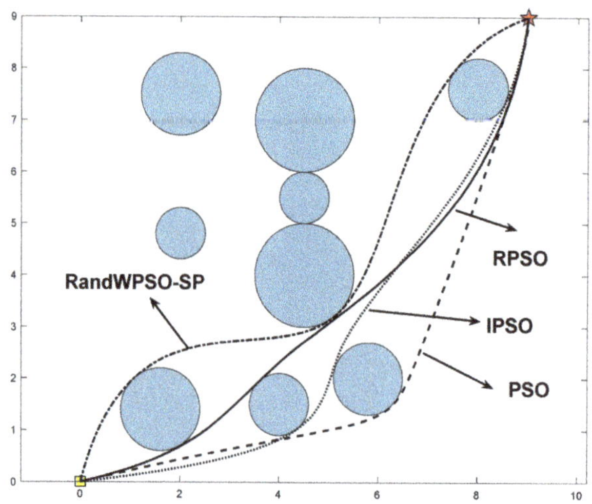

Figure 6. Comparison of path planning.

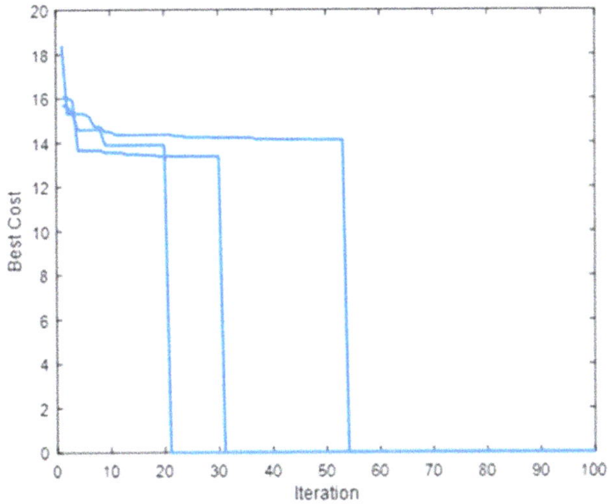

Figure 7. RDS-PSO iteration.

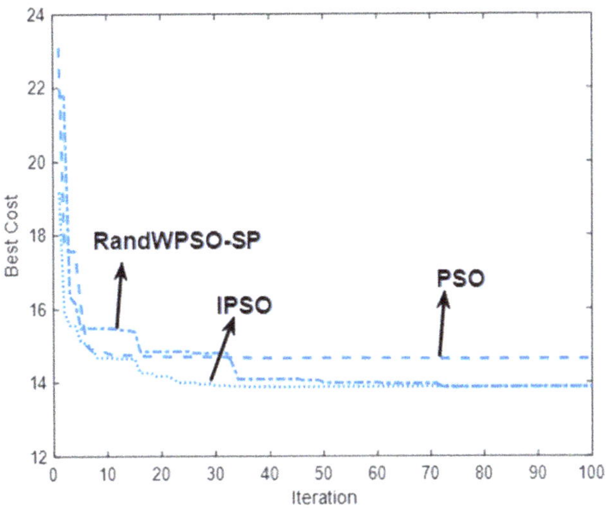

Figure 8. Iteration of three algorithms.

Table 2. Comparison of algorithm performance.

Algorithm	Longest Path	Shortest Path	Average Path	Average Time (s)	Accuracy
PSO	13.25	16.45	14.2	26.9	20%
IPSO	13.28	14.58	14.00	26.2	40%
RandWPSO-SP	13.29	15.24	14.03	25.76	50%
RDS-PSO	13.25	14.13	13.58	29	80%

3.4. Experiments in Map 3

Considering the diversity of actual obstacles, if all types of obstacles are expanded into circles, the feasible route may disappear, so this paper designed a third map for experimentation, as shown in Figures 9 and 10 below.

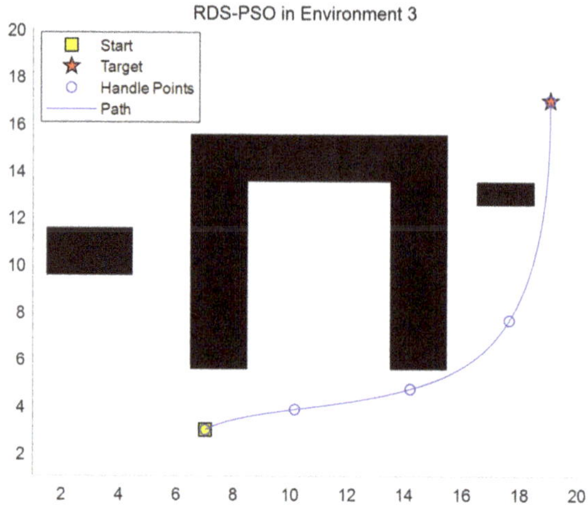

Figure 9. Path planning of RDS-PSO in map 3.

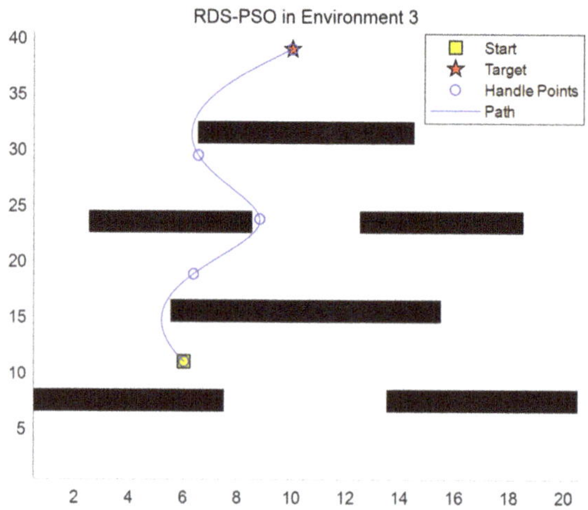

Figure 10. Path planning of RDS-PSO in map 3.

In experimental map 3, slender strips of various sizes are set up to evaluate the algorithm's universality and stability. Where-type barriers are used to see if the algorithm will fall into a state of local optimality. The final path fully avoids obstructions, is not "misled" by the center section of the gate, and chooses the best path, as can be seen from the planned route. The algorithm successfully avoided the obstacle, found the best path, and the path is also very smooth, as shown in Figure 10. Figure 10 enlarges the map range and sets up a continuous overlapping long bar obstacle. The starting point to the end point requires multiple turns, increasing the difficulty of planning. This report also confirms the search performance, ubiquity, and trustworthiness.

4. Conclusions

In this paper, an improved particle swarm algorithm combined with cubic spline interpolation is proposed to solve the robot path planning problem. For the "precociousness" in the basic PSO and some improved algorithms, the search ability is poor, it is easy to fall into local extremums, and it is difficult to jump out, resulting in problems such as search stagnation. First of all, the key parameters of PSO are improved, a new inverted S-type inertia weight and symmetric learning factor are introduced, and these three parameters are unified into one variable, which is convenient for practical application, improves the global optimization ability of the algorithm, and also improves the uniformity in the process of algorithm optimization, and enhances the search performance of the algorithm. At the same time, combined with the characteristics of the fast convergence speed of the particle swarm algorithm, a restart strategy is introduced, and when the algorithm search is stalled, it is reinitialized with random particles, which makes it easier for the algorithm to jump out of the local extremum, and also solves the problem of not being able to find a solution due to "precocious puberty". On this basis, the path nodes in the cubic spline interpolation are encoded as individual particles, so that the PSO and cubic spline interpolation method are combined with the robot path planning to plan a smooth path. An experimental comparison of four algorithms was carried out in two environments, and RDS-PSO was tested in complex environments, and the experimental results showed that the RDS-PSO improved algorithm in this paper had better solution performance under the same time, the shortest path of planning, the highest success rate, and the more stable algorithm, which proved the effectiveness and superiority of the improved algorithm in path planning problems.

Author Contributions: All of the authors contributed extensively to the work. H.X. proposed the key ideas; H.X. analyzed the key contents using a simulation and wrote the manuscript; L.L. obtained the financial support for the project leading to this publication; B.W., R.Z. and J.C. modified the manuscript. All authors have read and agreed to the published version of the manuscript.

Funding: This work was supported in part by the Natural Science Foundation of Fujian Province under Grant 2019J01773, in part by the Initial Scientific Research Fund of FJUT under Grant GY-Z12079, Grant GY-Z21036, and Grant GY-Z20067.

Acknowledgments: The authors would like to thank the anonymous reviewers for their valuable comments.

Conflicts of Interest: The authors declare no conflict of interest.

References

1. Eason, G.; Noble, B.; Sneddon, I.N. On certain integrals of Lipschitz-Hankel type involving products of bessel functions. *Philos. Trans. R. Soc.* **1955**, *247*, 529–551.
2. Yildirim, M.Y.; Rüştü, A. A Comparative Study of Optimization Algorithms for Global Path Planning of Mobile Robots. *Sakarya Univ. J. Sci.* **2021**, *25*, 417–428. [CrossRef]
3. Wang, H.; Yin, P.; Zheng, W.; Wang, H.; Zuo, J. Path planning of mobile robots based on improved A* algorithm and dynamic window method. *Robotics* **2020**, *42*, 346–353. [CrossRef]
4. Tan, B.; Luo, J.; Luo, Y.; Hu, C.; Zhuo, J.; Bai, Z.; Tian, J. Robot path planning for improved RRT algorithm. *J. Chongqing Univ.* **2022**, *25*, 1–13. Available online: http://kns.cnki.net/kcms/detail/50.1044.N.20220301.1410.005.html (accessed on 19 June 2022).
5. Liu, J.-S.; Ji, H.-Y.; Li, Y. Robot path planning based on improved bat algorithm and cubic spline interpolation. *Acta Autom. Sin.* **2021**, *47*, 1710–1719. [CrossRef]
6. Sun, H.; Hu, C.; Zhang, J. Deep Reinforcement Learning Methods for Motion Planning of Mobile Robots. *Control Decis.* **2021**, *36*, 1281–1292.
7. Wang, Y.; Jiang, X. Robot path planning using a hybrid grey wolf optimization algorithm. *Comput. Eng. Sci.* **2020**, *42*, 1294–1301.
8. Li, T.; Zhao, H. Path Optimization of Mobile Robot Based on Evolutionary Ant Colony Algorithm. *Control Decis.* **2022**. [CrossRef]
9. Xie, C.; Ying, L.I. Path planning of mobile robot based on improved algorithm. *J. Chongqing Univ.* **2021**, *44*, 140–148.
10. Eberhart, R.; Kennedy, J. Particle swarm optimization. In Proceedings of the IEEE International Conference on Neural Networks, Perth, WA, Australia, 27 November–1 December 1995; Volume 4, pp. 1942–1948.
11. Wang, X.; Wu, H.; Miao, Y.; Zhu, H. A Hybrid Routing Protocol Based on Naïve Bayes and Improved Particle Swarm Optimization Algorithms. *Electronics* **2022**, *11*, 869. [CrossRef]
12. Zhu, S.P.; Keshtegar, B.; Seghier, M.E.A.B.; Zio, E.; Taylan, O. Hybrid and enhanced PSO: Novel first order reliability method-based hybrid intelligent approaches. *Comput. Methods Appl. Mech. Eng.* **2022**, *393*, 114730. [CrossRef]
13. Tian, S.; Li, Y.; Kang, Y.; Xia, J. Multi-robot path planning in wireless sensor networks based on jump mechanism PSO and safety gap obstacle avoidance. *Future Gener. Comput. Syst.* **2021**, *118*, 37–47. [CrossRef]
14. Zhao, Q.; Li, C.; Zhu, D.; Xie, C. Coverage Optimization of Wireless Sensor Networks Using Combinations of PSO and Chaos Optimization. *Electronics* **2022**, *11*, 853. [CrossRef]

15. Kang, Y.; Jiang, C.; Qin, Y.; Ye, C. Robot Path Planning and Experiment with an Improved PSO Algorithm. *Robot* **2020**, *42*, 71–78. [CrossRef]
16. Panda, A.; Mallipeddi, R.; Das, S. Particle swarm optimization with a modified learning strategy and blending crossover. In Proceedings of the IEEE Symposium Series on Computational Intelligence (SSCI), Honolulu, HI, USA, 27 November–1 December 2017.
17. OuYang, H.; Quan, Y.; Gao, L.; Zou, D. Hierarchical path planning method based on mixed genetic particle swarm optimization algorithm. *J. Zhengzhou Univ.* **2020**, *41*, 34–40.
18. Song, B.; Wang, Z.; Zou, L. An improved PSO algorithm for smooth path planning of mobile robots using continuous high-degree Bezier curve. *Appl. Soft Comput.* **2021**, *100*, 106960. [CrossRef]
19. Miao, K.; Feng, Q.; Kuang, W. Particle Swarm Optimization Combined with Inertia-Free Velocity and Direction Search. *Electronics* **2021**, *10*, 597. [CrossRef]
20. Chen, G.; Jia, J.; Han, Q. Study on the Strategy of Decreasing Inertia Weight in Particle Swarm Optimization Algorithm. *J. Xi'an Jiaotong Univ.* **2006**, *40*, 53–56. [CrossRef]
21. Nan, J.; Wang, X. Particle swarm optimization algorithm with improved inertia weight. *J. Xi'an Polytech. Univ.* **2017**, *31*, 835–840. [CrossRef]
22. Zhao, Y.; Fang, Z. Particle swarm optimization algorithm with weight function's learning factor. *J. Comput. Appl.* **2013**, *33*, 2265–2268. [CrossRef]
23. Huberman, B.A.; Lukose, R.M. TadHogg. An Economics Approach to Hard Computational Problems. *Science* **1997**, *275*, 3. [CrossRef]
24. Chen, G.; Xie, X.; Xu, Y.; Jun, G.U. The construction of stochastic algorithm restart strategy and its application in TSP. *Chin. J. Comput. Sci.* **2002**, 514–519.
25. Li, X.; Wu, D.; Zhao, Z.; Wang, X.; Zhang, L. Path Planning Method for Indoor Robot Based on Improved PSO. *Comput. Meas. Control* **2020**, *28*, 206–211. [CrossRef]
26. Li, X.; Wu, D.; He, J.; Bashir, M.; Liping, M. An Improved Method of Particle Swarm Optimization for Path Planning of Mobile Robot. *J. Control Sci. Eng.* **2020**, *2020*, 3857894. [CrossRef]

electronics

Article

The Systems Approach and Design Path of Electronic Bidding Systems Based on Blockchain Technology

De Xu [1,2] and Qing Yang [3,*]

1 Intelligent Construction and Blockchain Collaborative Innovation Research Center, Jiangsu Open University, Nanjing 210000, China
2 Zhongru Information Technology Co., Ltd., Nanjing 210000, China
3 School of Architecture and Engineering, Jiangsu Open University, Nanjing 210000, China
* Correspondence: yangq@jsou.edu.cn

Abstract: The electronic tendering and bidding system has realized the digitalization, networking, and high integration of the whole process of tendering, bidding, bid evaluation, and contract, which has a wide range of applications. However, the trust degree, cooperation, and transaction efficiency of the parties involved in electronic bidding are low, and bidding fraud and collusion are forbidden repeatedly. Blockchain technology has the characteristics of decentralization, transparent transactions, traceability, non-tampering and forgery detection, and data security. This paper proposes a design path of an electronic bidding system based on blockchain technology, which aims to solve the efficiency, trust, and security of the electronic trading process. By building the underlying architecture platform of blockchain and embedding the business process of electronic bidding, this realizes the transparency, openness, and traceability during the whole process of electronic bidding. This paper uses qualitative and quantitative methods to prove the effectiveness of the system.

Keywords: blockchain technology; electronic bidding; system design

Citation: Xu, D.; Yang, Q. The Systems Approach and Design Path of Electronic Bidding Systems Based on Blockchain Technology. *Electronics* **2022**, *11*, 3501. https://doi.org/10.3390/electronics11213501

Academic Editor: Shinichi Yamagiwa

Received: 14 September 2022
Accepted: 24 October 2022
Published: 28 October 2022

1. Introduction

The construction industry plays an important role in the development of the social economy. However, the traditional bidding method has problems such as low information transparency, information asymmetry, and an opaque transaction process, which inhibit the development of the construction industry. Compared with the traditional bidding method, the electronic bidding (E-bidding) system is an essential transaction method in the current information era [1], which consolidates the process of bidding, tendering, bid evaluation, and contract-signing as an open network, and breaks through the limitation of time and space. Additionally, since the emergence of COVID-19 through the spread of Omicron, in the context of the global scale, it is of great practical significance to study how to realize the whole process of E-bidding online and how to ensure continuous economic activities and reduce personal contact.

Although the E-bidding system is widely used at present, the following problems still exist [2]. Firstly, there is no unified standard for E-bidding systems, which leads to poor real-time collaboration within different systems. Secondly, in the E-bidding system, it is difficult to ensure the identity authentication of users and data security, which is affected by network security. Finally, the E-bidding system has difficulties in the traceability of bidding participant behaviors. Namely, unfair situations often occur in transactions, but it is difficult for regulators to collect bidding fraud evidence and achieve supervision in real-time.

To solve the above problems, blockchain technology is brought into the E-bidding system. As a decentralized distributed ledger, blockchain co-records public data in chronological sequence, generates and updates data through distributed node consensus algorithm, and employs cryptography technology to ensure that data cannot be tampered with. Naturally, the blockchain enables collaboration without the authorization of a third-party, facilitates the construction of a highly credible transaction and supervision system with high security and reliability, and can trace all information in the transaction process to ensure transparency and fairness [3,4]. To this end, blockchain will become one of the most prevalent underlying protocols of the "Internet of Everything" and be applied in all fields of society, i.e., social governance, arbitration, auditing, smart city construction, etc. However, compared with other industries, it is the diverse and complicated transaction process that hinders the development of E-bidding in construction sectors. In addition, the application of E-bidding in the construction industry has lagged behind the manufacturing and retailing sectors [5], not to mention the adaptation of blockchain-based E-bidding. In addition, the research on blockchain-based E-bidding systems is limited to the preservation of information in each stage of bidding and does not consider how to avoid bidding fraud to maintain fair transactions. Thus, to promote further prosperity, decrease the large resource consumption, and improve the overall transaction process's efficiency and security, it is necessary to study blockchain-based E-bidding in the construction industry.

In this paper, we combine blockchain technology and an E-bidding system and propose a blockchain-based E-bidding system applied in the construction industry, which consists of a blockchain electronic transaction bidding system, a big data system, and a framework for mining evidence of bidding fraud. By virtue of a large amount of complex and frequently changing transaction information to handle, it is time-consuming and a great challenge for the E-bidding system to collect, process, and analyze the large-scale data. Hence, the introduction of "big data" technology into the blockchain-based E-bidding system will promote the interconnection and real-time sharing of information, as well as further optimize the market-based allocation of resources. In addition, bidding fraud detection is also an essential issue of concern in E-bidding. The "big data" analysis can quickly determine whether there is bidding fraud or collusion in the bidding process and provide fair digital "evidence" to assist the bidding administrative department to strengthen regulation of the entire bidding process and impose administrative penalties for violations, which is advantageous in improving the standardization, digitalization, and scientific level of bidding activities.

The main contributions of this work can be summarized as follows:

(1) This paper combines blockchain technology and an E-bidding system in the construction industry and designs a blockchain-based E-bidding framework to raise bidding efficiency and guarantee the fairness, impartiality, and transparency of transactions.
(2) On the basis of big data technology, a big data system (BDS) is designed to collect, handle, and analyze the data in the bidding process, which is convenient for maintaining transaction fairness and improving bidding efficiency.
(3) A bidding fraud evidence mining method is embedded in the big data system to mine fraud evidence and strengthen transaction supervision, which combines maximal frequent itemset mining, association rule mining, and binary support number calculation algorithms to boost operational efficiency.

The remainder of this paper is organized as follows. Section 2 offers the related work of E-bidding systems in the construction industry, application of blockchain, blockchain-based E-bidding systems in the construction industry, and big data system. Section 3 provides the proposed blockchain-based E-bidding systems. Section 4 shows the extensive experiments and results of the proposed method for electronic bidding. Section 5 presents the conclusion.

2. Related Work

2.1. Electronic Bidding System in the Construction Industry

The traditional project bidding field has gone through a long road of development under the norms of laws and regulations such as the "Tendering and Bidding Law" and the "Government Procurement Law", which have played an important role in unifying the rules of the bidding market and encouraging orderly competition in the market. The emergence and wide application of the internet is a revolution in industrial society; for the construction industry in the field of engineering bidding, the emergence and development of electronic bidding has also redefined the ways and methods of bidding by construction market entities and has played a positive role in further promoting a free, fair, just, and honest market environment. In recent years, the government and relevant industry organizations have supported and encouraged construction units and relevant market entities to carry out electronic bidding and bidding work, which has effectively promoted the application of electronic bidding. The electronic bidding system realizes business functions such as online bidding, bidding, bid evaluation, and contract management, reduces offline transaction costs, improves work efficiency, enhances the information management capabilities of governments and participating entities, and effectively promotes the digitalization, networking, and high integration of the whole bidding process. However, there are still some problems with the current electronic tendering, resulting in the application of electronic tendering still being quite limited. First, relevant laws and regulations lag behind, there are a lack of unified norms and standards, and it is difficult to promote. Second, there are many electronic bidding platforms, which are poorly compatible with each other, and the phenomena of administrative intervention and secret operations cannot be effectively prevented. Third, the security and stability of the electronic bidding platform need to be strengthened; if the data security and stability performance is not effectively guaranteed, it is very easy to enable the leakage of commercial secrets and malicious tampering of data information. The research on these problems is of great practical significance for the application of electronic bidding in the construction industry.

2.2. Application of Blockchain

Blockchain, sometimes known as a distributed shared ledger, is essentially a multi-participant, cooperatively maintained, continually growing distributed database system. Blockchain technology is very well liked by businesses and has been widely used because of its anonymous, decentralized, open and transparent, and tamper-evident characteristics. In the field of finance, when blockchain peer-to-peer (P2P) technology was applied to cross-border payments [6], the remittance becomes transparent, and transaction history data was traceable, providing security assurance for both the recipient and the remitter while also considerably enhancing efficiency and speed. In addition, with the application of blockchain in medical data privacy protection [7], medical data storage and access can be recorded and remain tamper-proof, which avoids unscrupulous individuals from using this information for fraud and blackmail. Also, the untamperable nature of blockchain renders the digital proof on the chain extremely believable, which may be utilized to create a new authentication mechanism in the areas of property rights [2], notarial services [8], and social welfare [9] and to raise the management standard of public service. Motivated by the compatibility of blockchain characteristics with trade process requirements, we attempt to integrate blockchain into the E-bidding system with its advantages of distribution, anonymity, transparency, and traceability to promote the reform and progress of the E-bidding system in the construction industry.

2.3. Blockchain-Based E-Bidding in the Construction Industry

Since the structure and technology of blockchain effectively ensure the authenticity and traceability of information, the research on the application of E-bidding systems in the construction industry has become popular in recent years. In 2017, motivated by the dynamic grouping of several companies in the projects, Turk et al. [10] introduced the P2P nature of the relationships in blockchain technology to establish a reliable infrastructure for information management throughout all stages of the building life-cycle. To improve the data reliability and verifiability and privacy of data transmission, Tso et al. [11] applied blockchain and smart contract technology and proposed the first decentralized electronic voting and bidding systems. In 2021, Sigalov et al. [12] combined Building Information Modeling (BIM) approaches with smart contracts to achieve automated billing, which enhances timely payment and guaranteed cash flow. Compared with these approaches, our method has higher operation efficiency and can mine bidding fraud evidence through big data technology, which will be later described in detail.

2.4. Big Data Technology

Big data technology has the following four characteristics: Volume, Variety, Value, and Velocity [13] when compared with traditional databases. With these advantages, after collecting and organizing the large-scale data, it is much easier and more practical to determine its potential laws and predict the development trend through intelligent analysis and data mining. This can assist people in decision-making [14], boost operational efficiency, and realize greater benefits. Therefore, there are many applications of big data technology in our daily life [15,16], such as finance [17,18], E-commerce [19], medical [20], and communication [21]. Moreover, it is data analysis that is the key point of big data technology, which usually uses data mining to acquire the diagnosis of anomalous data. In 2000, Pei et al. [22] proposed an efficient and scalable algorithm for frequent closed itemset mining with the use of a frequent pattern (FP) tree, which could provide a minimum description of abnormality. To reduce the computational complexity and memory usage, Halim et al. [23] presented a graph-based approach with storage of all relevant information to mine maximal frequent itemsets and prove its superiority. With only one access to the record of all frequent itemsets, it can significantly improve the execution efficiency of positive as well as negative association rule mining [24,25] and further increase the run-time efficiency of the whole process. Hence, we employ big data technology to assist in evaluating bidding activities, ensuring project quality, and boosting operational effectiveness while also providing reliable decision-making support for all types of transaction issues. Moreover, there is little research to study how to assist the blockchain-based E-bidding system through big data technology. Inspired by this, we integrate a big data system (BDS) into the E-bidding system in this work.

3. Method

3.1. Preliminaries of Blockchain

In this section, we introduce some preliminaries about the blockchain to which the traditional E-bidding system is adjusted.

3.1.1. Definition of Blockchain

Blockchain is generally considered as a decentralized, de-trusted, distributed, shared ledger system that combines blocked data, which includes transaction information, timestamps, and hash value in a chain chronologically and cryptographically [26]. From the view of data, blockchain can be interpreted as a distributed database that cannot be passively modified or forged. From the view of technic, blockchain is a distributed ledger technology integrated with various technologies, such as asymmetric cryptography [27], P2P network [28], and smart contracts [29].

3.1.2. Characteristics of Blockchain

A key characteristic of blockchain is that it is a distributed and decentralized system. While only one controller manages the completeness of data information in a centralized database [30], the term "distributed system" means that the content of transaction information can be stored and examined simultaneously by all participants, which makes it possible to maintain information integrity and trustworthiness without the need for authorization. The use of various distributed applications [31] is to achieve state change management, data storage, query validation, and control management. Therefore, blockchain has more obvious technical and management advantages compared with traditional centralized systems.

Additionally, using hash algorithms as encryption technology, the most prominent advantage of blockchain is its high level of security [32]. Since the information is all jointly owned in the blockchain, when viruses or hackers attack P2P-specific data, they cannot change or delete data at will. Secondly, a decentralized blockchain can minimize transaction costs to a maximum extent while having good technical scalability and improving transaction efficiency. Finally, blockchain, due to its openness nature, can improve the transparency and fairness of transactions, ensure security, and reduce regulatory costs. Although the access rights in the blockchain vary, almost all participants can access all the transaction records and information stored by the chain blocks anytime and anywhere [33]. All the above characteristics are summarized in Table 1.

Table 1. The characteristics of blockchain [3,4,6–9].

Characteristics	Description
Decentralization	Each node realizes information self-verification, self-transmission, and self-management.
Immutability	No one can modify the data without authorization once it has been written to the blockchain.
Security	All data on the chain are encrypted by hash operation, asymmetric encryption, private key, and other cryptographic methods.
Openness	All nodes in the chain can participate in the record maintenance of data.

3.1.3. Categories of Blockchain

The classification of blockchain is based on the degree of network openness and can be mainly classified as public, private, and industry blockchains [34,35], which is shown in Table 2. Concretely speaking, a public blockchain is a blockchain shared by any organization or individual that can operate and be confirmed on that blockchain, and other organizations or individuals can join it; a private blockchain is one in which the blockchain is used only internally for bookkeeping activities; a consortium blockchain is one in which some nodes are controlled by pre-selected nodes.

Table 2. The categories of blockchain.

Categories	Description	Scenarios	Trust Authority	Speed of Consensus
Public Blockchain	Anyone can operate and be confirmed.	Virtual Cryptocurrency	0	Slow
Private Blockchain	An organization controls the write access.	Only internally for bookkeeping activities.	1	Fast
Consortium Blockchain	Some nodes are controlled by pre-selected nodes.	Inter-institutional trade, settlement, or liquidation	≥ 1	Slightly Fast

3.1.4. Drawbacks of Blockchain

As mentioned before, the essential characteristic of blockchain, distribution, can not only verify all transaction information of participating subjects, effectively guaranteeing information authenticity and traceability [36], but also permit each node or user in the blockchain to enjoy the same equal and independent rights to supervise each other. Moreover, due to the Byzantine fault tolerance mechanism, the blockchain can function in an orderly fashion even when the system receives attacks. Thus, there are many well-known domestic and international projects based on blockchain, such as Bitcoin [37] and Ethereum [38], which rely on hardware arithmetic to reach consensus and have the advantage of high security. Although the application of blockchain is booming, it is undeniable that blockchain technology suffers from consensus mechanism security issues, block capacity, efficiency problems, and high hardware cost expenditure. To address the drawbacks of blockchain, this paper focuses on the block efficiency issue, as subsequently shown in Section 3.2.

3.2. Proposed E-Bidding System

3.2.1. System Structure of the Blockchain-Based Electronic Bidding System

The structure of our blockchain-based E-bidding system is composed mainly of three layers: the blockchain foundation layer, interface layer, and application layer, as shown in Figure 1 and Table 3. In addition, the big data system (BDS) is applied to assist the blockchain-based E-bidding system in providing reliable decision-making support for all types of transaction issues and mining the bidding fraud evidence.

(1) Blockchain foundation layer: To ensure the reliable operation of upper-layer bidding services, the blockchain foundation layer provides credible infrastructure for upper-layer architecture. Specifically, blockchain automatically executes the pre-defined smart contracts and triggers corresponding algorithms. Meanwhile, it implements the basic functions of data security sharing, such as on-chain data encryption, integrity assurance, and being untamperable.

(2) Interface layer: The interface layer plays a connecting role and provides an interface between the application layer and the blockchain layer, supporting JAVA-Software Development Kit (SDK), GO-SDK, etc. The SDK provides the blockchain address, private key generation, data signature, data uploading, data encryption, smart contract invocation, etc., and the data signature can support both the international and domestic cryptography standards.

(3) Application layer: The application layer is the gate to receive data and handles the business logic of bidding.

(4) Big data systems: BDS is employed to optimize the bidding process for vulnerabilities and avoid bidding fraud. Further, BDS collects data from all stages and can assist the decision-making for all types of transaction issues, while also boosting operational effectiveness and ensuring project quality.

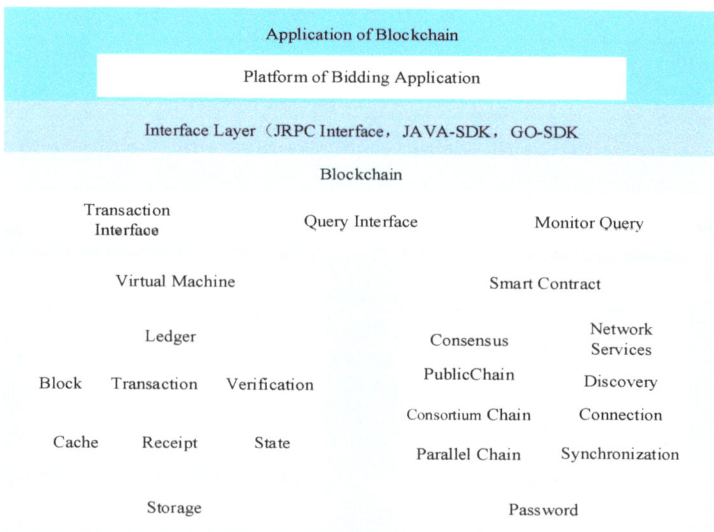

Figure 1. Architecture of the blockchain-based E-bidding system.

The main difference between the proposed system and the previous approach is whether blockchain technology and BDS are used. Therefore, we introduce only the blockchain foundation layer and BDS in detail in the following sections.

Table 3. The components of blockchain.

Components	Description
Blockchain foundation layer	Ensure the reliable operation of upper-layer bidding services.
Interface layer	Create a connection between the blockchain foundation layer and the application layer.
Application layer	The gate to receive data and handle the business logic of bidding.
Big data systems	Assist blockchain electronic bidding system to optimize the bidding process.

3.2.2. Structure of the Blockchain Foundation Layer

Blockchain records every key information in each segment, i.e., tenderer information, bid documents, evaluator information, the opening, evaluation, bidding determination, and contract signing. Various data need to be stored, including text, images, and documents, among which text information can be directly stored on the blockchain, while images and documents are usually stored with a hash value that easily suffers from being tampered from attackers. To address this issue, a distributed blockchain node system is the key component to ensure data security. The corresponding hash value will be changed if the original data on the chain is tampered with, which will lead to a data mismatch. This approach can not only solve the cost and efficiency problems of big data storage but also keep the data unchanged.

The blockchain node system consists of consensus nodes, supervisory nodes, and verification nodes, as shown in Figure 2 and Table 4. Specifically, the consensus node is involved in the consensus of the blocks in the business process, which is responsible for the security of the data; the supervisory nodes can conduct statistics on transaction behaviors, identify the true identity of users on the chain, review transactions, and when needed, the supervisory nodes can restrict transactions and freeze accounts by utilizing smart contracts of account management. Verification nodes, which are captured or released at any time, provide network resources as well as verify the validity of blocks, but they cannot become authentication nodes or super nodes.

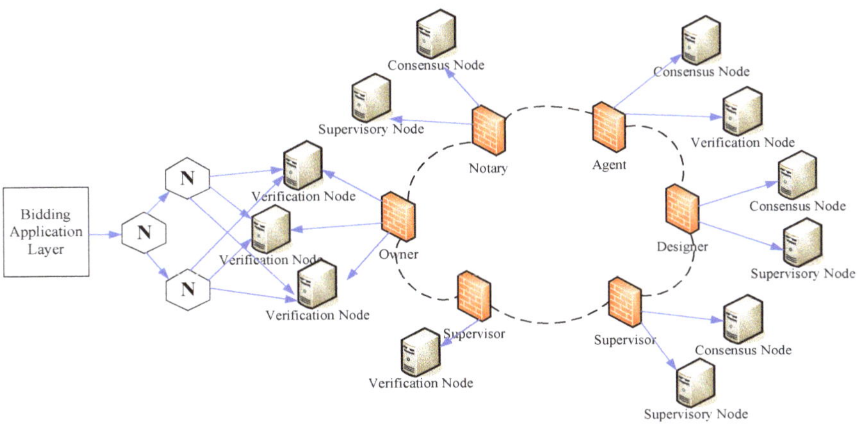

Figure 2. The relationship between nodes on the blockchain system.

Table 4. The description of nodes.

Types of Node	Description
Consensus nodes	Responsible for the security of the data.
Supervisory nodes	Supervise the process of transactions.
Verification nodes	Provide network resources as well as verify the validity of blocks.

The blockchain-based E-bidding system is designed as a consortium chain, which ensures that the traceable data on the chain cannot be tampered with. Moreover, the consensus mechanism of the consortium chain can tolerate node error rates up to one-third, which includes arbitrary node offline and malicious behaviors. Under this mechanism, each node executes the message that it has received most frequently to assure that the node reaches a consistent result; this algorithm is usually called the Byzantine fault tolerance mechanism [39] and is given in Algorithm 1. On the basis of this consortium chain, the consensus mechanism is divided into following parts: proposal phase, pre-selection phase, pre-submission phase, pre-submission waiting phase, submission phase, and block generation phase as shown in Figure 3.

Algorithm 1 Commit

```
Input: commitMsg
Output: ReplyMsg
if verifiedMsg(commitMsg) ! = true
    return error;
    end procedure
save commitMsg
if state prepared:
    return ReplyMsg;
    end procedure
return none
end procedure
```

(1) Proposal phase. The proposal node takes the transaction information out from the Mempool, packs it, and sends the proposal to other validation nodes. Then, the process enters in pre-selection phase.
(2) Pre-selection phase. Each validation node verifies whether the proposal is legitimate, such as whether the signature is authentic, whether the height is correct, etc. If the proposal passes the verification, it will be transmitted to a pre-selected state.
(3) Pre-submission phase. If each validation node receives pre-selected messages from more than 2/3 of the other nodes, the process moves on the pre-submission waiting phase.
(4) Pre-submission waiting phase. If each validation node receives pre-submission messages from more than 2/3 of the other nodes, the process goes to the submission phase.
(5) Submission phase. The consensus module sends the block to the smart contracts module, which is always regarded as an executor, for a specific execution. Then, when the execution succeeds, the block is stored in the blockchain and ingresses the next phase. After the contract signatory, transaction information is sent to the node's transaction Mempool module through the Remote Procedure Call (RPC) module while it is broadcasted to other nodes through the P2P module to ensure that the transactions of all nodes in the Mempool are consistent at the same time. In summary, the consensus module regularly pulls a list of transaction information from the Mempool, constructs blocks, performs a consensus mechanism, and sends blocks to the executor module to conduct the transactions, which is shown in Figure 4.
(6) Block generation phase. After the execution, the consensus module sends and writes the block to the Blockchain module. Then, the ledger broadcasts the block to other nodes through the P2P module. After receiving the block, nodes will verify and implement the transactions in the block again and store the block. This phase is shown in Figure 5.

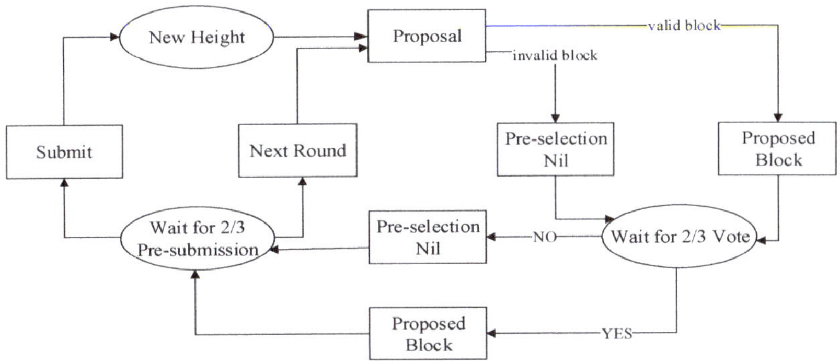

Figure 3. Consensus mechanism of consortium chain based on blockchain.

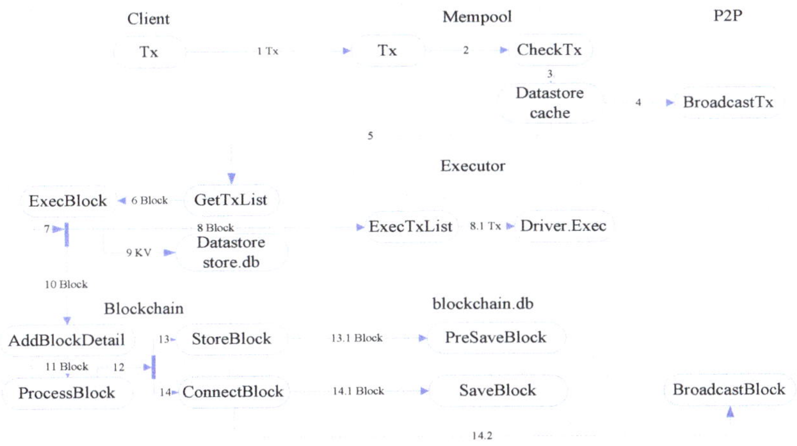

Figure 4. Execution of the smart contract module based on blockchain.

3.2.3. Process of the Big Data System

Big data technology analyzes the intrinsic linkage of information to quickly locate vulnerabilities in the E-bidding system. Therefore, BDS was designed to further improve the E-bidding system. More prosaically, BDS is organized into two parts: data collection and data analysis, and they will be described in detail in the following section.

Data collection: The main purpose of data collection is to extract valuable data from the entire bidding process, which provides the basis for subsequent analysis. There are three main types of data objects to be collected: data of the tender subject, data generated by the tender process, and evaluation information. Specifically, the data of the tender subject mainly include all types of information including enterprise information and tender information. These data allow a critical quality assessment of companies to limit the number of bidding participants and save running costs. Then, the data generated by the tender process become the main body of data analysis, including information about bid prices, anticipated prices, and expert evaluations, all of which are the most diverse, valuable, and largest part of the data collection phase. Moreover, evaluation information contains mainly contract evaluation and settlement audit information, which is used to supervise the legitimacy of bidding information.

Data analysis: Due to the complexity of large-scale data, it is a significant challenge to process and analyze these data. Thus, an association rule mining algorithm is applied to achieve efficient data analysis. In this stage, we use the frequent itemset mining method to detect the frequent closed itemsets and provide a minimum description of data fraud evidence, the number of which is between the maximum frequent itemsets and frequent itemsets. To reduce time complexity, we utilize an improved algorithm that mines the maximal frequent itemsets based on the FP tree and solves the problem of frequent itemset updating in bidding fraud data mining. Within the process of frequent itemset mining, the negative and positive association rule mining algorithm is executed, which is practical in solving the conflict between fraud evidence. In addition, the binary support number calculation method is applied to the simple logical operation of "yes" or "no" on the judgment operation of bidding fraud evidence so as to improve the execution efficiency of the algorithm. The progress of bidding fraud evidence mining is shown in Figure 6.

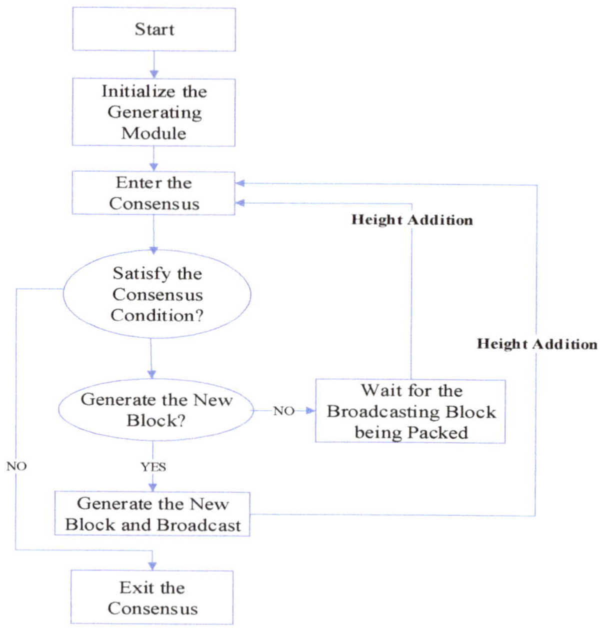

Figure 5. Demonstration of the block-generation process by a consensus algorithm.

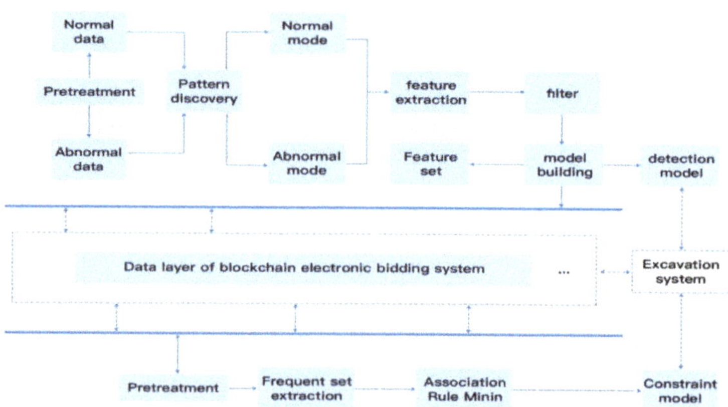

Figure 6. Evidence mining framework for bidding fraud.

3.2.4. Overall Processes of the Proposed System

This section describes the process of the blockchain-based E-bidding system in detail, which is divided into six stages: the registration of the tender and tenderer, signing up for the bidding, tender obtaining, tender submitting, tender opening, and the tender deciding and contract management stage, as shown in Figure 7. BDS is embedded into all phases of the E-bidding system and detects bidding fraud data in real-time.

(1) Registration of the Tender and Tenderer. On the blockchain, the corresponding account is assigned to the tenderer. Meanwhile, once uploaded to the chain, tenderers' basic information, such as credentials, credit, and performance, can be permanently stored and cannot be tampered with, and identity information is protected thanks to the blockchain's consensus mechanism. In addition, registration is an optional phase for the designed system, and the basic information of tenderers can be entered at the stage of potential tenderers if registration is not required.

(2) Signing up for the bidding. In this phase, the tenders post the information of specific bidding activity in the designed E-bidding system and this bid document will be stored in the blockchain. If necessary, the key material is encrypted for security. Moreover, the tender will verify the identity of the potential tenderers through blockchain and confirm the results. Each participant in the chain can get specified and reliable bid documents as credentials by using timestamps and produced hash values.

(3) Tender Obtaining Stage. Though the bid document is confidential, the potential tenderer can download and browse these documents to get more bidding details if they pay the bid document fee. Moreover, various previous successful cases are provided to these paid subscribers by the tender authority in the blockchain. Provided tenderers wish to join this bidding activity, they could download the specified bid documents and fill them in online or offline.

(4) Tender Submitting Stage. According to bid requirements and project characteristics, after tenderers complete the bid document, these bid documents will be uploaded to the E-bidding system before the deadline, and the system will automatically anchor the time-point and store the certificate. Due to the high volume of bid documents, a small amount of key information can be encrypted on the chain with specific digital signatures, and the large documents are hashed on the chain, while documents themselves are stored on the file server; this effectively avoids tampering and leakage of important information at the later stage, eliminates irregularities such as tenderer collusion, and ensures a fair and transparent bidding environment.

(5) Tender Opening Stage. Bid evaluators on the blockchain E-bidding system are given corresponding accounts and rights, and their personal data are made available to the public. The P2P and anonymity functions of blockchain can be used to implement P2P transactions, which ensure that remote evaluation of bids can do so impartially and without collusion or favoritism. Within a predetermined amount of time, after authenticating experts' identities on the chain using face or fingerprint recognition, their evaluation results according to the bid document will be stored on the chain.

(6) Tender Deciding and Contract Management Stage. Following the evaluation, the system authorizes the public key of the winning information based on the evaluation results and notifies the winner and the tender to sign the contract online. At the same time, the contract serial number, contract conditions, third-party certification of contract terms, contract subject, and contract filing are all written into the blockchain as witnesses. During the contract public period, any party or supervisory department with concerns about the bidding process can trace the original deposited data of the whole bidding process.

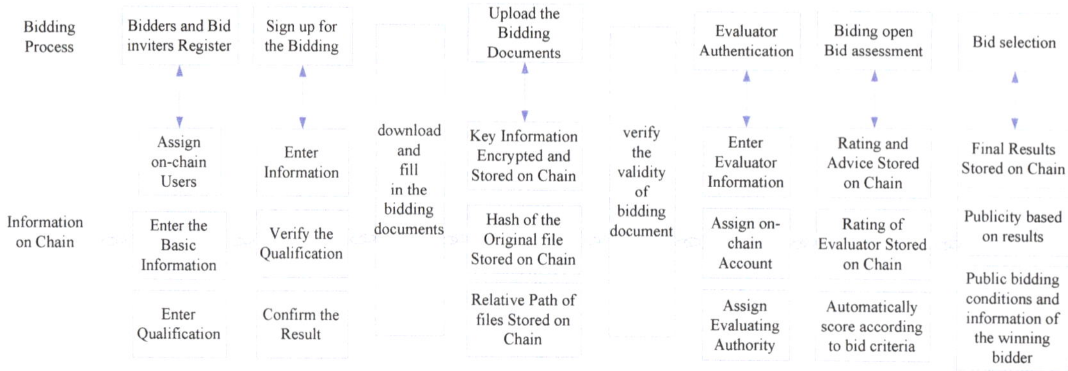

Figure 7. The process of blockchain-based E-bidding service.

4. Experiments

Our work has developed a decentralized electronic bidding framework based on blockchain technology and big data system and maintained the balance of algorithm complexity and performance to achieve transaction security and privacy protection. By handling bidding data in the big data system, an evidence mining framework for bidding fraud detection is designed, which is applied in the bidding system and has long-term significance for maintaining the fairness of the bidding environment. In this section, we first introduce the experimental settings and platform. We then conduct ablation experiments on the blockchain part to quantitatively evaluate the performance, subsequently compare two encryption algorithms in the proposed framework by designing quantitative and qualitative experiments to analyze the efficiency, test the computation cost of the proposed system, and finally, compare it with other blockchain-based E-bidding systems which are applied in different sectors.

4.1. Experimental Settings

We utilize four services as well as a CPU of Intel(R) Xeon(R) Platinum 8378 A and a RAM of 8 G to build the E-bidding system. The system is running on a 64-bit CentOS of version 7.9. As an open-source distributed ledger technology platform, Fabric not only has better performance in transaction processing and transaction confirmation delay but also realizes functions such as smart contracts and confidential transactions. Fabric is an open-source distributed ledger technology platform, and compared with the traditional public chain, it has better performance. Its most important feature is pluggability, and it can be configured to meet as diverse needs as possible. The underlying layer of Fabric consists of peers and orderer nodes that form a P2P network that interacts through Google's open-source RPC framework, gRPC. The middle is isolated using channel technology and each channel is an independent network with its own ledger. Fabric provides gRPC, API, and SDK for upper-layer applications, through which applications can access a variety of resources such as ledger, processing transactions, managing chain-code, registering events, and managing permissions [40]. Therefore, we conduct the experiments with Go language on Fabric, and the run-time calculations are obtained by using the computer system clock.

4.2. Ablation Experiment

Generally speaking, the metric of transactions per second is usually used to evaluate the performance of the blockchain. Thus, to validate the performance of the proposed method, we conduct an ablation experiment in terms of transactions per second. In five distinct sets of testing, the average throughput for the proposed system and the system without blockchain are compared in Figure 8. In addition, specific data are displayed in Table 5. Thanks to the parallel mechanism of blockchain, which allows the E-bidding system to implement several bidding activities at the same time, the transaction throughput of the proposed blockchain-based E-bidding system rises linearly with the number of transactions until it meets the peak at roughly 45 tps, at which point it starts to fall. Moreover, Figure 8 also demonstrates that the proposed methodology is much more effective than the system without blockchain. Specifically, the proposed system can process nearly 24 transactions per second while the system without blockchain can process only up to 11 transactions per second. That is, a system with blockchain technology can double the throughput of the original version method. From this point, it is clear how crucial blockchain technology is to transaction speed.

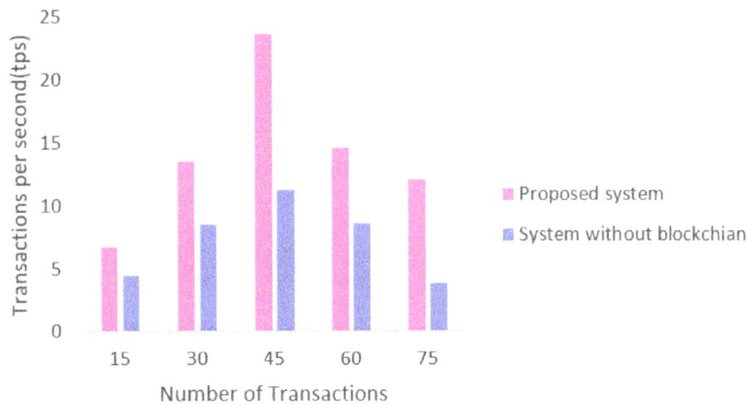

Figure 8. Average throughput comparison between proposed system and system without blockchain.

Table 5. Average throughput comparison of proposed system with/without blockchain.

No. of Transactions	Transactions Per Second (tps)	
	Proposed System	System without Blockchain
15	6.7	4.5
30	13.5	8.5
45	23.7	11.3
60	14.9	8.6
75	12.1	3.9

4.3. Performance Comparison of Encryption Algorithms

To a great extent, the efficiency of blockchain depends on the encryption algorithm [41]. Thus, in the proposed framework, we compared the two well-known encryption algorithms, elliptic curve cryptography (ECC) [42] and RSA [43], for time complexity and implementation of transaction validation. The relative pseudocode of ECC and RSA is given in Algorithms 2, 3, 4, and 5, respectively.

Algorithm 2 ECC encryption algorithm

Input: elliptic curve Ep(a, b), base point G, order n, random integer r, private key k, public key K, plaintext m
Output: ciphertexts c1 and c2
1: Select k (k < n)
2: Compute K = k ∗ G
3: Select r (r < n)
4: Compute c1 = m + r ∗ K
5: Compute c2 = r ∗ G
6: **return** c1, c2
7: **end procedure**

Algorithm 3 ECC decryption algorithm

Input: elliptic curve Ep(a, b), base point G, order n, random integer r, private key k, public key K, ciphertexts c1 and c2
Output: plaintext m
1: Compute M = c1 − k ∗ c2
2: Encode M
3: **return** M
4: **end procedure**

Algorithm 4 RSA encryption algorithm

Input: public key (x, y), plaintext m
Output: ciphertext c
1: Compute c = m^y mod x
2: **return** c
3: **end procedure**

Algorithm 5 RSA decryption algorithm

Input: public key (x, y), private key k, ciphertext c
Output: plaintext m
1: Compute m = c^k mod x
2: **return** m
3: **end procedure**

4.3.1. Time Complexity

Table 6 and Figure 9 certainly illustrate that ECC surpasses RSA in terms of time complexity. Even though both the corresponding time complexity of ECC and RSA tend to rise with the number of bits, the time complexity of ECC is consistently lower than that of RSA. The fundamental reason for this is that ECC, as opposed to RSA, better satisfies all the characteristics necessary to meet blockchain security requirements.

Table 6. Time complexity comparison of ECC and RSA.

Number	Time Complexity (ms)	
	ECC	RSA
1	3.5	13.8
2	3.8	15.2
3	4.6	17.3
4	5.2	18.9
5	5.8	19.7

4.3.2. Key Size, Encryption Time, and Decryption Time

On the basis of the comparison of ECC and RSA key size, encryption time, and decryption time shown in Table 7, we can observe that while ECC requires fewer bits, RSA has a similar level of protection. Concretely speaking, when RSA needs a 16,358-bit key to provide the resembled security level, ECC employs just a 622-bit key. Furthermore, though the encryption time of ECC is slower than the encryption time of RSA, ECC outperforms RSA in terms of efficiency when considering the decryption time as well. These outcomes are mainly because a shorter key leads to much less CPU and memory consumption as well as faster encryption and decryption time.

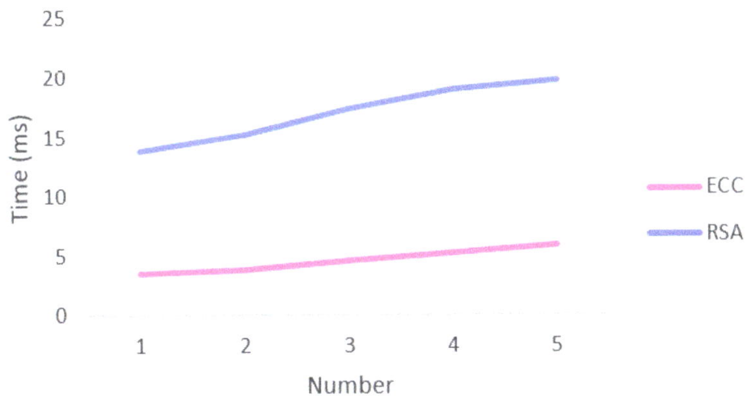

Figure 9. Time complexity comparison of algorithms.

Table 7. Performance comparison of ECC and RSA.

Key Size (Byte)		Encryption Time (s)		Decryption Time (s)	
ECC	RSA	ECC	RSA	ECC	RSA
178	1223	9.59	0.69	25.01	27.62
251	2362	61.23	0.82	25.98	121.38
297	3521	73.36	0.95	26.65	230.36
399	8353	100.26	1.24	35.01	313.67
622	16,358	121.35	1.62	47.91	455.61

To show the above trend more vividly, we illustrate the data of Table 7 in Figure 10, which also shows that the differences between ECC and RSA are more apparent as the key size grows and under the same degree of protection, RSA needs much more key size than ECC. As can be seen from Table 7, a robust ECC cryptosystem needs keys with a minimum key size of 178 bits. Therefore, we chose key sizes of 178 bits for ECC and 1223 bits for RSA as starting points in Figure 10. Afterward, Figure 10 presents the predominance of ECC.

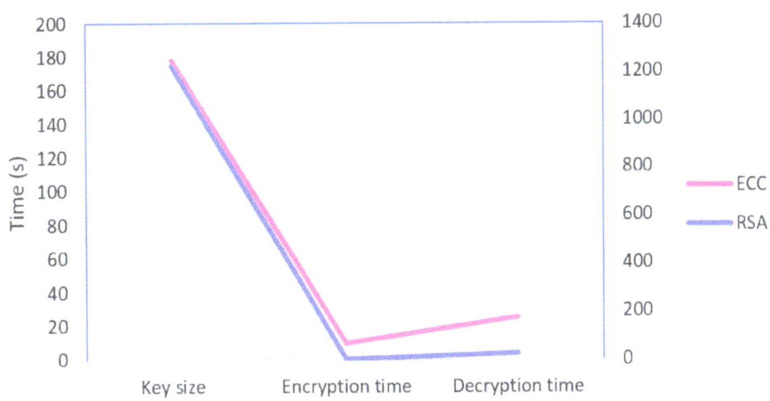

Figure 10. Comparison of ECC and RSA in terms of key size, encryption time, and decryption time.

4.4. Computation Cost

Additionally, six cases for various tenders with numerous amounts of bids are tested. Table 8 and Figure 11 reveal that even with 41 tenders and 70 bids, the computation cost is only 72.353 ms, which indicates that the adoption of big data technology can substantially decrease the large resource consumption and enhance the effectiveness of the proposed system. As a result, high performance can be attained by implementing our framework.

Table 8. Computation cost of the blockchain.

No. of Case	Tenders	Bids	Computation Cost (ms)
Case 1	9	10.5	12.12
Case 2	13.65	20.7	25.27
Case 3	19.36	34.98	39.79
Case 4	25.78	49.71	56.25
Case 5	36.352	65.57	62.291
Case 6	41.695	70.39	72.353

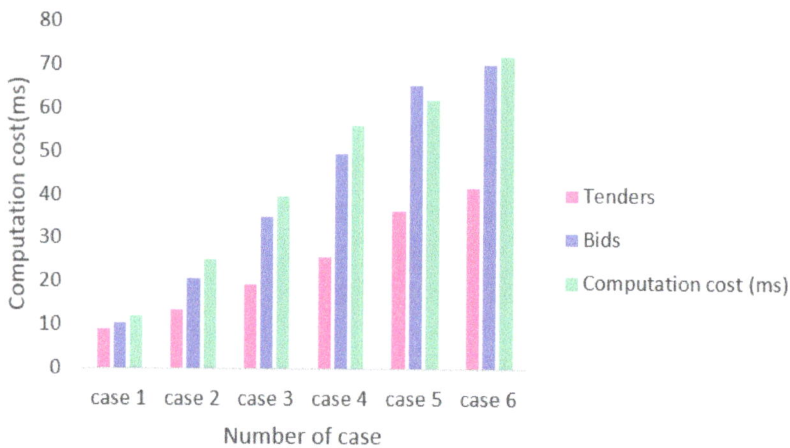

Figure 11. Computation cost with various tenders and bids.

4.5. Comparisons of Security with The-State-of-the-Art Methods

To evaluate the performance of the proposed bidding system, four popular systems are utilized, namely, those of Chen [44], Nair [45], Johnson [46], and Wang [47]. Additionally, six metrics that are necessary for an E-voting system are adopted to exhibit comprehensive comparisons, i.e., completeness, anonymity, fairness, eligibility, rationality, and non-repeatability. Specifically, the meanings of these metrics are provided as follows.

Completeness: Completeness is when each person can check whether the bidding information is correct.

Anonymity: Anonymity ensures that no internal or external attackers can know the identity and transactions of other people.

Fairness: A technology or protocol that does not discriminate against the honest and correctly participating members is said to be fair.

Eligibility: Eligibility means that only those with legal qualifications have access to the system to protect the fairness of the voting or bidding process.

Rationality: Rationality denotes that no internal or external attackers have the opportunity to maliciously tamper with other people's bidding, thereby ensuring the legitimacy of the voting process.

Non-repeatability: Non-repeatability denotes that each operation is done only once.

As can be seen in Table 9, our system has more comprehensive security than several systems, which is based on the following merits. (1) In our system, every node verifies whether the new data is correct through the existing data on the blockchain. Due to the great difficulty in tampering with existing information and the closeness of the blockchain system, completeness is ensured. (2) Our system ensures the traceability and anonymity of data through an encryption algorithm. (3) We use a decentralized consensus mechanism that makes each member's encrypted identity and bidding information public to other members for verification, which can also reflect fairness, to a great extent. (4) Our big data system could filter out malicious bidding attacks. Furthermore, in the registration stage and bidding information transferring procedures, our system verifies and encrypts the identity information of tender and tenderers to guarantee eligibility. (5) In our scheme, if individuals want to tamper with the information of a block on the blockchain, they must lead a new branch from the block and create a new chain that exceeds the length of the original chain, which is computationally impossible. (6) In our system, the blockchain prevents double-bidding by timestamping groups of transactions and then broadcasting them to all of the nodes in the system. As operations are time-stamped on the blockchain and mathematically related to the previous ones, they are irreversible and impossible to tamper with.

Table 9. Comparison of security properties.

Method	Completeness	Anonymity	Fairness	Eligibility	Rationality	Non-Repeatability
Chen's [44]	✓	✓	✗	✓	✓	✓
Nair's [45]	✗	✗	✓	✗	✓	✗
Johnson's [46]	✗	✓	✓	✓	✓	✓
Wang's [47]	✓	✓	✓	✓	✓	✗
Ours	✓	✓	✓	✓	✓	✓

In summary, our proposed bidding system is very beneficial for improving security, data traceability, and cooperation.

5. Conclusions

This paper proposes an implementation path and method of blockchain technology to solve the existing problems of electronic bidding system, which provides a realistic solution for solving the design standardization of electronic bidding platforms, system security and stability, and traceability and storage of bidding process. In addition, through this paper, practitioners related to electronic bidding can understand the latest research trends and technological innovation methods of blockchain technology in this field and become familiar with the main problems and technical paths solved by blockchain technology in electronic bidding. At present, there is still a gap in research in this field at home and abroad, and this paper is of great significance for blockchain technology to empower the industrialization, industrialization, and digitalization of construction, and promote the transformation and upgrading of the construction industry.

Author Contributions: Conceptualization, D.X. and Q.Y.; methodology, D.X.; software, D.X.; validation, D.X. and Q.Y.; formal analysis, D.X.; investigation, D.X.; data curation, D.X.; writing—original draft preparation, D.X.; writing—review and editing, Q.Y.; visualization, Q.Y.; supervision, D.X.; project administration, D.X.; funding acquisition, D.X. All authors have read and agreed to the published version of the manuscript.

Funding: This research was funded by the 2020 Science and Technology Plan Project of the Ministry of Housing and Urban-Rural Development, under Project Number 2020-K-061.

Data Availability Statement: Not applicable.

Acknowledgments: Gao Baojian of Jiangsu Construction Engineering Group Co., Ltd. who also contributed to this article. We would like to express our gratitude to Baojian Gao.

Conflicts of Interest: The authors declare no conflict of interest.

References

1. Chen, B.; Li, X.; Xiang, T.; Wang, P. SBRAC: Blockchain-based sealed-bid auction with bidding price privacy and public verifiability. *J. Inf. Secur. Appl.* **2022**, *65*, 103082. [CrossRef]
2. Wang, Y.-C.; Chen, C.-L.; Deng, Y.-Y. Authorization mechanism based on blockchain technology for protecting museum-digital property rights. *Appl. Sci.* **2021**, *11*, 1085. [CrossRef]
3. Munim, Z.H.; Balasubramaniyan, S.; Kouhizadeh, M.; Hossain, N.U.I. Assessing blockchain technology adoption in the Norwegian oil and gas industry using Bayesian Best Worst Method. *J. Ind. Inf. Integr.* **2022**, *28*, 100346. [CrossRef]
4. Gorkhali, A.; Chowdhury, R. Blockchain and the evolving financial market: A literature review. *J. Ind. Integr. Manag.* **2022**, *7*, 47–81. [CrossRef]
5. Laryea, S.; Ibem, E.O. Patterns of Technological Innovation in the Use of e-Procurement in Construction. *Electron. J. Inf. Technol. Constr.* **2014**, *19*, 104–125.
6. Qiu, T.; Zhang, R.; Gao, Y. Ripple vs. SWIFT: Transforming cross border remittance using blockchain technology. *Procedia Comput. Sci.* **2019**, *147*, 428–434. [CrossRef]
7. Wang, B.; Li, Z. Healthchain: A Privacy Protection System for Medical Data Based on Blockchain. *Future Internet* **2021**, *13*, 247. [CrossRef]
8. Song, G.; Kim, S.; Hwang, H.; Lee, K. Blockchain-based notarization for social media. In Proceedings of the 2019 IEEE International Conference on Consumer Electronics (ICCE), Las Vegas, NV, USA, 11–13 January 2019; pp. 1–2.
9. Hsu, C.-S.; Tu, S.-F.; Huang, Z.-J. Design of an E-voucher system for supporting social welfare using blockchain technology. *Sustainability* **2020**, *12*, 3362. [CrossRef]
10. Turk, Ž.; Klinc, R. Potentials of blockchain technology for construction management. *Procedia Eng.* **2017**, *196*, 638–645. [CrossRef]
11. Tso, R.; Liu, Z.-Y.; Hsiao, J.-H. Distributed E-voting and E-bidding systems based on smart contract. *Electronics* **2019**, *8*, 422. [CrossRef]
12. Sigalov, K.; Ye, X.; König, M.; Hagedorn, P.; Blum, F.; Severin, B.; Hettmer, M.; Hückinghaus, P.; Wölkerling, J.; Groß, D. Automated payment and contract management in the construction industry by integrating building information modeling and blockchain-based smart contracts. *Appl. Sci.* **2021**, *11*, 7653. [CrossRef]
13. , A.C. Big Data, 4v: Volume, velocity, variety, value. *Monit. Public Opin. Econ. Soc. Chang.* **2015**, 156–159.
14. Duan, Y.; Edwards, J.S.; Dwivedi, Y.K. Artificial intelligence for decision making in the era of Big Data–evolution, challenges and research agenda. *Int. J. Inf. Manag.* **2019**, *48*, 63–71. [CrossRef]
15. Duan, L.; Xiong, Y. Big data analytics and business analytics. *J. Manag. Anal.* **2015**, *2*, 1–21. [CrossRef]
16. Javaid, M.; Haleem, A.; Singh, R.P.; Suman, R. Significant applications of big data in Industry 4.0. *J.-Dustrial Integr. Manag.* **2021**, *6*, 429–447.
17. Hassani, H.; Huang, X.; Silva, E. Banking with blockchain-ed big data. *J. Manag. Anal.* **2018**, *5*, 256–275. [CrossRef]
18. Hasan, M.; Popp, J.; Oláh, J. Current landscape and influence of big data on finance. *J. Big Data* **2020**, *7*, 1–17. [CrossRef]
19. Painuly, S.; Sharma, S.; Matta, P. Big Data Driven E-Commerce Application Management System. In Proceedings of the 2021 6th International Conference on Communication and Electronics Systems (ICCES), Coimbatre, India, 8–10 July 2021; pp. 1–5.
20. Wu, J.; Wang, J.; Nicholas, S.; Maitland, E.; Fan, Q. Application of big data technology for COVID-19 prevention and control in China: Lessons and recommendations. *J. Med. Internet Res.* **2020**, *22*, e21980.
21. Du, M. Application of information communication network security management and control based on big data technology. *Int. J. Commun. Syst.* **2022**, *35*, e4643. [CrossRef]
22. Pei, J.; Han, J.; Mao, R. CLOSET: An efficient algorithm for mining frequent closed itemsets. In Proceedings of the ACM SIGMOD Workshop on Research Issues in Data Mining and Knowledge Discovery, Paris, France, 13 June 2004; pp. 21–30.
23. Halim, Z.; Ali, O.; Ghufran Khan, G. On the Efficient Representation of Datasets as Graphs to Mine Maximal Frequent Itemsets. *IEEE Trans. Knowl. Data Eng.* **2021**, *33*, 1674–1691. [CrossRef]
24. Lin, Z.; Jianli, Z. Frequent Item Sets and Association Rules Mining Algorithm Based on Floyd Algorithm. *J. Comput. Theor. Nanosci.* **2015**, *12*, 2574–2578. [CrossRef]
25. Bagui, S.; Dhar, P.C. Positive and negative association rule mining in Hadoop's MapReduce environment. *J. Big Data* **2019**, *6*, 75. [CrossRef]
26. Zheng, Z.B.; Xie, S.A.; Dai, H.N.; Chen, X.P.; Wang, H.M. An Overview of Blockchain Technology: Architecture, Consensus, and Future Trends. In Proceedings of the IEEE 6th International Congress on Big Data (BigData Congress), Honolulu, HI, USA, 25–30 June 2017; pp. 557–564.
27. Zhai, S.P.; Yang, Y.Y.; Li, J.; Qiu, C.; Zhao, J.M. Research on the Application of Cryptography on the Blockchain. In Proceedings of the International Conference on Computer Information Science and Application Technology (CISAT), Daqing, China, 7–9 December 2019.
28. Donet, J.A.D.; Perez-Sola, C.; Herrera-Joancomarti, J. The Bitcoin P2P Network. In Proceedings of the 18th International Conference on Financial Cryptography and Data Security (FC), Christ Church, Barbados, 3–7 March 2014; pp. 87–102.
29. Watanabe, H.; Fujimura, S.; Nakadaira, A.; Miyazaki, Y.; Akutsu, A.; Kishigami, J. Blockchain Contract: A Complete Consensus using Blockchain. In Proceedings of the IEEE 4th Global Conference on Consumer Electronics GCCE, Osaka, Japan, 27–30 October 2022; pp. 577–578.

30. Cole, R.; Stevenson, M.; Aitken, J. Blockchain technology: Implications for operations and supply chain management. *Supply Chain Manag.-Int. J.* **2019**, *24*, 469–483. [CrossRef]
31. Kuo, T.T.; Kim, H.E.; Ohno-Machado, L. Blockchain distributed ledger technologies for biomedical and health care applications. *J. Am. Med. Inform. Assoc.* **2017**, *24*, 1211–1220. [CrossRef]
32. Li, Q. Research on e-commerce user information encryption technology based on Merkle hash tree. In Proceedings of the 2019 International Conference on Robots & Intelligent System (ICRIS), Haikou, China, 15–16 June 2019; pp. 365–369.
33. Zhao, W.S.; Liu, K.; Ma, K. Design of Student Capability Evaluation System Merging Blockchain Technology. In Proceedings of the International Conference on Computer Information Science and Application Technology (CISAT), Daqing, China, 7–9 December 2019.
34. Yang, R.; Wakefield, R.; Lyu, S.N.; Jayasuriya, S.; Han, F.L.; Yi, X.; Yang, X.C.; Amarasinghe, G.; Chen, S.P. Public and private blockchain in construction business process and information integration. *Autom. Constr.* **2020**, *118*. [CrossRef]
35. Zhang, R.; Xue, R.; Liu, L. Security and privacy on blockchain. *ACM Comput. Surv. (CSUR)* **2019**, *52*, 1–34. [CrossRef]
36. Lee, H.A.; Kung, H.H.; Udayasankaran, J.G.; Kijsanayotin, B.; Marcelo, A.B.; Chao, L.R.; Fisu, C.Y. An Architecture and Management Platform for Blockchain-Based Personal Health Record Exchange: Development and Usability Study. *J. Med. Internet Res.* **2020**, *22*. [CrossRef] [PubMed]
37. Tschorsch, F.; Scheuermann, B. Bitcoin and Beyond: A Technical Survey on Decentralized Digital Currencies. *IEEE Commun. Surv. Tutor.* **2016**, *18*, 2084–2123. [CrossRef]
38. Hu, X.Y.; Zhu, C.; Tong, Z.Q.; Gao, W.J.; Cheng, G.; Li, R.D.; Wu, H.; Gong, J. Identifying Ethereum traffic based on an active node library and DEVp2p features. *Future Gener. Comput. Syst.-Int. J. Escience* **2022**, *132*, 162–177. [CrossRef]
39. Li, Y.; Qiao, L.; Lv, Z. An optimized byzantine fault tolerance algorithm for consortium blockchain. *Peer-Peer Netw. Appl.* **2021**, *14*, 2826–2839. [CrossRef]
40. Zhao, H.Q.; Zhang, L.L. An architecture evolution algorithm for Fabric blockchain application software. *Software* **2020**, *41*, 1–10.
41. Sarfaraz, A.; Chakrabortty, R.K.; Essam, D.L. A tree structure-based improved blockchain framework for a secure online bidding system. *Comput. Secur.* **2021**, *102*, 102147. [CrossRef]
42. Chandel, S.; Cao, W.; Sun, Z.; Yang, J.; Zhang, B.; Ni, T.-Y. A multi-dimensional adversary analysis of RSA and ECC in blockchain encryption. In Proceedings of the Future of Information and Communication Conference, San Francisco, CA, USA, 14–15 March 2019; pp. 988–1003. [CrossRef]
43. Rivest, R.L.; Shamir, A.; Adleman, L. A method for obtaining digital signatures and public-key cryptosystems. *Commun. ACM* **1978**, *21*, 120–126. [CrossRef]
44. Chen, Y.; Tso, R. *An E-Voting System Based on Oblivious Signatures*; National Chengchi University: Taipei, Taiwan, 2013.
45. Nair, D.G.; Binu, V.; Kumar, G.S. An improved e-voting scheme using secret sharing based secure multi-party computation. *arXiv* **2015**, arXiv:1502.07469.
46. Johnson, N.; Jones, B.M.; Clendenon, K. E-voting in america: Current realities and future directions. In Proceedings of the International Conference on Social Computing and Social Media, Vancouver, BC, Canada, 9–14 July 2017; pp. 337–349. [CrossRef]
47. Wang, D.; Zhao, J.; Mu, C. Research on blockchain-based e-bidding system. *Appl. Sci.* **2021**, *11*, 4011. [CrossRef]

electronics

MDPI

Article

A More Effective Zero-DCE Variant: Zero-DCE Tiny

Weiwen Mu, Huixiang Liu, Wenbai Chen * and Yiqun Wang

School of Automation, Beijing Information Science and Technology University, Beijing 100101, China
* Correspondence: chenwb03@126.com

Abstract: The purpose of Low Illumination Image Enhancement (LLIE) is to improve the perception or interpretability of images taken in low illumination environments. This work inherits the work of Zero-Reference Deep Curve Estimation (Zero-DCE) and proposes a more effective image enhancement model, Zero-DCE Tiny. First, the new model introduces the Cross Stage Partial Network (CSPNet) into the original U-net structure, divides basic feature maps into two parts, and then recombines it through the structure of cross-phase connection to achieve a richer gradient combination with less computation. Second, we replace all the deep separable convolutions except the last layer with Ghost modules, which makes the network lighter. Finally, we introduce the channel consistency loss into the non-reference loss, which further strengthens the constraint on the pixel distribution of the enhanced image and the original image. Experiments show that compared with Zero-DCE++, the network proposed in this work is more lightweight and surpasses the Zero-DCE++ method in some important image enhancement evaluation indexes.

Keywords: image enhancement; cross stage partial network; zero-reference; Ghost module

Citation: Mu, W.; Liu, H.; Chen, W.; Wang, Y. A More Effective Zero-DCE Variant: Zero-DCE Tiny. *Electronics* **2022**, *11*, 2750. https://doi.org/10.3390/electronics11172750

Academic Editor: Manohar Das

Received: 29 July 2022
Accepted: 30 August 2022
Published: 1 September 2022

1. Introduction

Due to the interference of equipment and environmental factors such as insufficient lighting and limited exposure time, the final image is often taken in a suboptimal environment, which is affected by the backlight, uneven lighting, and low light interference, resulting in the aesthetic quality of these images being impacted, which is unsatisfactory for higher-level tasks such as cell classification [1] and semantic segmentation in the process of robot arm grasping [2]. Therefore, the enhancement of low-light-level images is a research field worth exploring.

Traditional low light level enhancement methods include the histogram equalization method [3] and the Retinex model method [4]. Based on the histogram equalization method, the gray value of pixels in the image is changed by gray operation, so that the transformed image histogram is more uniform and the gray is clearer than the original image, to achieve the purpose of image enhancement. The method based on the Retinex model considers that the image data acquired by the human eyes depends on the incident light and the reflection of the object's surface. Usually, the incident light component can be obtained after filtering the original image signal, and then the reflection component can be solved through the mathematical relationship between the three variables to obtain the purpose of image enhancement. Although these traditional algorithms can achieve the effect of image enhancement, it is difficult to suppress the noise information generated in the process of image enhancement, resulting in the poor usability of the enhanced image.

With the development of deep learning, learning-based methods have been applied to image enhancement, including supervised learning (SL), reinforcement learning (RL), un-supervised learning (UL), zero sample learning (ZSL), and semi-supervised learning (SSL). Unsupervised learning and zero sample learning can directly learn from unlabeled samples, and the model can learn more generalized feature expressions from data. The model training in this work inherits the series work of Zero-DCE [5,6], which is different from the methods based on image reconstruction [7–13]. Through the constraint of non-reference loss (the non-reference loss here refers to the loss function used by the algorithm that does not use labeled data for training), the model can be well generalized to the test set data after training through the unlabeled data set.

This work mainly inherits the work of Zero-DCE++ [6] and proposes a more lightweight model for low-light-level image enhancement. The model can deal with pictures under various lighting conditions, including uneven lighting and weak lighting. Compared with the original method, the new model can become further lightweight while improving its performance. Our contributions are summarized below.

- The CSPNet structure is introduced into the original U-net structure, which can reduce the amount of computation and achieve a richer gradient combination. At the same time, except for the last layer, the Ghost module is used to replace the depth separable convolution, which further reduces the size of the image enhancement model.
- The channel consistency loss is introduced into the non-reference loss: using KL divergence to enhance the consistency between the original image and the enhanced image on the difference between channels.

Section 2 introduces the overall architecture of Zero-DCE Tiny and the non-reference loss function used. Section 3 introduces the parameter setting of the Ablation Experiment and the comparison of relevant experimental results. Finally, this work compares the new method with Zero-DCE and Zero-DCE++ methods in sensory and quantitative aspects and tests the effect of each method in the downstream application.

2. Related Works

In this section, we mainly focus on the relevant work of zero sample learning in the field of image enhancement and summarize some commonly used model lightweight methods.

2.1. Zero Sample Learning for Image Enhancement

Zhang et al. [14] proposed a zero-order learning method that uses Exposure Correction Network (ExCNet) for backlight image restoration. It first uses a depth network to estimate the S-curve. Zhu et al. [15] proposed a three-branch CNN, called RRDNet, to repair the underexposed image by decomposing the input image into illuminance, reflectivity, and noise. Several kinds of loss functions are specially designed to drive zero-order learning. Zhao et al. [16] performed Retinex decomposition through a neural network and then used the RetinexDIP model based on Retinex to enhance low illumination images. Inspired by deep image priority (DIP) [17], RetinexDIP takes randomly sampled white noise as input, generates reflection components and illumination components through Retinex decomposition, and then uses the obtained illumination components for image enhancement. The training process uses some losses as constraints, such as reflection loss. Liu et al. [18] proposed a new principled framework to search for a lightweight priority architecture for low-light-level images in real scenes by injecting knowledge of low-light-level images. Zero-DCE [5] regards light enhancement as a curve estimation task for images. It takes low-light images as input and generates high-order curves as its output. These curves are used to adjust the input dynamic range at the pixel level to obtain an enhanced image. In addition, a fast and lightweight version called Zero-DCE++ [6] is proposed. Because the mapping from image to curve only needs a lightweight network, it realizes fast estimation.

2.2. Model Lightweight Method

Howard et al. [19] proposed the MobileNet network. In this network, the depth separable convolution is used to replace the ordinary convolution for the first time. The depth separable convolution is mainly composed of the depthwise convolution and the pointwise convolution. The depthwise convolution uses convolution to check the input features and convolute them respectively according to the channel to obtain the spatial information of the features, and the pointwise convolution uses 1×1 to obtain the information between different channels in the feature and achieve the lightweight effect through this combination method. In the ShuffleNet [20], the feature map obtained by group convolution was randomly and uniformly scrambled in deep separable convolution on the channel, and then a group convolution operation was carried out to replace the pointwise convolution operation, which also solved the problem of the lack of information exchange between different groups in the training process, as well as maintained the feature extraction ability of the neural network while reducing the weight. Han K et al. [21] proposed the GhostNet to solve the problem of traditional convolution containing a large amount of redundant information when extracting features. First, conventional convolution is performed with fewer convolution check inputs to obtain output features with fewer channels. After linear transformation of these features, the ghost feature map is obtained, and then the final feature map is obtained by splicing with the output features. Chien Yao Wang et al. [22] proposed the CSPNet to solve the incompatibility between deep separable convolution technology and some industrial IC designs. This network not only realizes richer gradient combinations but also reduces the calculation of the model.

3. Materials and Methods

3.1. Overall Architecture

This work inherits the method of image enhancement in the Zero-DCE++ paper [6], learns the mapping curve from a weak light image to a strong light image through a convolution neural network, and then uses the learned mapping curve to iteratively adjust the pixels of the original image for many times to achieve the purpose of adjusting the image in a large dynamic range. It is assumed that the enhancement curve parameter map $A_n(x)$ obtained through network learning is related to the coordinates of pixels. A corresponding enhancement curve will be applied to each pixel on the original image. The expression of the designed image enhancement is shown in Equation (1):

$$LE_n(x) = LE_{n-1}(x) + A_n(x)LE_{n-1}(x)(1 - LE_{n-1}(x)) \tag{1}$$

where I represents the input image and n is the number of iterations. In this work, n is set to 8, which can achieve the relatively best image enhancement results. $LE_n(x)$ is an enhanced version of the last enhanced image $LE_{n-1}(x)$, and $A_n(x)$ is a curve parameter mapping that has the same size as the given image. The process of image enhancement using Zero-DCE Tiny is shown in Figure 1.

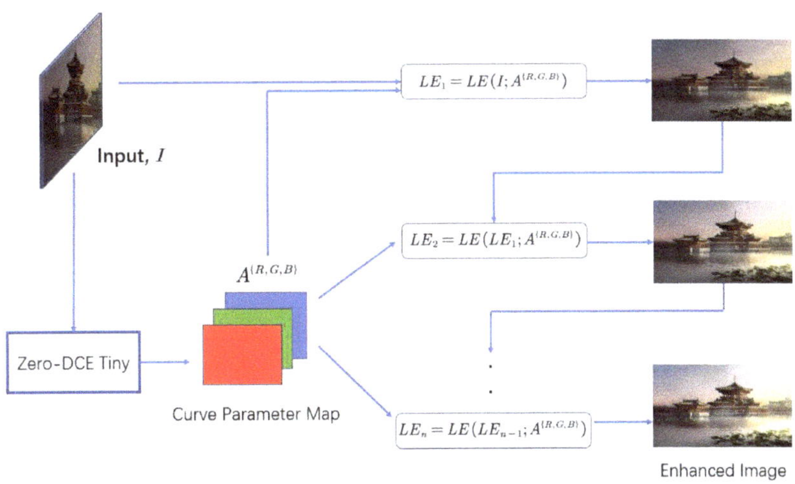

Figure 1. Overall structure diagram.

3.2. DCE-Net Tiny

The original DCE-Net [5] used a simple CNN composed of seven convolution layers. It has a U-net structure. In the first six convolution layers, each convolution layer consists of 32 convolution layers, the kernel size is 3 × 3 of which stride is 1, followed by the ReLU activation function. The last convolution layer consists of 32 convolution layers with a size of 3 × 3 of which stride is 1, followed by the Tanh activation function, which generates 24 curve parameter mappings for eight iterations, in which each iteration generates three curve parameter mappings for three channels (i.e., RGB channels). The downsampling and batch normalization layers that destroy the relationship between adjacent pixels are discarded. Later, in Zero-DCE++ [6], the ordinary convolution processing was replaced by the deep separable convolution to reduce the amount of computation. Wherein the size of the depthwise convolution kernel is 3 × 3, the stride is 1, and when the pointwise convolution kernel size is 1 × 1, the stride is 1. At the same time, the output layer only generates 3 curve parameter maps and then reuses them in different iteration stages. This will reduce the risk of oversaturation.

The reason for choosing this U-net structure is that the U-net structure can effectively integrate multi-scale features, which are very important to achieve satisfactory low illumination enhancement. However, layer hopping connections used in U-net networks may introduce redundant feature information into the final results. Therefore, we need to design a network that effectively combines shallow features and deep features to achieve the purpose of being lightweight but effective. Inspired by CSPNet [22] and GhostNet [21], we designed the model shown in Figure 2 to replace the previous model structure.

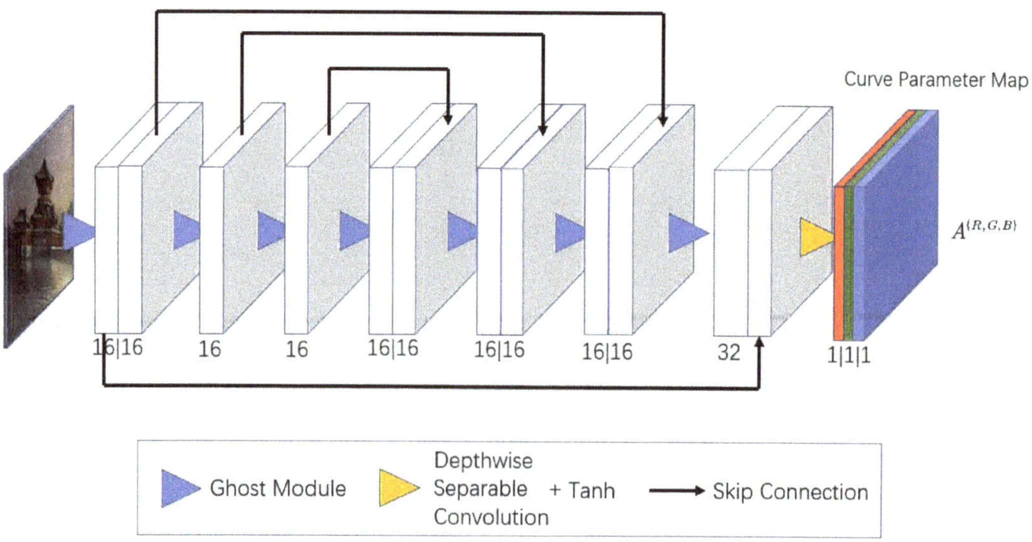

Figure 2. Zero-DCE Tiny structure diagram.

In the new structure, the basic feature maps are split into two parts through the channel. The former is directly connected to the output layer, and the latter will act as the DCE-net [5]. With the exception of the last layer, which still uses deep separable convolution, other layers are replaced by Ghost modules. As shown in Figure 3, the Ghost module first uses 1 × 1 convolution to condense the input feature map to achieve cross-channel feature extraction. After obtaining the condensed feature, it uses a 3 × 3 convolution kernel to convolute layer-by-layer to obtain an additional feature map. Finally, it stacks the 1 × 1 convolution result and the layer-by-layer convolution result to obtain the final feature map. The feature map obtained in the two steps is processed by the ReLU activation function. In

Zero-DCE Tiny, the last convolutional layer is still followed by the Tanh activation function, and the input is iterated 8 times through the curve parameter map to generate the final enhanced image.

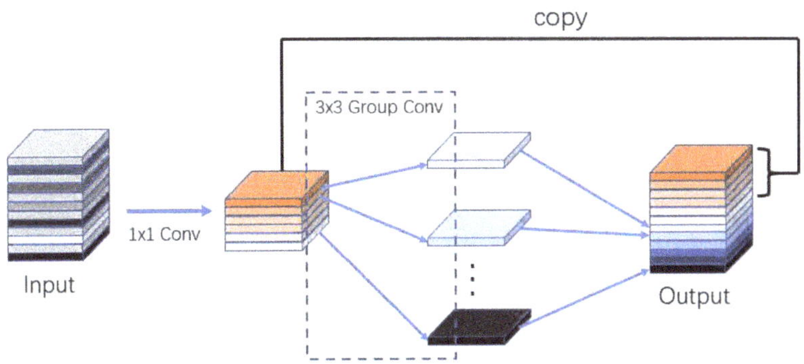

Figure 3. Ghost module sketch map.

The new network structure uses the Ghost modules to replace the depthwise separable convolution operation, which greatly improves the utilization efficiency of the feature map and reduces the amount of calculation. At the same time, introducing the CSPNet structure realizes richer feature fusion and strengthens the learning ability of the network. Moreover, the final amount of calculation is further reduced due to the segmentation of the base feature map.

3.3. Non-Reference Loss Functions

This work inherits the non-reference loss function used in the Zero-DCE++ paper [6]. Spatial consistency loss is mainly used to maintain the difference between the adjacent areas between the input image and its enhanced version and to encourage the spatial consistency of the enhanced image.

$$L_{spa} = \frac{1}{K}\sum_{i=1}^{K}\sum_{j\in\Omega(i)}(|(Y_i - Y_j)| - |I_i - I_j|)^2 \tag{2}$$

where K is the number of the local region and $\Omega(i)$ represents a collection of adjacent areas centered at the region i. As shown in Figure 4, Y and I are the average pixel value of the local region in the enhanced image and the original image. Our local region is set to 4×4.

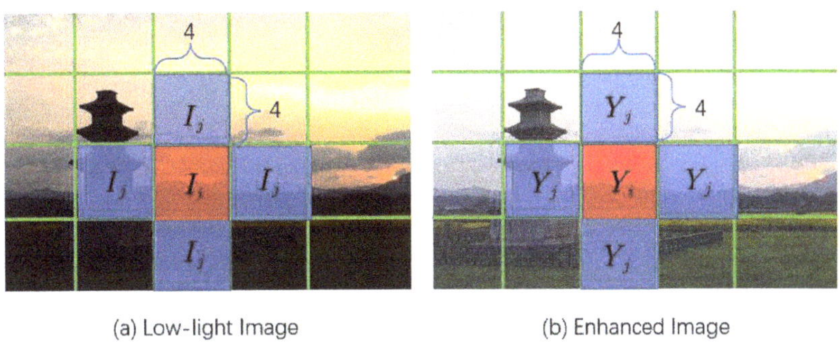

Figure 4. Mapping of spatial consistency loss. Subfigure (**a**,**b**) respectively show the setting of local regions in the original image and the enhanced image.

Exposure Control Loss is used to control the exposure level. $L_{\exp}$ can be expressed as:

$$L_{\exp} = \frac{1}{M}\sum_{k=1}^{M}|Y_k - E| \tag{3}$$

where M is the number of nonoverlapping local regions of size 16×16, the average pixel value of a local region in the enhanced version is denoted as Y, and E indicates a good exposure level.

The loss of color constancy is mainly used to reduce color deviation in enhanced images, which can be expressed as:

$$L_{col} = \sum_{\forall(p,q)\in\varepsilon}(J^p - J^q)^2,\ \varepsilon = \{(R,G),(R,B),(G,B)\} \tag{4}$$

where J^p denotes the pixel average value of the p channel in the enhanced image, and (p,q) represents a pair of channels.

The loss of illumination smoothness keeps the adjacent pixel values monotonous, thus avoiding overexposure and underexposure, which can be expressed as:

$$L_{tv_A} = \frac{1}{N}\sum_{n=1}^{N}\sum_{c\in\xi}\left(|\nabla_x A_n^c| + |\nabla_y A_n^c|\right)^2, \xi = \{R,G,B\} \tag{5}$$

where N is the number of iterations, and ∇_x and ∇_y represent the horizontal and vertical gradient operations, respectively.

Inspired by the spatial consistency loss, this work proposes a new non-reference loss: channel consistency loss. As a new loss, channel consistency loss mainly enhances the consistency between the original image and the enhanced image in the channel pixel difference through *KL* divergence, and suppresses the generation of noise information and invalid features to improve the image enhancement effect. The channel consistency loss can be expressed as:

$$L_{kl} = KL[R - B||R' - B'] + KL[R - G||R' - G'] + KL[G - B||G' - B'] \tag{6}$$

In this work, R, G, and B represent the color channels of the original image, R', G' and B' represent the three-color channels of the enhanced image, and *KL* divergence is used to represent the difference between the two distributions. If the difference between the two is small, the *KL* divergence is small. When the two distributions are consistent, the *KL* divergence value is 0.

The total loss can be expressed as:

$$L_{total} = W_{spa}L_{spa} + W_{\exp}L_{\exp} + W_{col}L_{col} + W_{tv_A}L_{tv_A} + W_{kl}L_{kl} \tag{7}$$

where W_{spa}, $W_{\exp}$, W_{col}, and W_{kl} are the weights of the losses.

4. Results

To be consistent with the previous work [5,6], we also used 360 multiple exposure sequences from Part 1 of the SICE dataset [23] as our training dataset. We randomly divided 3022 images with different exposure levels in the Part 1 [23] subset into two parts (2422 images for training and 600 images for validation). The images were resized to $512 \times 512 \times 3$. We implemented our framework on RTX3060 GPU using PyTorch. The batch size is 8. We used a Gaussian function with a mean of 0 and a standard deviation of 0.02 to initialize the convolutional neural network and used the Adam optimizer to optimize the network. The Adam optimizer uses default parameters and a constant learning rate. The weights W_{spa}, $W_{\exp}$, W_{col}, and W_{kl} were set to 1, 10, 5, 1600, and 5 to balance the loss ratio. Network training 100 rounds in total.

We used some public datasets for testing, including LIME [24] (10 images), and DICM [25] (64 images). In addition, we also collected a total of 2300 low light/normal light images on the part2 subset of the SICE dataset as the test dataset, and all images were adjusted to $1200 \times 900 \times 3$.

4.1. Ablation Study

4.1.1. Ablation Study of Each Loss

We performed ablation experiments on each loss function; the results are shown in Figure 5. As shown in Figure 5c, lack of spatial consistency loss L_{spa} reduces the image contrast, for example, the part of the cloud in the image. As shown in Figure 5d, lack of exposure control loss L_{exp} causes image enhancement invalid. As shown in Figure 5e, When the loss of color constancy L_{col} is discarded, serious color projection occurs. Finally, as shown in Figure 5f, removing the light smoothness loss L_{tv_A} leads to obvious artifacts.

Figure 5. Ablation study of each loss. Subfigure (**a**) shows the original input, subfigure (**b**) shows the enhanced image result through Zero-DCE Tiny method, subfigure (**c**–**f**) respectively show the image enhancement results after removing spatial consistency loss, exposure control loss, color consistency loss and illumination smoothness loss.

We added the channel consistency loss to the original version of the non-reference loss function and performed ablation experiments. Figure 6 compares the sensory results of the test image: After adding the loss of spatial consistency, the enhanced image is more natural and the overall contrast distribution of the image is more balanced. As shown in Figure 6, the house is less affected by the halo, and the details are clearer.

4.1.2. Ablation Study of Backbone Network

For the new backbone network, we introduce the CSPNet network structure and set the number of feature maps in the base layer to 32. We divide the basic feature maps into two parts. The former is directly connected to the output layer, and the latter will act as the DCE-net [5]. At the same time, we replace all depth separable convolutions outside the last layer with the Ghost module.

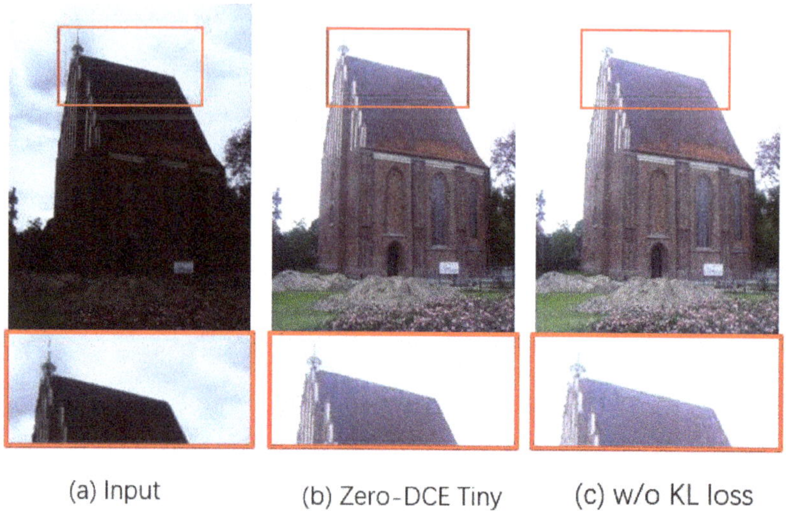

Figure 6. Sensory comparison of kl loss ablation experiment. Subfigure (**a**) shows the original input, subfigure (**b**) shows the enhanced image result through Zero-DCE Tiny method, subfigure (**c**) shows the image enhancement result after removing channel consistency loss.

Table 1 shows the original network and three network parameters. We divide the network structure into three types: "Only CSPNet structure", "Only Ghost module", and "Zero-DCE Tiny". We mainly compare five parameters, namely the number of network parameters (Total params), the amount of memory required for node reasoning (Total memory), the number of floating-point operations (Total Flops), the amount of multiplication and addition required for network reasoning (Total Madd), and the sum of memory read and write (Total MemR + W). It can be seen from Table 1 that Zero-DCE Tiny has achieved a lighter effect on multiple indicators. However, since the Ghost module uses a large number of group convolutions, resulting in more memory occupancy, the "Total Madd" and "Total MemR + W" metrics are slightly higher than "Only CSPNet structure". However, the experiments show that "Only CSPNet structure" will lead to a poor image enhancement effect, so we finally choose to obtain better image enhancement performance at the cost of certain memory occupation.

Table 1. Parameter comparison of the backbone network; the parameter is computed for an image of size 256 × 256 × 3.

Method	Total Params	Total Memory	Total Flops	Total MAdd	Total MemR + W
Zero-DCE	79,416	62.00 MB	10.38 GFlops	5.21 GMAdd	143.05 MB
Zero-DCE++	10,561	129.50 MB	1.32 GFlops	694.22 MMAdd	283.04 MB
Only CSPNet structure	5153	93.50 MB	632.68 MFlops	339.8 MMAdd	199.02 MB
only Ghost module	5331	112.75 MB	689.96 MFlops	361.96 MMAdd	273.52 MB
Zero-DCE Tiny	2731	104.75 MB	353.37 MFlops	190.51 MMAdd	215.51 MB

4.1.3. Ablation Study of Input Size

We provide input of different sizes for Zero-DCE Tiny. Table 2 summarizes the statistical relationship between enhanced performance and input size. We also show some results by modifying the size of the network input image, as shown in Figure 7. As shown in Figure 7 and Table 2, the downsampling input size has no significant impact on the enhanced performance, but significantly saves computing costs. As shown in Table 2, 6 × ↓ obtained the highest average PSNR value, but because 12 × ↓ is better in model efficiency, we use it as the default configuration for the new network.

Table 2. Effect of different input image resolutions on image enhancement. The FLOPs (in G) are computed for an image of size 1200 × 900 × 3. "number × ↓" indicates the times of downsampling the input image. The test image is from the part2 dataset of SCIE.

Metrics	Original Resolution	2 × ↓	4 × ↓	6 × ↓	12 × ↓	20 × ↓	50 × ↓
PSNR	16.14	16.22	16.38	16.45	16.42	15.95	15.02
FLOPs	5.82	1.46	0.355	0.158	0.039	0.014	0.002

(a) input (b) original resolution (c) 2 × ↓ (d) 12 × ↓ (e) 50 × ↓

Figure 7. Ablation study of input image size. Subfigure (**a**) shows the original input, subfigure (**b**) shows the enhanced image result when the image resolution is not changed, subfigure (**c**–**e**) show the image enhancement results after downsampling the input image.

4.2. Benchmark Evaluations

In this section, we compare the new method with the classical benchmark models in qualitative and quantitative experiments. Finally, the new image enhancement method's gain effect on object detection in the dark is tested.

4.2.1. Visual and Perceptual Comparisons

We selected some classical benchmark methods to compare them with our methods for visual and perceptual comparisons. The new method chooses Zero-DCE Tiny as the backbone network, and adds the spatial consistency loss to the non-reference loss for training and testing. Figure 8 shows the enhanced image effects of some test images obtained by different methods under the same conditions. We tested three CNN-based methods (RetinexNet [9], LightenNet [26], MBLLEN [8]) and one GAN-based method (EnlightenGAN [27]) to replicate the results using open-source code.

Figure 8 shows the results of our tests on the SICE dataset. For outdoor scenes, the LightenNet, the MBLLEN, and the EnlightenGAN find it difficult to achieve clear enhancement results for difficult backlight areas, such as the face part. For RetinexNet, there are many overexposure cases in the image, including the face part, with poor overall sensory effects. For indoor scenes, MBLLEN performs well visually, but it is too smooth, which may filter out the detailed features of the original image. For RetinexNet, the noise information in the image is amplified, resulting in a poor enhancement effect. For EnlightenGAN, the enhanced image shows a certain color deviation. For the Zero-DCE series methods, the effects of Zero-DCE and Zero-DCE tiny methods are very close. Compared with Zero-DCE++, the enhancement effect of the face region is better.

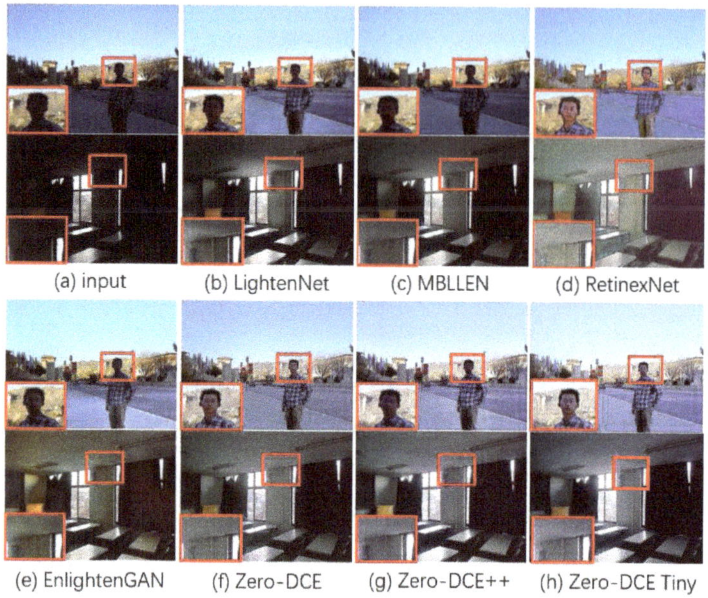

Figure 8. A visual comparison among the results generated by different methods. Subfigure (**a**) shows the original input, subfigure (**b–h**) respectively show the enhanced image results through LightenNet [26], MBLLEN [8], RetinexNet [9], EnlightenGAN [27], Zero-DCE [5], Zero-DCE++ [6] and Zero-DCE Tiny methods.

In the experiment, we found that in the Zero-DCE series of methods, the image enhancement effect of Zero-DCE Tiny is softer, as shown in Figure 9. For areas with strong sunlight, the roof part and the cross part in the enhanced image of Zero-DCE Tiny are clearer. At the sensory level, it shows that the new method is conducive to suppressing the problem of excessive local exposure.

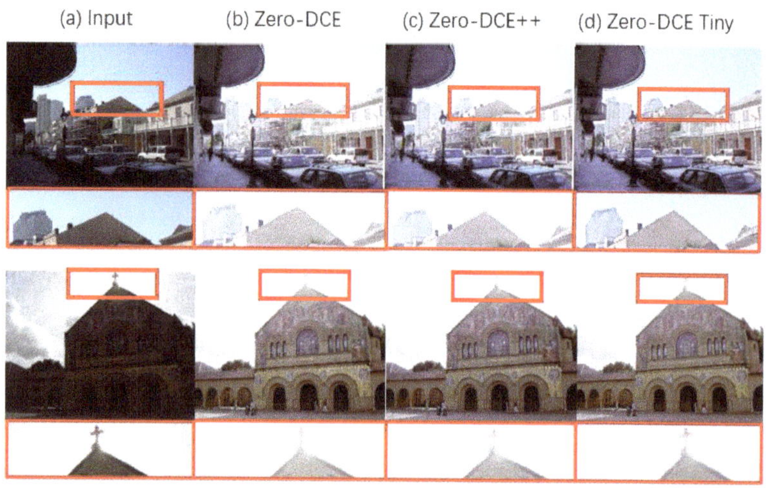

Figure 9. Visual comparisons among the results generated by the Zero-DCE series of methods. Subfigure (**a**) shows the original input, subfigure (**b–d**) respectively shows the enhanced image results through Zero-DCE [5], Zero-DCE++ [6] and Zero-DCE Tiny methods.

The part2 subset of SCIE dataset is also used to compare different methods. The comparison results are shown in Figure 10. The LightenNet has obvious light spots in the wall area, and RetinexNet, EnlightenGAN, and Zero-DCE++ all have different degrees of color deviation. The image enhancement results of MBLLEN are dark, whereas the results of Zero-DCE and Zero-DCE Tiny are very close. The image enhancement result obtained by Zero-DCE Tiny is closer to the natural situation in color and contrast, and the sensory effect is better.

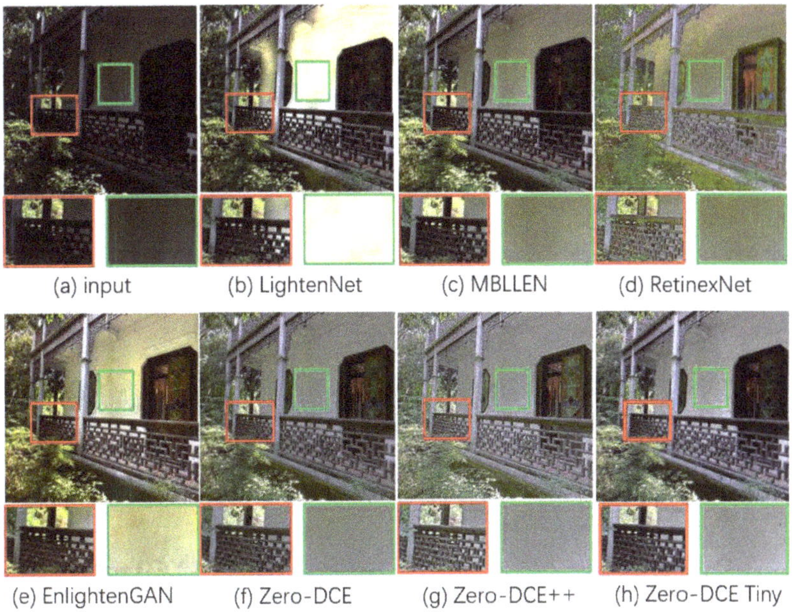

Figure 10. Visual comparison of Part 2 subset of SCIE dataset. Subfigure (**a**) shows the original input, subfigure (**b–h**) respectively show the enhanced image results through LightenNet [26], MBLLEN [8], RetinexNet [9], EnlightenGAN [27], Zero-DCE [5], Zero-DCE++ [6] and Zero-DCE Tiny methods.

4.2.2. Quantitative Comparisons

Table 3 shows the quantitative comparison of several image enhancement methods. We compared three image enhancement indicators on the part2 test dataset [23]: peak signal to noise ratio (PSNR), structural similarity (SSIM), and mean absolute error (MAE), where the SSIM value represents the similarity between the results and the real results in terms of structural characteristics. The PSNR value (in the case of a low MAE value) indicates that the results obtained are closer to the actual situation.

Table 3. Comparison of image enhancement indexes.

Metrics	PSNR	SSIM	MAE
MBLLEN	15.02	0.52	119.14
RetinexNet	15.99	0.53	104.81
LightenNet	13.17	0.55	140.92
EnlighenGAN	16.21	0.59	102.78
Zero-DCE	**16.57**	0.59	**98.78**
Zero-DCE++	16.42	0.58	102.87
Zero-DCE Tiny	16.50	**0.61**	102.52

It can be seen from Table 3 that by introducing a new backbone network and channel consistency loss, the PSNR index and SSIM index are improved compared with Zero-

DCE++ (when the MAE value is low), wherein the SSIM index even exceeds Zero-DCE. It shows that the loss of channel consistency helps improve the structural consistency of the original image and the enhanced image. At the same time, from Tables 1 and 4, we know that compared with Zero-DCE++, our network is more lightweight and the reasoning speed is more friendly to practical applications. At the same time, due to the reduction of the number of parameters, during our training, it only takes 35 min to train the model with a single RTX3060 graphics card. So, it is also very friendly to the second training of developers. In general, the new model is a more efficient image enhancement model that achieves lightweight while maintaining a good image enhancement effect.

Table 4. Model running speed comparison.

Metrics	Runtime (s)	Total Params	Platform
MBLLEN	13.9949	450,171	TensorFlow (GPU)
RetinexNet	0.1200	555,205	TensorFlow (GPU)
LightenNet	25.7716	29,532	MATLAB (CPU)
EnlighenGAN	0.0078	8,636,675	PyTorch (GPU)
Zero-DCE	0.0025	79,416	PyTorch (GPU)
Zero-DCE++	0.0012	10,561	PyTorch (GPU)
Zero-DCE Tiny	**0.0008**	**2731**	PyTorch (GPU)

4.2.3. Object Detection in the Dark

To test the gain effect of the improved image enhancement algorithm in the downstream application, we selected the object detection task in the low light environment to test the new algorithm. We mainly tested on the ExDark dataset [28], which was built specifically for low-light-level image recognition tasks. The ExDark dataset consists of 7363 low-light images which are marked as 12 object classes. We only use its test dataset, take Zero-DCE Tiny as the preprocessing step, and then use the pretrain ResNet50 classifier through the ImageNet. In the weak light test set, Zero-DCE Tiny was used as pretreatment to improve the classification accuracy from 22.02% (top-1) and 39.46% (top-5) to 27.86% (top-1) and 44.86% (top-5) after enhancement. This provides side evidence that image enhancement using Zero-DCE Tiny not only produces pleasant visual effects but also provides richer image details for downstream applications, which is conducive to improving the application effect of downstream applications.

5. Discussion

The new model Zero-DCE Tiny proposed in this paper is a further lightweight product of the Zero-DCE series models. The comprehensive results of multiple test datasets show that the new model can deal with low-light images in various scenarios well. In Furthermore, compared with the Zero-DCE++ version, the efficiency of the model is further improved. Shorter reasoning time and lower training cost make the new model more friendly to practical applications; this will promote the application of the deep learning image enhancement model in real life, such as the night vision instrument. More importantly, the upstream benefits of image enhancement will benefit downstream applications, so that image processing algorithms such as image detection and semantic segmentation can better cope with images in complex environments.

6. Conclusions

We propose a new backbone network Zero-DCE Tiny to replace Zero-DCE++ for low illumination image enhancement. It can use zero reference images for end-to-end training. At the same time, compared with the original method, the backbone network used in this paper not only enhances the feature fusion but also reduces the amount of computation and memory consumption. This paper also tests the new non-reference loss to verify the effectiveness of channel consistency loss in improving image contrast balance. The results show that the new image enhancement method can better balance the image enhancement effect and the lightweight level of the model. This will further promote the application of the deep learning model in the field of image enhancement. However, the method proposed in this work also has some problems to be solved. For example, although this image enhancement model enhances the fusion of features, it inevitably introduces noise and redundant information. Therefore, there is still much room for improvement in the effect of image enhancement. In the future, we will try more noise suppression methods to retain the semantic information in the original image and enhance the promotion of image enhancement to downstream applications.

Author Contributions: Conceptualization, W.M. and H.L.; methodology, W.M.; software, W.M.; validation, W.M.; formal analysis, W.M.; investigation, W.M.; resources, W.C.; data curation, Y.W.; writing—original draft preparation, W.M.; writing—review and editing, W.M. and H.L.; visualization, W.M.; supervision, W.C. All authors have read and agreed to the published version of the manuscript.

Funding: This research was funded by [the Major Project of Scientific and Technological Innovation 2030] grant number [2021ZD0113603], [Natural Science Foundation of Beijing Municipal] grant number [4202026], [the Qin Xin Talents Cultivation Program of Beijing Information Science and Technology University] grant number [QXTCP A202102], [the R&D Program of Beijing Municipal Education Commission] grant number [KM202011232023].

Conflicts of Interest: The authors declare no conflict of interest.

References

1. Iqbal, M.S.; Ahmad, I.; Bin, L.; Khan, S.; Rodrigues, J.J.P.C. Deep learning recognition of diseased and normal cell representation. *Trans. Emerg. Telecommun. Technol.* **2020**, *32*, e4017. [CrossRef]
2. Ainetter, S.; Fraundorfer, F. End-to-end Trainable Deep Neural Network for Robotic Grasp Detection and Semantic Segmentation from RGB. In Proceedings of the 2021 IEEE International Conference on Robotics and Automation (ICRA), Xi'an, China, 30 May–5 June 2021.
3. Abdullah-Al-Wadud, M.; Kabir, H.; Dewan, M.A.A.; Chae, O. A Dynamic Histogram Equalization for Image Contrast Enhancement. *Int. Conf. Consum. Electron.* **2007**, *53*, 593–600. [CrossRef]
4. Wang, S.; Zheng, J.; Hu, H.; Li, B. Naturalness Preserved Enhancement Algorithm for Non-Uniform Illumination Images. *IEEE Trans. Image Process.* **2013**, *22*, 3538–3548. [CrossRef] [PubMed]
5. Guo, C.; Li, C.; Guo, J.; Loy, C.C.; Hou, J.; Kwong, S.; Cong, R. Zero-Reference Deep Curve Estimation for Low-Light Image Enhancement. In Proceedings of the IEEE/CVF Conference on Computer Vision and Pattern Recognition (CVPR), Seattle, WA, USA, 14–19 June 2020.
6. Li, C.; Guo, C.; Loy, C.C. Learning to Enhance Low-Light Image via Zero-Reference Deep Curve Estimation. *arXiv* **2021**, arXiv:2103.00860. [CrossRef] [PubMed]
7. Lore, K.G.; Akintayo, A.; Sarkar, S. LLNet: A Deep Autoencoder Approach to Natural Low-light Image Enhancement. *Pattern Recognit.* **2015**, *61*, 650–662. [CrossRef]
8. Lv, F.; Lu, F.; Wu, J.; Lim, C. MBLLEN: Low-Light Image/Video Enhancement Using CNNs. *Br. Mach. Vis. Conf.* **2018**, *220*, 4.
9. Wei, C.; Wang, W.; Yang, W.; Liu, J. Deep Retinex Decomposition for Low-Light Enhancement. *arXiv* **2018**, arXiv:1808.04560.
10. Zhang, Y.; Zhang, J.; Guo, X. Kindling the Darkness: A Practical Low-light Image Enhancer. In Proceedings of the 27th ACM International Conference on Multimedia, Nice, France, 21–25 October 2019.
11. Ren, W.; Liu, S.; Ma, L.; Xu, Q.; Xu, X.; Cao, X.; Du, J.; Yang, M. Low-Light Image Enhancement via a Deep Hybrid Network. *IEEE Trans. Image Process.* **2019**, *28*, 4364–4375. [CrossRef] [PubMed]
12. Lim, S.; Kim, W.J. DSLR: Deep Stacked Laplacian Restorer for Low-Light Image Enhancement. *IEEE Trans. Multimed.* **2021**, *23*, 4272–4284. [CrossRef]
13. Zhang, Y.; Guo, X.; Ma, J.; Liu, W.; Zhang, J. Beyond Brightening Low-light Images. *Int. J. Comput. Vis.* **2021**, *129*, 1013–1037. [CrossRef]
14. Zhang, L.; Zhang, L.; Liu, X.; Shen, Y.; Zhang, S.; Zhao, S. Zero-Shot Restoration of Back-lit Images Using Deep Internal Learning. In Proceedings of the 27th ACM International Conference on Multimedia, Nice, France, 21–25 October 2019.
15. Zhu, A.; Zhang, L.; Shen, Y.; Ma, Y.; Zhao, S.; Zhou, Y. Zero-Shot Restoration of Underexposed Images via Robust Retinex Decomposition. In Proceedings of the 2020 IEEE International Conference on Multimedia and Expo (ICME), London, UK, 6–10 July 2020.
16. Zhao, Z.; Xiong, B.; Wang, L.; Ou, Q.; Yu, L.; Kuang, F. RetinexDIP: A Unified Deep Framework for Low-light Image Enhancement. *IEEE Trans. Circuits Syst. Video Technol.* **2021**, *32*, 1076–1088. [CrossRef]
17. Ulyanov, D.; Vedaldi, A.; Lempitsky, V. Deep Image Prior. *Int. J. Comput. Vis.* **2017**, *128*, 1867–1888. [CrossRef]

18. Liu, R.; Ma, L.; Zhang, J.; Fan, X.; Luo, Z. Retinex-inspired Unrolling with Cooperative Prior Architecture Search for Low-light Image Enhancement. In Proceedings of the IEEE/CVF Conference on Computer Vision and Pattern Recognition (CVPR), Nashville, TN, USA, 20–25 June 2021.
19. Howard, A.; Zhu, M.; Chen, B.; Kalenichenko, D.; Wang, W.; Weyand, T.; Andreetto, M.; Adam, H. MobileNets: Efficient Convolutional Neural Networks for Mobile Vision Applications. Computer Vision and Pattern Recognition. *arXiv* **2017**, arXiv:1704.04861.
20. Zhang, X.; Zhou, X.; Lin, M.; Sun, J. ShuffleNet: An Extremely Efficient Convolutional Neural Network for Mobile Devices. In Proceedings of the IEEE Conference on Computer Vision and Pattern Recognition (CVPR), Salt Lake City, UT, USA, 18–23 June 2018.
21. Han, K.; Wang, Y.; Tian, Q.; Guo, J.; Xu, C.; Xu, C. GhostNet: More Features from Cheap Operations. In Proceedings of the IEEE/CVF Conference on Computer Vision and Pattern Recognition (CVPR), Long Beach, CA, USA, 15–20 June 2019.
22. Wang, C.; Liao, H.M.; Yeh, I.-H.; Wu, Y.; Chen, P.; Hsieh, J. CSPNet: A New Backbone that can Enhance Learning Capability of CNN. In Proceedings of the IEEE/CVF Conference on Computer Vision and Pattern Recognition (CVPR) Workshops, Long Beach, CA, USA, 15–20 June 2019.
23. Cai, J.; Gu, S.; Zhang, L. Learning a Deep Single Image Contrast Enhancer from Multi-Exposure Images. *IEEE Trans. Image Process.* **2018**, *27*, 2049–2062. [CrossRef] [PubMed]
24. Guo, X.; Li, Y.; Ling, H. LIME: Low-light image enhancement via illumination map estimation. *IEEE Trans. Image Process.* **2017**, *26*, 982–993. [CrossRef] [PubMed]
25. Lee, C.; Lee, C.; Kim, C.-S. Contrast enhancement based on layered difference representation. *IEEE Trans. Image Process.* **2012**, *22*, 965–968.
26. Li, C.; Guo, J.; Porikli, F.; Pang, Y. LightenNet: A convolutional neural network for weakly illuminated image enhancement. *Pattern Recognit. Lett.* **2018**, *104*, 15–22. [CrossRef]
27. Jiang, Y.; Gong, X.; Liu, D.; Cheng, Y.; Fang, C.; Shen, X.; Yang, J.; Zhou, P.; Wang, A.Z. EnlightenGAN: Deep light enhancement without paired supervision. *arXiv* **2019**, arXiv:1906.06972. [CrossRef] [PubMed]
28. Loh, Y.P.; Chan, C.S. Getting to Know Low-light Images with The Exclusively Dark Dataset. *Comput. Vis. Image Underst.* **2018**, *178*, 30–42. [CrossRef]

electronics

MDPI

Article

YKP-SLAM: A Visual SLAM Based on Static Probability Update Strategy for Dynamic Environments

Lisang Liu [1,2], Jiangfeng Guo [1,2,*] and Rongsheng Zhang [1,2]

1 School of Electronic, Electrical Engineering and Physics, Fujian University of Technology, Fuzhou 350118, China
2 National Demonstration Center for Experimental Electronic Information and Electrical Technology Education, Fujian University of Technology, Fuzhou 350118, China
* Correspondence: 2201905130@smail.fjut.edu.cn

Abstract: Visual simultaneous localization and mapping (SLAM) algorithms in dynamic scenes can incorrectly add moving feature points to the camera pose calculation, which leads to low accuracy and poor robustness of pose estimation. In this paper, we propose a visual SLAM algorithm based on object detection and static probability update strategy for dynamic scenes, named YKP-SLAM. Firstly, we use the YOLOv5 target detection algorithm and the improved K-means clustering algorithm to segment the image into static regions, suspicious static regions, and dynamic regions. Secondly, the static probability of feature points in each region is initialized and used as weights to solve for the initial camera pose. Then, we use the motion constraints and epipolar constraints to update the static probability of the feature points to solve the final pose of the camera. Finally, it is tested on the TUM RGB-D dataset. The results show that the YKP-SLAM algorithm proposed in this paper can effectively improve the pose estimation accuracy. Compared with the ORBSLAM2 algorithm, the absolute pose estimation accuracy is improved by 56.07% and 96.45% in low dynamic scenes and high dynamic scenes, respectively, and the best results are almost obtained compared with other advanced dynamic SLAM algorithms.

Keywords: Visual SLAM; dynamic scene; YOLOv5; K-means clustering; probability update

Citation: Liu, L.; Guo, J.; Zhang, R. YKP-SLAM: A Visual SLAM Based on Static Probability Update Strategy for Dynamic Environments. *Electronics* **2022**, *11*, 2872. https://doi.org/10.3390/electronics11182872

Academic Editor: Giovanni Ramponi

Received: 24 June 2022
Accepted: 5 September 2022
Published: 11 September 2022

Publisher's Note: MDPI stays neutral with regard to jurisdictional claims in published maps and institutional affiliations.

1. Introduction

Simultaneous localization and mapping (SLAM) is to estimate camera pose and build a map of the environment simultaneously during motion from sensor data collected by the robot. After decades of development, some very mature SLAM algorithms have emerged, such as PTAM [1], LSD-SLAM [2], DSO [3], ORB-SLAM2 [4], and VINS Mono [5], which are basically based on the assumption of static environments. However, in practical applications of robotics, motion scenes are more common than static scenes, and most application scenes encounter dynamic objects, e.g., pedestrians, vehicles, animals, etc. Dynamic objects can introduce anomalous "outliers" that disrupt the normal correspondence between image features, resulting in significant drift in camera pose. Some optimization algorithms, such as random sample consensus [6] (RANSAC) and graph optimization, can filter out a small number of weak dynamic features in the environment as outliers. These methods can achieve certain results for low-speed motion with a small number of outliers. Though, they are not able to process dynamic features very well for high-speed complex motion scenes, and the visual SLAM system might fail to track and localize. Therefore, it is particularly important to study SLAM algorithms in dynamic environments.

In order to solve the visual SLAM problem in a dynamic environment, the traditional method is to eliminate dynamic objects through geometric constraints and set a threshold according to the size of the reprojection error to distinguish static objects from dynamic objects. However, this method has two problems. (1) The method cannot distinguish the residuals caused by moving objects from those caused by mis-matching. (2) The segmentation threshold is difficult to set; if the threshold set is too large, the static features will be mis-rejected, and if the segmentation threshold set is too small, it is difficult to completely reject the dynamic features in the environment. Therefore, the method is more suitable for a low dynamic environment. Additionally, in a high dynamic environment, the accuracy of dynamic feature detection is low, and the accuracy of pose estimation is poor.

In recent years, with the development of computer vision and deep learning, semantic constraints have been widely applied to visual SLAM problems in dynamic environments. The semantic constraint approach mainly applies semantic segmentation and target detection to obtain semantic information in the environment. By identifying and removing potential dynamic objects, the performance of visual SLAM in dynamic scenes can be greatly improved. The semantic segmentation algorithm can provide fine pixel-level object masks, but its real-time performance is poor. The improvement of segmentation accuracy and robustness often comes at the cost of huge computational cost. Even then, the segmentation boundary of an object cannot be very accurate. The target detection algorithm can quickly obtain the object frame of an object with low computational cost, but it cannot obtain accurate object boundaries, and if the features in the dynamic object frame are directly removed, it will lead to the false removal of some static features. Moreover, there are three problems with semantic constraints. (1) The actual motion is stationary, however, the algorithm cannot judge a semantic prior is a dynamic object or not, which may lead to the false removal of some static features. (2) It can only handle known objects labeled in the training set of the network but may still fail in the face of unknown moving objects, which leads to the missed detection of some dynamic features. (3) It deletes all dynamic features of semantic information discrimination and does not calculate the pose. This will lead to a reduction in constraints in pose calculation, knowing that dynamic features can still provide weak constraints for pose calculation. If it is deleted directly, it will lead to a decrease in the accuracy of pose estimation.

To address the above problems, in order to improve the pose estimation accuracy and robustness of the SLAM system in a dynamic environment, this paper proposes a YKP-SLAM algorithm in a dynamic environment. On the basis of ORBSLAM2, YKP-SLAM adds three major processes: YOLOv5 target detection, improved K-means clustering, and probability updating strategy. Our experiments prove that the YKP-SLAM algorithm can effectively reduce the tracking error and improve the accuracy and robustness of the SLAM system, both in a slow-moving dynamic environment and in a fast-moving dynamic environment.

The main contributions of this paper are as follows:

(1) We incorporate the lightweight YOLOv5 object detection algorithm into the SLAM system, which can quickly and accurately provide accurate semantic priors for subsequent operations.

(2) A K-means clustering algorithm specifically for depth images is proposed, which can select the number of clusters adaptively and can segment dynamic object contours from dynamic object frames quickly and accurately.

(3) A method for initializing static probability is proposed. The image is divided into three regions by combining YOLOv5 and improved K-means clustering. Then, the initial poses are solved by probability initialization of feature points in each region separately. More accurate initial poses are provided for the subsequent motion constraints and polar constraints.

(4) A probability update strategy based on motion constraints and epipolar constraints is proposed. Probability updates are performed for all feature points in the image. Then, all feature points are added to the pose calculation to solve the final pose.

2. Related Work

2.1. Dynamic SLAM Based on Traditional Method

Traditional dynamic SLAM algorithms are mainly based on geometric constraints to filter out dynamic feature points in the environment. For example, Zou [7] et al. project feature points from the previous frame onto the current frame and calculate the 2D reprojection error of matching points with the current frame and classify feature points into static and dynamic feature points according to the magnitude of the reprojection error. Wang [8] et al. detected the matched outlier points in two adjacent frames by epipolar constraint and then fused the clustering information of the depth map provided by RGB-D cameras to identify the moving targets in the scene. Dai [9] et al. proposed a static object geometry prior method in a feature-based SLAM framework. The algorithm utilizes the connectivity of map points to separate moving objects from the static background, thus reducing the impact of moving objects on the pose estimation.

In addition to geometric constraints, optical flow methods are also used to distinguish dynamic and static features. For example, Klappstein [10] et al. defined the likelihood of "moving objects in the scene" based on the motion metric calculated by optical flow. Fang [11] et al. improved the optical flow method to detect dynamic targets based on point matching techniques and uniform sampling strategies and introduced a Kalman filter to enhance detection and tracking. FlowFusion [12] estimated the optical flow of two adjacent frames through a PWC-Net [13] network, and at the same time, estimated the camera pose based on the intensity and depth of the two adjacent frames and then used the estimated optical flow and camera motion to compute the 2D scene flow and finally used the 2D scene flow for dynamic feature segmentation.

2.2. Dynamic SLAM Based on Semantic Constraints

In recent years, deep-learning-based image semantic segmentation and target recognition have been widely used, and the detection methods have evolved greatly in terms of efficiency and accuracy. Many researchers have tried to solve the dynamic SLAM problem by removing potential dynamic objects through semantic tagging or target detection preprocessing. For example, Yang [14] et al. used the target detection network Faster R-CNN [15] to detect dynamic objects and then performed geometric matching with the current frame and keyframes to determine whether they are dynamic objects. Yu [16] et al. proposed the DS-SLAM algorithm, combining a semantic segmentation network and optical flow method to provide a semantic representation of octree maps, thus reducing the dynamic objects. The DynaSLAM proposed by Bescos [17] et al. uses a combination of multi-view geometry and Mask RCNN [18] to detect and filter dynamic targets. ZHANG Jinfeng [19] et al. used the target detection network YOLOv3 [20] to filter dynamic feature points in the scene, which effectively reduced the trajectory error of the SLAM system. Zhong [21] et al. proposed Detect-SLAM combined with the target detection network SSD [22] to identify dynamic targets, such as pedestrians and vehicles, in the environment as a priori dynamic targets and then filter the feature points on the a priori dynamic target to improve its localization accuracy. Blitz-SLAM [23] obtains the mask of the object by BlitzNet [24], then completes the mask by depth information, and finally classifies the static feature points and dynamic feature points by epipolar constraints.

3. Materials and Methods

3.1. System Architecture

The algorithm framework of YKP-SLAM is shown in Figure 1. Based on ORBSALM2, we added the YOLOv5 target detection algorithm and the improved K-means clustering algorithm to the fore-end and added a complete probability update strategy to the back-end pose calculation. The algorithmic flow of YKP-SLAM can be described as follows. Firstly, the RGB image is detected by YOLOv5 target detection algorithm to obtain the dynamic object frame, and at the same time, the ORB [25] feature points are extracted from the RGB image. Secondly, the depth values of the pixel points are clustered within the dynamic object frame by the improved K-means clustering algorithm combined with the depth image. The results of YOLOv5 target detection and K-means clustering are used to segment the image into static regions, suspicious static regions, and dynamic regions, initialize the static probability of feature points within each region, and add them as weights to the camera pose estimation to calculate the initial camera pose T_{cw1}. Finally, the static probability of feature points is updated by the motion constraint and the epipolar constraint, and the second stage pose T_{cw2} and the final pose T_{cw} of the camera are solved, respectively.

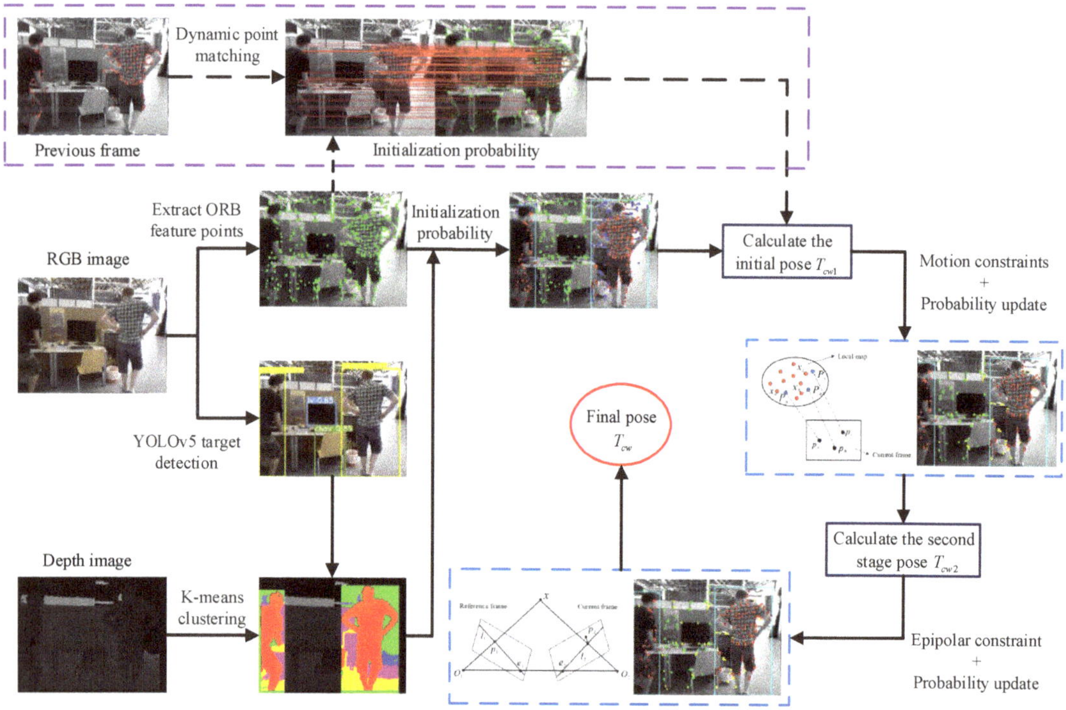

Figure 1. The algorithmic framework of YKP-SLAM. In the image, green points represent static points, blue points represent suspicious static points, red points represent dynamic points, and yellow points represent points where the probability changes.

Of course, we also considered the failure of the YOLOv5 algorithm. When YOLOv5 fails, the dynamic object frame cannot be obtained. Then, at this time, we perform feature matching between the feature points in the current frame and the dynamic feature points in the previous frame. Mark the successfully matched feature points of the current frame as dynamic feature points, and mark the remaining feature points as static feature points. The only difference from a normal operation is that the characteristic points are divided into three categories in a normal operation, and the characteristic points in a fault operation are divided into two categories. The subsequent static probability initialization method and probability update strategy are the same. The feature point classification process of YOLOv5 fault runtime is shown in the purple dashed box in Figure 1.

3.2. *YOLOv5 Target Detection*

You Only Look Once (YOLO) is a regression-based target detection algorithm. It is the pioneering work of the one-stage method. It was released by Ultralytics on 10 June 2020. It is one of the most widely used target detection algorithms. It solves target detection as a regression problem and directly obtains the bounding box position and classification of the predicted object from an input image. It ensures the accuracy while taking into account the real-time performance and achieves very good speed and accuracy. YOLOv5 proposes a total of 4 network models: YOLOv5s, YOLOv5m, YOLOv5l, and YOLOv5x. The network structure of the four models is the same; the difference is that the depth_multiple and width_multiple parameters can be used to control the depth of the model and the number of convolution kernels, respectively. Among them, YOLOv5s is the network with the smallest network depth and the smallest feature map width. It occupies only 7.5 M of memory. Its detection speed on TeslaP100 reaches 140FPS, which fully meets real-time performance. The other three are continuously deepened and widened on this basis, with improved accuracy and slower speed.

In order to meet the real-time nature of the SLAM system, the fastest YOLOv5s algorithm is adopted, which is embedded in the fore-end of the SLAM system, to perform target detection on each RGB image passed by the camera and obtain the bounding box position of the object and its category. In the bounding box, the people and animals are located as dynamic object boxes DB. The target detection results of YOLOv5s are shown Figure 2. The yellow frame in Figure 2 is the dynamic object box. It can be seen from the figure that whether the person is on the front, side, back, or only half of the body is exposed, YOLOv5 can be accurately framed.

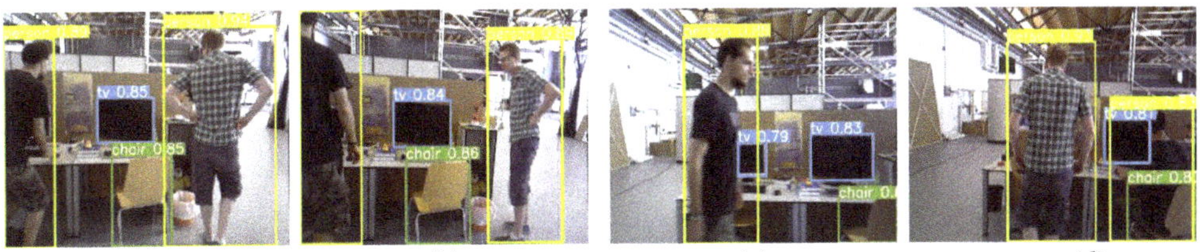

(a) (b) (c) (d)

Figure 2. YOLOv5 target detection results. (**a–d**) represents the detection results of the YOLOv5 algorithm in several different scenes.

3.3. Improved Adaptive K-means Clustering Algorithm

Although the YOLOv5 target detection algorithm can quickly and accurately locate the bounding boxes of dynamic objects, it cannot obtain an accurate dynamic object mask. Therefore, this paper proposes an adaptive K-means clustering segmentation algorithm based on depth images, which can segment dynamic objects from the dynamic object box DB quickly and accurately.

The K-means algorithm is an unsupervised clustering algorithm, which is easy to implement and runs fast. However, the traditional K-means clustering algorithm pre-specifies the number of clusters and randomly initializes the cluster centers according to experience, which is likely to cause too many iterations of the algorithm or misclassification. Since the number of clusters is artificially set in advance, the direct application of the traditional K-means clustering algorithm to depth image clustering will have the following two problems:

(1) If the number of clusters set is too large, a complete dynamic objects would be divided into multiple categories, which might cause incomplete segmentation of dynamic objects.

(2) If the number of clusters set is too small, the dynamic objects cannot be separated from the static background.

In order to solve the above problems, an improved adaptive K-means algorithm is proposed in this paper. The algorithm can automatically generate the optimal number of clusters and the initial cluster centers, so that dynamic objects can be segmented from the static background more quickly and accurately. The steps of the improved K-means algorithm are as follows:

(1) Take out the depth image IDB_i in the dynamic object frame DB and count the total number of pixels M and the maximum pixel depth $D_{\max}$ in IDB_i.

(2) Solve the histogram of the depth image IDB_i and divide the data of the histogram into k segments:

$$k = \frac{D_{\max}}{T} \tag{1}$$

where T is the segmentation threshold, whose size can determine the number of clusters. Since the depths of dynamic objects do not change much in the two adjacent frames, we first use the depth mean D_p of dynamic feature points in the previous frame as the prior of the depth value of dynamic objects in the current frame. Then, the ratio λ of the number of pixels in the dynamic object in the previous frame to the number of pixels in the dynamic object frame is calculated. Finally, find the neighborhood $U(D_p, \delta) = \{x \mid D_p - \delta < x < D_p + \delta\}$ of point D_p in the histogram of the depth image IDB_i, so that the number of pixels in the neighborhood is equal to λM; then, the size of the segmentation threshold T is the range of the neighborhood.

$$T = 2\delta \tag{2}$$

(3) We take k as the number of clusters for subsequent K-means clustering and take the maximum depth value of each piece of data as the initial cluster center for each category.

(4) The K-means clustering algorithm obtains a depth image segmentation graph based on the number of clusters calculated in step (3) and the initial cluster centers.

Since the depth values of dynamic objects do not change too much within the two adjacent frames, the depth mean value D_p of dynamic features in the previous frame is used as a criterion, and then, the pixel depth mean value of each cluster in the dynamic object box is solved, and the cluster with a pixel depth mean value closest to the depth mean value of dynamic points in the previous frame is marked as a dynamic region; the other clusters in the dynamic object box are marked as suspicious static regions, and the regions outside the dynamic object box are marked as static regions. The whole dynamic region classification process is shown in Figure 3.

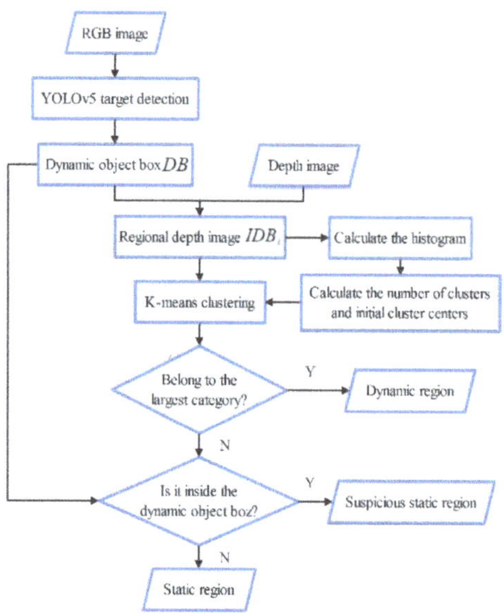

Figure 3. Schematic diagram of dynamic region division.

The results of K-means clustering are shown in Figure 4. From the figure, we can see that the improved K-means clustering algorithm proposed in this paper can segment people from the background completely and does not lead to mis-segmentation.

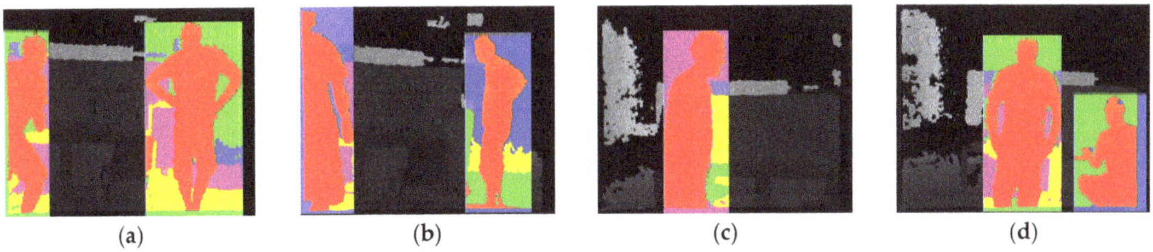

Figure 4. Improved K-means clustering, where each color represents one class. The red region is the dynamic region, and the other colored regions are suspicious static regions. (a–d) represents the clustering results of the improved K-means clustering algorithm in several different scenes.

3.4. Initialize the Static Probability and Calculate the Initial Camera Pose

In this paper, the YOLOv5 target detection algorithm and the improved adaptive K-means clustering algorithm are used to segment the image into dynamic regions, suspicious static regions, and static regions. In order to obtain a more accurate initial pose, the feature points in different regions are assigned static probability initial values of

$$\textit{Static probability}\begin{cases} \omega_a = 0 & \textit{Dynamic region} \\ \omega_b = 0.5 & \textit{Suspicious static region} \\ \omega_c = 1 & \textit{Static region} \end{cases} \tag{3}$$

These initial static probabilities are then used as weights for the pose calculation, and the initial pose T_{cw1} for the current frame is calculated according to the weighted minimization reprojection error.

The structure of the camera pose T_{cw1} is

$$\mathrm{SE}(3) = \left\{ T_{cw1} = \begin{bmatrix} R_{cw1} & t_{cw1} \\ 0^T & 1 \end{bmatrix} \in \mathbb{R}^{4\times4} \mid R_{cw1} \in \mathrm{SO}(3), t_{cw1} \in \mathbb{R}^3 \right\} \tag{4}$$

where R_{cw1} is the rotation matrix, and t_{cw1} is the translation vector.

T_{cw1} can be solved by Equation (5).

$$\begin{aligned} T_{cw1} = \mathrm{argmin}(&\sum_{a=1}^{N_a} \|KT_{cw1}x_a - p_a\|^2_{\Sigma_1} + \\ &\sum_{b=1}^{N_b} \|KT_{cw1}x_b - p_b\|^2_{\Sigma_2} + \\ &\sum_{c=1}^{N_c} \|KT_{cw1}x_c - p_c\|^2_{\Sigma_3}) \end{aligned} \tag{5}$$

Among them

$$\begin{aligned} \Sigma_1 &= \omega_a \times n \times E \\ \Sigma_2 &= \omega_b \times n \times E \\ \Sigma_3 &= \omega_c \times n \times E \end{aligned} \tag{6}$$

Where, p_a, p_b, p_c are the 2D pixel point coordinates of dynamic feature points, suspicious static points, and static points in the current frame, respectively, while x_a, x_b, x_c are the coordinates of their corresponding matching 3D map points. $\Sigma_1, \Sigma_2, \Sigma_3$ is the information matrix of feature points in each region, n is the number of layers of the image pyramid where the current feature point is located, and E is the unit matrix of 3×3. N_a, N_b, N_c are the numbers of dynamic feature points, suspicious static points, and static points in the current frame, respectively.

3.5. Probability Update Based on Motion Constraints

The traditional geometric method distinguishes dynamic points and static points by the size of the reprojection error and sets the threshold value and judges the points with a reprojection error larger than the threshold value as dynamic points and those smaller than the threshold value as static points. The threshold size of this method is difficult to set, which can easily lead to mis-segmentation of dynamic and static points. Therefore, this paper proposes a new segmentation method that uses the motion distance of the a priori dynamic point p_a judged by the front-end of the SLAM system (YOLOv5 and K-means) as a scale to update the static probability of the suspicious static point p_b and static point p_c. The schematic diagram of the motion constraint is shown in Figure 5.

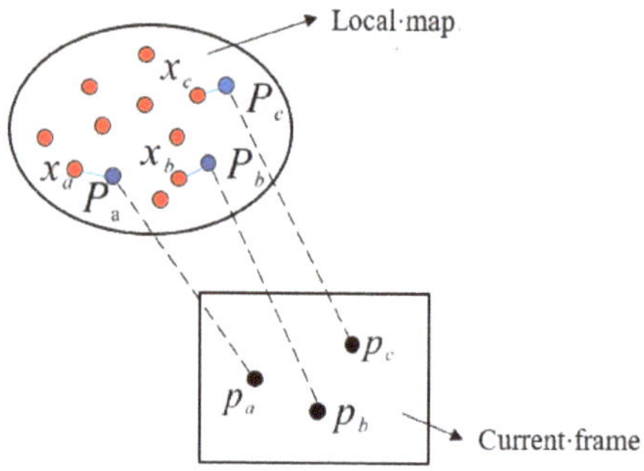

Figure 5. Schematic diagram of motion constraints, where the ellipse represents the local map, the rectangle represents the current frame, the red point inside the ellipse represents the local map point, the blue point represents the 3D point of the current frame feature point back-projected to the world coordinate system, and the line between the red point and the blue point represents the motion distance of the feature point.

We now know the initial pose T_{cw1} and the camera internal reference K of the current frame, and we can also directly obtain the depth information Z of the feature points through the depth camera. Then, we first back-project the dynamic point p_a in the current frame to the world coordinate system to obtain the 3D point coordinate P_a in the world coordinate system.

$$P_a = \begin{bmatrix} X_a \\ Y_a \\ Z_a \end{bmatrix} = T_{wc1} K p_a \tag{7}$$

Calculate the square value L_a of the movement distance between the back-projection point P_a and the corresponding map point x_a:

$$L_a = (X_a - X_a{}')^2 + (Y_a - Y_a{}')^2 + (Z_a - Z_a{}')^2 \tag{8}$$

where $\begin{bmatrix} X_a{}' & Y_a{}' & Z_a{}' \end{bmatrix}^T$ are the 3D point coordinates of the map point x_a.

Similarly, the squares of the motion distances of the suspicious static point p_b and the static point p_c can be solved as L_b and L_c, respectively.

Then, solve the mean μ_L and variance S_L of the square of the motion distance of the dynamic point p_a in the current frame:

$$\mu_L = \frac{\sum\limits_{a=1}^{N_a} L_a}{N_a} \tag{9}$$

$$S_L = \sqrt{\frac{\sum_{a=1}^{N_a} (L_a - \mu_L)^2}{N_a}} \quad (10)$$

By comparing the motion distance of the suspicious static point p_b, static point p_c, and dynamic point p_a to update their static probability, this paper designs a sigmoid function to calculate the static probability of each suspicious static point p_b and static point p_c as follows:

$$\omega_{b1} = \frac{1}{1 + \exp(\alpha(\frac{L_b - \mu_L}{S_L}))} \quad (11)$$

$$\omega_{c1} = \frac{1}{1 + \exp(\alpha(\frac{L_c - \mu_L}{S_L}))} \quad (12)$$

where α is a coefficient greater than 0.

Update the static probability of each feature point in each region in combination with the initial static probability:

$$\begin{aligned} \omega_a &= \omega_a \\ \omega_b &= \omega_b \times \omega_{b1} \\ \omega_c &= \omega_c \times \omega_{c1} \end{aligned} \quad (13)$$

Based on the updated static probability of the feature points, the static probabilities are brought into Equation (5) to calculate the camera pose T_{cw2} in the second stage.

3.6. Probability Update Based on Epipolar Constraint

As shown in Figure 6, O_1, O_2 is the camera optical center at the moment of the current frame and reference frame, respectively, and p_1, p_2 is a pair of matching points between the current frame and reference frame. x is the map point corresponding to the p_1 point on the reference frame, and the projection point of this point on the current frame should be located on the polar line l_2 if the point is stationary, or not on the polar line if it is moving. In this paper, the static probability of the feature points is updated based on the distance from point p_2 to the polar line l_2.

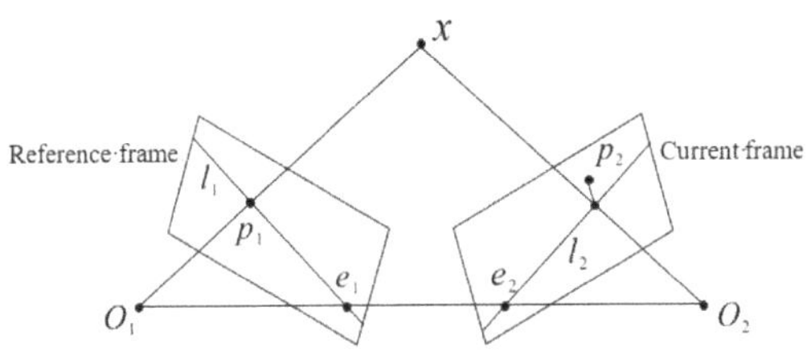

Figure 6. Schematic diagram of epipolar constraint.

Through the current frame camera pose T_{cw2} and the reference frame camera pose T_{cwr} solved in the second stage, the rotation matrix and translation matrix t_{2r} between the two frames can be solved:

$$R_{2r} = R_{cw2} \times R_{cwr}^{-1} \quad (14)$$

$$t_{2r} = -R_{cw2} \times R_{cwr}^{-1} \times t_{cwr} + t_{cw2} \quad (15)$$

Among them, R_{cw2} and t_{cw2} are the rotation matrix and translation matrix of the current frame, respectively, and R_{cwr} and t_{cwr} are the rotation matrix and translation matrix of the reference frame.

Fundamental matrix F

$$F = K^{-T} (t_{2r})^{\wedge} R_{2r} K^{-1} \quad (16)$$

Solve the polar equation corresponding to the feature point on the reference frame to the current frame according to the fundamental matrix. The polar equation is expressed as

$$\begin{bmatrix} A & B & C \end{bmatrix}^T = F\begin{bmatrix} u_1 & v_1 & 1 \end{bmatrix} \quad (17)$$

$\begin{bmatrix} u_1 & v_1 & 1 \end{bmatrix}$ is the homogeneous coordinate of the reference frame feature point p_1.

Calculate the square of the polar distance from the feature point of the current frame to the corresponding polar line:

$$H = \frac{(Au_2 + Bv_2 + C)^2}{A^2 + B^2} \quad (18)$$

$\begin{bmatrix} u_2 & v_2 & 1 \end{bmatrix}$ is the homogeneous coordinate of the current frame feature point p_2.

According to the above Equations (16)–(18), the polar distance H_a, H_b, H_c of the dynamic point, suspicious static point, and static point of the current frame can be calculated, respectively.

Calculate the mean μ_H and variance S_H of the polar distance of the dynamic points, as with the motion constraints:

$$\mu_H = \frac{\sum_{a=1}^{N_a} H_a}{N_a} \quad (19)$$

$$S_H = \sqrt{\frac{\sum_{a=1}^{N_a} (H_a - \mu_H)^2}{N_a}} \quad (20)$$

By comparing the polar distance of the suspicious static point p_b, static point p_c, and dynamic point p_a to update their static probability

$$\omega_{b2} = \frac{1}{1 + \exp(\beta(\frac{H_b - \mu_H}{S_H}))} \quad (21)$$

$$\omega_{c2} = \frac{1}{1 + \exp(\beta(\frac{H_c - \mu_H}{S_H}))} \quad (22)$$

where β is a coefficient greater than 0.

Update the final static probability of the feature points in each region using the static probability of epipolar constraints:

$$\begin{aligned} \omega_a &= \omega_a \\ \omega_b &= \omega_b \times \omega_{b2} \\ \omega_c &= \omega_c \times \omega_{c2} \end{aligned} \quad (23)$$

The final camera pose T_{cw} can be calculated from the final static probability of the feature points and Equation (5).

4. Experiments and Analysis

In order to evaluate the performance of the YKP-SLAM algorithm, this paper uses the public TUM RGB-D dataset [26] to conduct the experiments. The TUM dataset is produced by the University of Munich, Germany, and uses a Kinect sensor to capture information at a rate of 30 HZ with an image resolution of 640 $*$ 480 and uses a high-precision motion capture system VICON with an inertial measurement system while acquiring image data. The camera position and pose data are acquired in real time, which can be approximated as the real positional data of the RGB-D camera. In this paper, we mainly use eight dynamic scene sequences from the TUM RGB-D dataset for experiments, which are divided into two categories: walking and sitting. The sitting dataset series are low dynamic scenes, in which two people are sitting in front of a table and chatting, with low motion. The walking dataset series are high dynamic scenes, in which two people are walking in front of or around a table, with high motion. For each type of dataset series, the camera motion is also divided into four states, where static means the camera is at rest, xyz means the camera is moving along the spatial X-Y-Z axis in translation, rpy means the camera is rotating in a flip angle, pitch angle, and yaw angle, and halfsphere means the camera is moving along the trajectory of a hemisphere with a diameter of 1 m.

The experiments were run on a server with Ubuntu 18.04, a GeForce RTX 3060 graphics card with 12 GB of video memory, a 7-core Intel(R) Xeon(R) CPU, and 20 GB of RAM.

4.1. Comparison with ORBSLAM2

Since the YKP-SLAM algorithm proposed in this paper is improved on the basis of ORBSLAM2, a comparison experiment with ORBSLAM2 is conducted first. In this paper, the absolute trajectory error (ATE) and relative pose error (RPE) [26] are adopted to evaluate algorithm accuracy. The absolute trajectory error is the direct difference between the estimated and real poses, which can reflect the algorithm accuracy and global consistency of the trajectory very intuitively. The relative trajectory error contains the relative translation error and relative rotation error, which are directly measured by the odometer. The experimental results are shown in Tables 1 and 2, where RMSE denotes the root mean square error, Mean denotes the mean error, and Std denotes the standard deviation.

Table 1. Comparison of absolute trajectory error (ATE) between ORB-SLAM2 and YKP-SLAM.

Sequences	ORB-SLAM2/m			YKP-SLAM/m			Improvement/%		
	RMSE	Mean	Std	RMSE	Mean	Std	RMSE	Mean	Std
sitting_xyz	0.0111	0.0093	0.0059	0.0072	0.0065	0.0033	35.14	30.11	44.07
sitting_half	0.0437	0.0360	0.0247	0.0153	0.0132	0.0076	64.99	63.33	69.23
sitting_static	0.0128	0.0120	0.0046	0.0052	0.0043	0.0028	59.38	64.17	39.13
sitting_rpy	0.0358	0.0293	0.0205	0.0268	0.0237	0.0126	25.13	19.11	38.53
walking_xyz	0.5185	0.4420	0.2711	0.0147	0.0130	0.0068	97.16	97.06	97.49
walking_half	0.5820	0.4571	0.3603	0.0245	0.0220	0.0107	95.79	95.19	97.03
walking_static	0.2742	0.2286	0.1514	0.0063	0.0056	0.0026	97.70	97.55	98.28
walking_rpy	1.5320	1.4262	0.5594	0.0702	0.0489	0.0514	95.41	96.57	90.81

Table 2. Comparison of relative pose error (RPE) between ORB-SLAM2 and YKP-SLAM.

Sequences	ORB-SLAM2/m			YKP-SLAM/m			Improvement/%		
	RMSE	Mean	Std	RMSE	Mean	Std	RMSE	Mean	Std
sitting_xyz	0.0148	0.0126	0.0077	0.0079	0.0070	0.0038	46.62	44.44	50.65
sitting_half	0.0227	0.0121	0.0192	0.0137	0.0108	0.0084	39.64	10.74	56.25
sitting_static	0.0180	0.0169	0.0063	0.0058	0.0055	0.0031	67.78	67.46	50.79
sitting_rpy	0.0256	0.0208	0.0148	0.0232	0.0171	0.0151	9.38	17.79	−2.27
walking_xyz	0.0382	0.0303	0.0233	0.0139	0.0116	0.0076	63.61	61.72	67.38
walking_half	0.0452	0.0317	0.0322	0.0196	0.0148	0.0128	56.64	53.31	60.25
walking_static	0.0473	0.0291	0.0373	0.0072	0.0062	0.0031	84.78	78.69	91.69
walking_rpy	0.0429	0.0316	0.0291	0.0317	0.0218	0.0239	26.11	31.01	17.97

The improvement rates in the table are calculated as follows:

$$\eta = \left(1 - \frac{\beta}{\alpha}\right) \times 100\% \tag{24}$$

where η represents the algorithm improvement rate, β represents the experimental results of the YKP-SLAM algorithm, and α represents the experimental results of the ORBSLAM2 algorithm.

Tables 1 and 2 show the quantitative evaluation of the errors, from which it can be seen that in the low dynamic scene sitting dataset series, the average improvement of the RMSE of absolute and relative trajectory errors of the YKP-SLAM algorithm compared with the ORBSLAM2 algorithm is 46.16% and 40.86%, respectively. The average improvement of the RMSE of absolute and relative trajectory errors of this algorithm over ORBSLAM2 is 96.52% and 57.79%, respectively, in the walking data set series of high dynamic scenes, which shows that the YKP-SLAM algorithm has a great improvement over the traditional ORBSLAM2 algorithm in both low and high dynamic scenes. The trajectory accuracy is greatly improved in both low and high dynamic scenes.

Figures 7 and 8 show the absolute trajectory error distributions of the ORBSLAM2 algorithm and the YKP-SLAM algorithm under the low dynamic sequences s_xyz, s_half and the high dynamic sequences w_xyz, w_half, respectively. Figures 9 and 10 show the comparison of the estimated trajectory and the real trajectory of the ORBSLAM2 algorithm and the YKP-SLAM algorithm under the low dynamic sequences s_xyz, s_half and the high dynamic sequences w_xyz, w_half, respectively. It can be seen that under the low dynamic sequences s_xyz and s_half, the absolute trajectory error of the YKP-SLAM algorithm is slightly smaller than that of the ORBSLAM2 algorithm, and the estimated trajectory is closer to the real trajectory than the ORBSLAM2 algorithm. Under the high dynamic sequences w_xyz and w_half, the absolute pose error of the YKP-SLAM algorithm is smaller than that of the ORBSLAM2 algorithm, and the estimated trajectory is still very close to the real trajectory, while the estimated trajectory of the ORBSLAM2 algorithm is far away from the real trajectory. This proves that the YKP-SLAM algorithm can effectively improve the pose estimation accuracy of the SLAM system in low dynamic and high dynamic scenes.

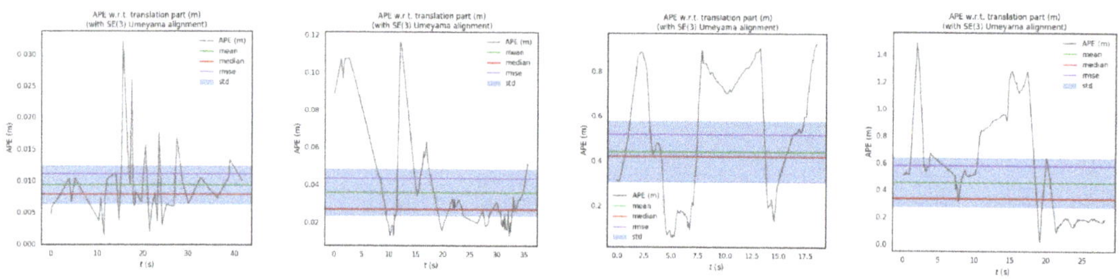

(a) (b) (c) (d)

Figure 7. Absolute trajectory error distribution of ORBSLAM2 algorithm. (**a**) s_xyz. (**b**) s_half. (**c**) w_xyz. (**d**) w_half.

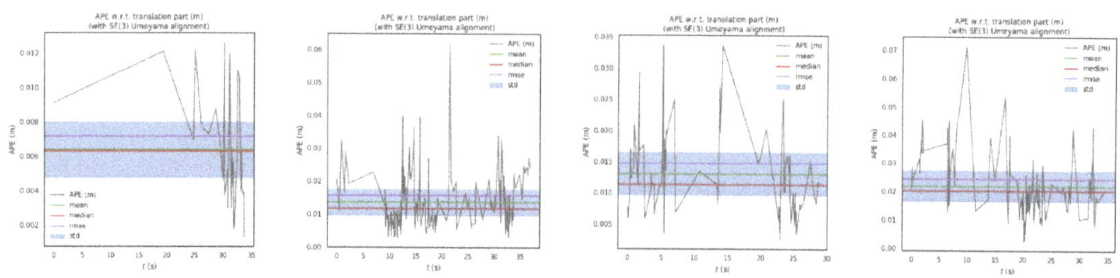

(a) (b) (c) (d)

Figure 8. Absolute trajectory error distribution of YKP-SLAM algorithm. (**a**) s_xyz. (**b**) s_half. (**c**) w_xyz. (**d**) w_half.

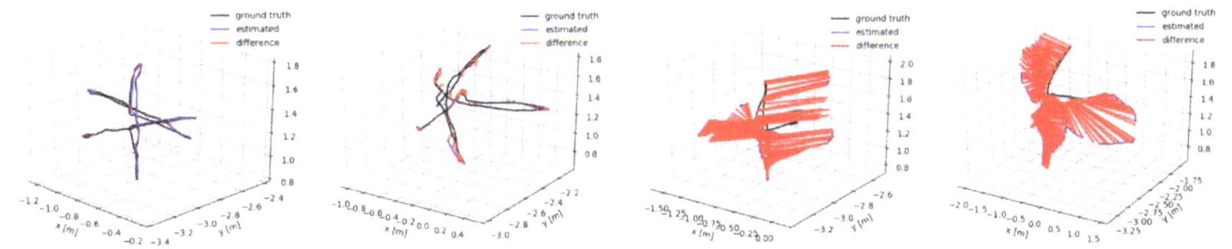

(**a**) (**b**) (**c**) (**d**)

Figure 9. Comparison of estimated trajectory and real trajectory of ORBSLAM2 algorithm. The colored line is the estimated trajectory, and the gray line is the real trajectory. (**a**) s_xyz. (**b**) s_half. (**c**) w_xyz. (**d**) w_half.

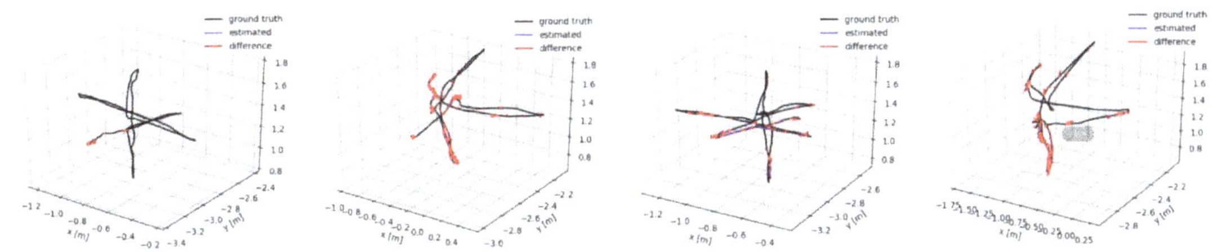

(**a**) (**b**) (**c**) (**d**)

Figure 10. Comparison of estimated trajectory and real trajectory of YKP-SLAM algorithm. The colored line is the estimated trajectory, and the gray line is the real trajectory. (**a**) s_xyz. (**b**) s_half. (**c**) w_xyz. (**d**) w_half.

4.2. Comparison with Advanced Dynamic SLAM Algorithms

In order to verify the superiority of the YKP-SLAM algorithm, DS-SLAM [16], DynaSLAM [17], and Blitz-SLAM [23] are selected for comparison experiments with YKP-SLAM in this paper. The root mean square error RMSE and variance Std in the absolute trajectory error are selected as the evaluation metrics for verification. The experimental results are shown in Table 3, where the bold font indicates the best results. Among them, the DS-SLAM and DynaSLAM codes were open sourced as well as the experimental data, while the Blitz-SLAM algorithm code was not open sourced. As can be seen from the table, the YKP-SLAM algorithm achieves almost the best results compared to the other dynamic SLAM algorithms, both in high dynamic scenes and in low dynamic scenes. The performance is slightly worse under the s_rpy and w_rpy data sets, which is caused by the fact that the camera motion is too large at this time, making the YOLOv5 target detection results less accurate.

Table 3. Comparison of absolute trajectory error (ATE) between YKP-SLAM algorithm and other dynamic SLAM algorithms.

Sequences	DS-SLAM/m		DynaSLAM/m		Blitz-SLAM/m		YKP-SLAM/m	
	RMSE	Std	RMSE	Std	RMSE	Std	RMSE	Std
sitting_xyz	0.0187	0.0119	0.0135	0.0063	0.0148	0.0069	**0.0072**	**0.0033**
sitting_half	0.0162	**0.0061**	0.0193	0.0084	0.0160	0.0076	**0.0153**	0.0076
sitting_static	0.0065	0.0033	0.0085	0.0051	/	/	**0.0052**	**0.0028**
sitting_rpy	**0.0266**	0.0153	0.0865	0.0516	/	/	0.0268	**0.0126**
walking_xyz	0.0247	0.0186	0.0176	0.0086	0.0153	0.0078	**0.0147**	**0.0068**
walking_half	0.0303	0.0159	0.0273	0.0130	0.0256	0.0126	**0.0245**	**0.0107**
walking_static	0.0081	0.0036	0.0067	0.0031	0.0102	0.0052	**0.0063**	**0.0026**
walking_rpy	0.4442	0.2350	0.0389	0.0237	**0.0356**	**0.0220**	0.0702	0.0514

4.3. Ablation Experiment

In order to verify the effectiveness of the improved K-means clustering algorithm and probability update strategy proposed in this paper, we conduct ablation experiments, and the experimental results are shown in Table 4 The bold font indicates the best results, and the underlined ones represent the second best results.

Table 4. Comparison of absolute trajectory error of ablation experiment.

Sequences	Y-SLAM/m		YK-SLAM/m		YKP-SLAM/m	
	RMSE	Std	RMSE	Std	RMSE	Std
sitting_xyz	0.0168	0.0079	<u>0.0129</u>	<u>0.0068</u>	**0.0072**	**0.0033**
sitting_half	0.0858	0.0178	<u>0.0189</u>	<u>0.0084</u>	**0.0153**	**0.0076**
sitting_static	<u>0.0072</u>	<u>0.0035</u>	0.0079	<u>0.0032</u>	**0.0052**	**0.0028**
sitting_rpy	0.0481	0.0376	<u>0.0384</u>	<u>0.0221</u>	**0.0268**	**0.0126**
walking_xyz	<u>0.0181</u>	<u>0.0105</u>	0.0212	0.0111	**0.0147**	**0.0068**
walking_half	<u>0.0292</u>	0.0144	0.0301	<u>0.0135</u>	**0.0245**	**0.0107**
walking_static	<u>0.0079</u>	<u>0.0034</u>	0.0080	0.0035	**0.0063**	**0.0026**
walking_rpy	<u>0.0962</u>	<u>0.0625</u>	0.1457	0.0701	**0.0702**	**0.0514**

In Table 4, Y-SLAM refers to the direct elimination of feature points within the dynamic object frame by YOLOv5 target detection; YK-SLAM is the combination of YOLOv5 and improved K-means clustering to eliminate feature points within the dynamic object; YKP-SLAM is the proposed algorithm.

The comparison between Y-SLAM and YK-SLAM shows that the performance of YK-SLAM is better than Y-SLAM in the low dynamic environment, which is due to the fact that the number of dynamic points is smaller in the low dynamic environment. In contrast, Y-SLAM eliminates all the points in the dynamic object frame and deletes some static points by mistake, resulting in a reduction in constraints in the pose calculation, thus causing a decrease in pose accuracy. The performance of Y-SLAM is better than that of YK-SLAM in the high dynamic environment, which is due to the higher number of dynamic points and larger dynamic amplitude in the high dynamic environment. The area of the dynamic object frame is larger than that of the dynamic object, which allows Y-SLAM to reject more dynamic points and thus make its pose accuracy more accurate. YKP-SLAM with the addition of the probability update strategy achieves the best results in both low and high dynamic scenes. This is due to the fact that the probability update strategy assigns appropriate static probabilities to static and dynamic points and then adds all points to the pose calculation, which does not lead to either false deletion of static points or missed detection of dynamic points.

4.4. Real-Time Analysis

Real-time performance is one of the important evaluation indicators of SLAM systems. As shown in Table 5, in order to measure the real-time performance of the YKP-SLAM algorithm proposed in this paper, we test each module of the YKP-SLAM algorithm and the ORBSALM2 algorithm, respectively, under the highly dynamic "walking_xyz" sequence. In the table, A represents the YOLOv5 target detection module, B represents the ORB feature extraction module, C represents the improved K-means clustering module, D represents the probability update module, and E represents the normal tracking calculation pose module. Among them, the YOLOv5 target detection module and the ORB feature extraction module in the YKP-SLAM algorithm are run in parallel. The results show that the YOLOv5 target detection module cost less time than the ORB feature extraction module; that is to say, there is no need to wait for the detection results of YOLOv5 after the ORB feature extraction is completed. Therefore, in the case of sufficient computing power, adding the YOLOv5 module will not increase the system time. The average total time per frame of ORBSLAM2 and YKP-SLAM is 48.20ms and 62.05ms, respectively; that is, the running speed reaches 20

Fps and 16 Fps, respectively. Overall, YKP-SLAM basically meets the real-time performance of SLAM while ensuring accuracy in dynamic environments.

Table 5. The average running time of each module.

Algorithm	A/ms	B/ms	C/ms	D/ms	E/ms	Total Time/ms
ORBSLAM2	/	19.28	/	/	28.92	48.20
YKP-SLAM	15.46	19.28	7.33	6.52	28.92	62.05

5. Conclusions

In this paper, a YKP-SLAM algorithm in dynamic environment is proposed. The algorithm first segments the whole current frame image by YOLOv5 target detection algorithm and improved K-means clustering algorithm and assigns a priori static probability to each feature point according to the segmentation result. The a priori static probability is used as the weight to calculate the initial camera pose, and then, the static probability of the feature points is updated according to the motion constraint and the epipolar constraint to solve the final camera pose. The algorithm in this paper is verified under the TUM dataset. Compared with the ORBSLAM2 algorithm, the accuracy and robustness of this algorithm are greatly improved in both low and high dynamic scenes. Compared with the other SLAM algorithms in dynamic scenes, the YKP-SLAM algorithm also achieves almost the best localization accuracy. In future work, we will propose a dense semantic map construction method in dynamic scenes based on the existing one and make full use of the advantages of localization accuracy in high dynamic scenes and the semantic information provided by YOLOv5 to realize path planning and obstacle avoidance in dynamic scenes.

Author Contributions: Conceptualization, L.L. and J.G.; methodology, J.G.; software, J.G.; validation, L.L., J.G. and R.Z.; formal analysis, L.L.; investigation, R.Z.; resources, J.G.; data curation, J.G.; writing—original draft preparation, L.L. and J.G.; writing—review and editing, J.G.; visualization, R.Z.; supervision, L.L.; project administration, L.L.; funding acquisition, L.L. All authors have read and agreed to the published version of the manuscript.

Funding: This research was supported in part by the Natural Science Foundation of Fujian Province under Grant 2022H6005 and 2022J01952, in part by the Initial Scientific Research Fund of FJUT under Grant GY-Z12079, Grant GY-Z21036, and Grant GY-Z20067.

Data Availability Statement: Not applicable.

Acknowledgments: The authors are grateful to the editors and the anonymous reviewers for their insightful comments and suggestions.

Conflicts of Interest: The authors declare no conflict of interest.

References

1. Klein, G.; Murray, D. Parallel tracking and mapping for small AR workspaces. In Proceedings of the 2007 6th IEEE and ACM International Symposium on Mixed and Augmented Reality, Nara, Japan, 13–16 November 2007; pp. 225–234.
2. Engel, J.; Schöps, T.; Cremers, D. LSD-SLAM: Large-scale direct monocular SLAM. In *Computer Vision—ECCV 2014, Proceedings of the European Conference on Computer Vision, Zurich, Switzerland, 6–12 September 2014*; Springer: Cham, Switzerland, 2014; pp. 834–849.
3. Engel, J.; Koltun, V.; Cremers, D. Direct sparse odometry. *IEEE Trans. Pattern Anal. Mach. Intell.* **2017**, *40*, 611–625. [CrossRef] [PubMed]
4. Mur-Artal, R.; Tardós, J.D. Orb-slam2: An open-source slam system for monocular, stereo, and RGB-D cameras. *IEEE Trans. Robot.* **2017**, *33*, 1255–1262. [CrossRef]
5. Qin, T.; Li, P.; Shen, S. Vins-mono: A robust and versatile monocular visual-inertial state estimator. *IEEE Trans. Robot.* **2018**, *34*, 1004–1020. [CrossRef]
6. Fischler, M.A.; Bolles, R.C. Random sample consensus: A paradigm for model fitting with applications to image analysis and automated cartography. *Commun. ACM* **1981**, *24*, 381–395. [CrossRef]
7. Zou, D.; Tan, P. Coslam: Collaborative visual slam in dynamic environments. *IEEE Trans. Pattern Anal. Mach. Intell.* **2012**, *35*, 354–366. [CrossRef] [PubMed]

8. Wang, R.; Wan, W.; Wang, Y.; Di, K. A new RGB-D SLAM method with moving object detection for dynamic indoor scenes. *Remote Sens.* **2019**, *11*, 1143. [CrossRef]
9. Dai, W.; Zhang, Y.; Li, P.; Fang, Z.; Scherer, S. RGB-D SLAM in dynamic environments using point correlations. *IEEE Trans. Pattern Anal. Mach. Intell.* **2020**, *44*, 373–389. [CrossRef] [PubMed]
10. Klappstein, J.; Vaudrey, T.; Rabe, C.; Wedel, A.; Klette, R. Moving object segmentation using optical flow and depth information. In *Advances in Image and Video Technology, Proceedings of the Pacific-Rim Symposium on Image and Video Technology, Tokyo, Japan, 13–16 January 2009*; Springer: Berlin/Heidelberg, Germany, 2009; pp. 611–623.
11. Fang, Y.; Dai, B. An improved moving target detecting and tracking based on optical flow technique and Kalman filter. In Proceedings of the 2009 4th International Conference on Computer Science & Education, Nanning, China, 25–28 July 2009; pp. 1197–1202.
12. Zhang, T.; Zhang, H.; Li, Y.; Nakamura, Y.; Zhang, L. Flowfusion: Dynamic dense RGB-D SLAM based on optical flow. In Proceedings of the 2020 IEEE International Conference on Robotics and Automation (ICRA), Paris, France, 15 September 2020; pp. 7322–7328.
13. Sun, D.; Yang, X.; Liu, M.Y.; Kautz, J. PWC-Net: CNNs for optical flow using pyramid, warping, and cost volume. In Proceedings of the IEEE Conference on Computer Vision and Pattern Recognition, Salt Lake City, UT, USA, 18–22 June 2018; pp. 8934–8943.
14. Yang, S.; Wang, J.; Wang, G.; Hu, X.; Zhou, M.; Liao, Q. Robust RGB-D SLAM in dynamic environment using faster R-CNN. In Proceedings of the 2017 3rd IEEE International Conference on Computer and Communications (ICCC), Chengdu, China, 13–16 December 2017; pp. 2398–2402.
15. Ren, S.; He, K.; Girshick, R.; Sun, J. Faster R-CNN: Towards real-time object detection with region proposal networks. *Adv. Neural Inf. Process. Syst.* **2015**, *28*. Available online: https://proceedings.neurips.cc/paper/2015/file/14bfa6bb14875e45bba028a21ed38046-Paper.pdf (accessed on 23 June 2022). [CrossRef] [PubMed]
16. Yu, C.; Liu, Z.; Liu, X.J.; Xie, F.; Yang, Y.; Wei, Q.; Fei, Q. DS-SLAM: A semantic visual SLAM towards dynamic environments. In Proceedings of the 2018 IEEE/RSJ International Conference on Intelligent Robots and Systems (IROS), Madrid, Spain, 1–5 October 2018; pp. 1168–1174.
17. Bescos, B.; Fácil, J.M.; Civera, J.; Neira, J. DynaSLAM: Tracking, mapping, and inpainting in dynamic scenes. *IEEE Robot. Autom. Lett.* **2018**, *3*, 4076–4083. [CrossRef]
18. He, K.; Gkioxari, G.; Dollár, P.; Girshick, R. Mask R-CNN. In Proceedings of the IEEE International Conference on Computer Vision, Venice, Italy, 22–29 October 2017; pp. 2961–2969.
19. Zhang, J.; Shi, C.; Wang, Y. SLAM method based on visual features in dynamic scene. *Comput. Eng.* **2020**, *46*, 95–102.
20. Redmon, J.; Farhadi, A. Yolov3: An incremental improvement. *arXiv* **2018**, arXiv:1804.02767.
21. Zhong, F.; Wang, S.; Zhang, Z.; Chen, C.; Wang, Y. Detect-SLAM: Making object detection and SLAM mutually beneficial. In Proceedings of the 2018 IEEE Winter Conference on Applications of Computer Vision (WACV), Lake Tahoe, NV, USA, 12–15 March 2018; pp. 1001–1010.
22. Liu, W.; Anguelov, D.; Erhan, D.; Szegedy, C.; Reed, S.; Fu, C.-Y.; Berg, A.C. SSD: Single shot multibox detector. In *Computer Vision—ECCV 2016, Proceedings of the European Conference on Computer Vision, Amsterdam, The Netherlands, 11–14 October 2016*; Springer: Cham, Switzerland, 2016; pp. 21–37.
23. Fan, Y.; Zhang, Q.; Tang, Y.; Liu, S.; Han, H. Blitz-SLAM: A semantic SLAM in dynamic environments. *Pattern Recognit.* **2022**, *121*, 108225. [CrossRef]
24. Dvornik, N.; Shmelkov, K.; Mairal, J.; Schmid, C. BlitzNet: A real-time deep network for scene understanding. In Proceedings of the IEEE International Conference on Computer Vision, Venice, Italy, 22–29 October 2017; pp. 4154–4162.
25. Rublee, E.; Rabaud, V.; Konolige, K.; Bradski, G. ORB: An efficient alternative to SIFT or SURF. In Proceedings of the 2011 International Conference on Computer Vision, Barcelona, Spain, 6–13 November 2011; pp. 2564–2571.
26. Sturm, J.; Engelhard, N.; Endres, F.; Burgard, W.; Cremers, D. A benchmark for the evaluation of RGB-D SLAM systems. In Proceedings of the 2012 IEEE/RSJ International Conference on Intelligent Robots and Systems, Vilamoura-Algarve, Portugal, 7–12 October 2012; pp. 573–580.

MDPI
St. Alban-Anlage 66
4052 Basel
Switzerland
Tel. +41 61 683 77 34
Fax +41 61 302 89 18
www.mdpi.com

Electronics Editorial Office
E-mail: electronics@mdpi.com
www.mdpi.com/journal/electronics

www.ingramcontent.com/pod-product-compliance
Lightning Source LLC
Chambersburg PA
CBHW061616220926
51850CB00005B/518
* 9 7 8 3 0 3 6 5 7 9 4 1 2 *